Springer Proceedings in Mathematics & Statistics

Volume 56

For further volumes:
http://www.springer.com/series/10533

Springer Proceedings in Mathematics & Statistics

This book series features volumes composed of select contributions from workshops and conferences in all areas of current research in mathematics and statistics, including OR and optimization. In addition to an overall evaluation of the interest, scientific quality, and timeliness of each proposal at the hands of the publisher, individual contributions are all refereed to the high quality standards of leading journals in the field. Thus, this series provides the research community with well-edited, authoritative reports on developments in the most exciting areas of mathematical and statistical research today.

Hervé Abdi • Wynne W. Chin
Vincenzo Esposito Vinzi • Giorgio Russolillo
Laura Trinchera
Editors

New Perspectives in Partial Least Squares and Related Methods

Springer

Editors
Hervé Abdi
School of Behavioral & Brain Sciences
The University of Texas at Dallas
Richardson, TX, USA

Wynne W. Chin
Department of Decision
and Information Systems
University of Houston
Houston, TX, USA

Vincenzo Esposito Vinzi
ESSEC Business School of Paris
Cergy-Pontoise Cedex, France

Giorgio Russolillo
CNAM, Paris, France

Laura Trinchera
Rouen Business School
Rouen, France

ISSN 2194-1009 ISSN 2194-1017 (electronic)
ISBN 978-1-4614-8282-6 ISBN 978-1-4614-8283-3 (eBook)
DOI 10.1007/978-1-4614-8283-3
Springer New York Heidelberg Dordrecht London

Library of Congress Control Number: 2013946412

Printed on acid-free paper

Springer is part of Springer Science+Business Media (www.springer.com)

Foreword

This book contains write-ups of presentations at PLS 2012 which was the 7th meeting in the series of PLS conferences and chaired by Professors Wynne Chin and David Francis at the University of Texas at Houston. Here PLS stands for partial least squares projection to latent structures. It is an approach for modeling relations between data matrices of different types of variables measured on the same set of objects (cases, individuals, chemical or biological samples, businesses, etc.).

The series was started in Paris in 1999 by Michel Tenenhaus, Vincenzo Vinci, et al., with two objectives: (a) to present interesting developments and applications of the PLS approach and (b) bring together PLS users from a wide range of fields in social, economic, natural sciences, and engineering.

It was a privilege and a joy to attend this 7th conference, listening to interesting talks, experiencing Houston and the hospitality of Wynne Chin and his team, including a fascinating visit to NASA to hear the story of the American space program.

In the tradition of the PLS-*nn* meetings, we were entertained by a great variety of PLS developments (e.g., PLS metamodels, variable selection, sparse PLS regression, distance-based PLS, significance vs. reliability, nonlinear PLS, and much more) and applications on a wide range of data, from the traditional econometric/economic data to data from genomics, brain images, epidemiology, and chemical spectroscopy. All this gave rise to numerous interesting discussions and new scientific contacts spanning the world.

It is now about 50 years since Herman Wold, my dear late father, started to look into multivariate analysis and its potential for analyzing and interpreting multidimensional econometric data. He started with principal component analysis (PCA), in computer science often called singular value decomposition (SVD). PCA can be seen as the simplest PLS model with a single block **X**. One of Herman's first accomplishments in this field was to make PCA handle missing data, and he soon realized that PCA was a wonderful tool to reduce a block of data (a matrix) to something much smaller. This could then, with some modifications, be used as building blocks in complex schemes of information flow, so-called PLS path models. The PLS path models grew in different directions as summarized in the second volume of the book *Systems under Indirect Observation* (edited by KG Jöreskog and H Wold, Amsterdam, North Holland 1982).

Around 1975 some chemists–Bruce Kowalski in Seattle, and myself in Umeå, Sweden—started to be interested in PLS-modeling. Bruce successfully tried PLS path modeling on water samples from the "Trout Creek" area in Colorado, with the same 11 ions measured at 5 sites along the creek. Myself (Svante), I investigated the simplest two-block PLS model and its use for multiple regression problems with one or several responses (**y** or **Y**). This turned out to be a powerful approach. Suddenly we could do regression-like modeling with arbitrary many X-variables even for data sets with a small number of observations. And PLS provided a model also of the X-space, greatly facilitating the chemical (or biological, or, etc.) interpretation of the results and prediction of new events. In collaboration with Harald Martens in Oslo, this was further developed to an approach of multivariate calibration where the whole spectra of samples were related to concentrations of interest, instead of the traditional calibration using data at a single wavelength.

The two-block PLS, or PLS regression, became the core of a toolbox of data mining, long before the latter term was coined, and today this has expanded to PLS-discriminant analysis, time-series PLS, hierarchical PLS, PLS-trees, three-way PLS, batch-PLS, nonlinear PLS, and more.

So, before 1999, we had two apparently dissimilar PLS camps, one in social-economic science with complex multiblock PLS path models and one in chemistry-biology and engineering limiting themselves to the simplest two-block PLS model (PLS regression). Then, at the first PLS-*nn* conference in Paris in 1999, these two camps were brought together, discussed the PLS approach in its different flavors, and considered interesting applications in various fields. The present PLS 12 conference had the same format and scope. The expansion on the natural science side is noticeable, especially what concerns biological applications, but the integration of the two approaches is still limited (see, however, the article by Löfstedt, Hanafi, and Trygg). Given the close connection between the two PLS approaches, this is somewhat strange, and with time we can hope that this division will disappear, much helped by this series of PLS-*nn* conferences.

Some interesting connections between the two approaches are seen by considering the relations between:

(a) The simplest two-block PLS model
(b) A hierarchical PLS model where matrix **X** is split into blocks—a "star-formed" path model where each **X**-block is connected to **Y** and only to **Y**
(c) A PLS path model, where the blocks in (b) are arranged in a path structure, all applied to the same data, (**X**, **Y**), and using the same Mode A for the estimation of all score vectors (LVs)

Going from (a) to (b) corresponds to installing restrictions on the model since implicit interactions between variables in different blocks are forced to be zero. Similarly, going from (b) to (c) installs further restrictions since only a few of the inner relations between the score vectors have coefficients different from zero.

Hence, by comparing the amount of explained and cross-validated variances for the **Y**-block (R^2 and Q^2, respectively) for the three models, one will have Model (a) be "better" than Model (b), which in turn will be "better" than Model (c). This occurs because a nonrestricted model always fits data better than a restricted one, but not necessarily predicts better. If the differences in fit or predictivity are large, this indicates a misspecification of Model (c) and/or of Model (b), which can then be diagnosed further by means of the residual variances of blocks and individual variables, for both fitted and cross-validated and other residuals.

So, to conclude this brief foreword, I wish once more to thank and congratulate the organizers to an excellent and most interesting conference. And second, I would like to see—hopefully already at the next PLS-*nn* conference (i.e., PLS 2014 in Paris)—more cases where different types of PLS models (e.g., path models, hierarchical PLS regression, and ordinary PLS regression) are run on the same data, including discussions of similarities and differences.

Thanks a million.

Umeå, Sweden Svante Wold

Preface

In 1999 the first meeting dedicated to partial least squares methods (abbreviated as PLS and also, sometimes, expanded as *projection to latent structures*) took place in Paris. Other meetings in this series took place in various cities, and in 2012, from the 19th to the 22nd of May, the seventh meeting of the partial least squares (PLS) series took place for the first time in the United States (in Houston, Texas). This *première* was a superb success with roughly 120 authors presenting 44 papers during these 4 days. These contributions were all very impressive by their quality and by their breadth. They covered the multiple dimensions of partial least squares-based methods, ranging from partial least squares regression and correlation to component-based path modeling, regularized regression, and subspace visualization. In addition several of these papers presented exciting new theoretical developments. This diversity was also expressed in the large number of domains of application presented in these papers: brain imaging, genomics, chemometrics, marketing, management, and information systems to name only but a few.

After the conference, we decided that a large number of the papers presented in the meeting were of such an impressive high quality and originality that they deserved to be made available to a wider audience and we asked the authors of the best papers if they would like to prepare a revised version of their paper. Most of the authors contacted shared our enthusiasm, and the papers that they submitted were then read and commented by anonymous reviewers, revised, and finally edited for inclusion in this volume. These 22 papers (including three invited contributions from our keynote speakers), included in *New perspectives in Partial Least Squares and Related Methods*, provide a comprehensive overview of the current state of the most advanced research related to PLS and cover all domains of PLS and related domains.

Each paper was overviewed by one editor who took charge of having the paper reviewed and edited (Hervé was in charge of the papers of Martens et al., Marcoulides and Chin, Beaton et al., Ciampi et al., Krishnan et al., Le Floch et al., Kovacecic et al., and Churchill et al.; Wynne was in charge of the papers of Chin et al., Cepeda et al., Murray et al., and Newman et al.; Vincenzo was in charge of the papers of Magidson and Aluja-Banet et al.; Giorgio was in charge of the papers of Sharma and Kim, Liu et al., Löfstedt et al., and Eslami et al.; Laura was in charge of the papers of Mehmood and Snipen, Farooq et al., and Martinez-Ruiz and Aluja-Banet). The final production of the LaTeXversion of the book was mostly the work of Laura, Giorgio, and Hervé.

We are particularly grateful to Professor Svante Wold—one of the creators of PLS—who opened the seventh PLS meeting with an outstanding opening keynote that presented an overview of the field from its creation to its most recent developments and wrote the foreword presenting this book. We are also particularly grateful to our (anonymous) reviewers for their help and dedication.

Finally, this meeting would not have been possible without the generosity, help, and dedication of several persons, and we would like to specifically thank John Antel, Thierry Fahmy, David Francis, Michele Hoffman, Jennifer James, Yong Jin Kim, Ken Nieser, Blair Stauffer, Doug Steel Sarah J. Sweaney, Reza Vaezi, and Sean Woodward.

Richardson, TX, USA — Hervé Abdi
Houston, TX, USA — Wynne W. Chin
Cergy-Pontoise, France — Vincenzo Esposito Vinzi
Paris, France — Giorgio Russolillo
Rouen, France — Laura Trinchera

Contents

Part I Keynotes

Contributors

Hervé Abdi
School of Behavioral and Brain Sciences, The University of Texas at Dallas, Richardson, TX, USA

Tomas Aluja-Banet
Universitat Politecnica de Catalunya, Barcelona, Spain

Elizabeth Anderson-Fletcher
University of Houston, Houston, TX, USA

Carmen Barroso
University of Seville, Seville, Spain

Derek Beaton
School of Behavioral and Brain Sciences, The University of Texas at Dallas, Richardson, TX, USA

Stéphanie Bougeard
Department of Epidemiology, French Agency for Food, Environmental and Occupational Health Safety (Anses), Ploufragan, France

Gabriel Cepeda
University of Seville, Seville, Spain

Wynne W. Chin
Department of Decision and Information Systems, C. T. Bauer College of Business, University of Houston, Houston, TX, USA

Nathan Churchill
Baycrest Center, Rotman Research Institute, Toronto, ON, Canada

Antonio Ciampi
Department of Epidemiology, Biostatistics, and Occupational Health, Montréal, Canada

Edouard Duchesnay
CEA, Saclay, France

Aida Eslami
Department of Epidemiology, French Agency for Food, Environmental and Occupational Health Safety (Anses), Ploufragan, France

Vincenzo Esposito Vinzi
ESSEC Business School, Cergy Pontoise Cedex, France

Omer Farooq
EUROMED, Marseilles, France

Francesca Filbey
School of Behavioral and Brain Sciences, The University of Texas at Dallas, Richardson, TX, USA

Edith Le Floch
CEA, Saclay, France

Vincent Frouin
CEA, Saclay, France

George O. Gamble
University of Houston, Houston, TX, USA

Arne B. Gjuvsland
IMT (CIGENE), Norwegian University of Life Sciences, Ås, Norway

Vincent Guillemot
CEA, Saclay, France

Mohamed Hanafi
Sensometrics and Chemometrics Laboratory, Nantes, France

Alfred O. Hero
Electrical Engineering and Computer Science Department, University of Michigan, Ann Arbor, MI, USA,

Kevin H. Kim
School of Education and Joseph M. Katz Graduate School of Business, University of Pittsburgh, Pittsburgh, PA, USA

Yong Jin Kim
Global Service Management, Sogang University, Seoul, Korea

Achim Kohler
Centre for Integrative Genetics, Norwegian University of Life Sciences, Ås, Norway

Natasa Kovacevic
Baycrest Center, Rotman Research Institute, Toronto, ON, Canada

Nikolaus Kriegeskorte
MRC Cognition and Brain Sciences Unit, University of Cambridge, Cambridge, UK

Anjali Krishnan
Institute of Cognitive Science, University of Colorado Boulder, Boulder, CO, USA

Aurélie Labbe
Department of Epidemiology, Biostatistics, and Occupational Health, Montréal, Canada

Giuseppe Lamberti
Universitat Politecnica de Catalunya, Barcelona, Spain

Gunhee Lee
Sogang Business School, Sogang University, Seoul, Korea

Tzu-Yu Liu
Electrical Engineering and Computer Science Department, University of Michigan, Ann Arbor, MI, USA

Tommy Löfstedt
Department of Chemistry, Computational Life Science Cluster (CLiC), Umeå University, Umeå, Sweden

Jay Magidson
Statistical Innovations Inc, Belmont, MA, USA

George A. Marcoulides
Research Methods and Statistics, Graduate School of Education and Interdepartmental Graduate Program in Management, A. Gary Anderson Graduate School of Management, University of California, Riverside, CA, USA

Alba Martínez-Ruiz
Universidad Católica de la Santísima Concepción, Concepción, Chile

Silvia Martelo
University of Seville, Seville, Spain

Harald Martens
IMT (CIGENE), Norwegian University of Life Sciences, Ås, Norway
Nofima Ås, Ås, Norway

Anthony R. McIntosh
Baycrest Center, Rotman Research Institute, Toronto, ON, Canada

Tahir Mehmood
Biostatistics, Department of Chemistry, Biotechnology and Food Sciences, Norwegian University of Life Sciences, Ås, Norway

Chantal Mérette
Faculty of Medicine, Department of Psychiatry and Neurosciences, Université Laval, Quebec City, Canada

Dwight Merunka
IAE and Cergam, University Aix-Marseille, Aix-Marseille, France

EUROMED Marseilles School of Management, Marseille, France

Michael J. Murray
University of Houston, Houston, TX, USA

Michael R. Newman
University of Houston, Houston, TX, USA

Stig W. Omholt
IMT (CIGENE), Norwegian University of Life Sciences, Ås, Norway

Jaime Ortega
University of Seville, Seville, Spain

Erik Plahté
IMT (CIGENE), Norwegian University of Life Sciences, Ås, Norway

Jean-Baptiste Poline
CEA, Saclay, France

El Mostafa Qannari
Sensometrics and Chemometrics Laboratory, ONIRIS, LUNAM University, Nantes, France

Giorgio Russolillo
Conservatoire National des Arts et Métiers, Paris, France

Gastón Sánchez
CTEG, Unité de Recherche de Sensométrie et Chimiométrie (USC INRA), ONIRIS, Nantes, France

Pratyush N. Sharma
Joseph M. Katz Graduate School of Business, University of Pittsburgh, Pittsburgh, PA, USA

Lars Snipen
Biostatistics, Department of Chemistry, Biotechnology and Food Sciences, Norwegian University of Life Sciences, Ås, Norway

Robyn Spring
Baycrest Center, Rotman Research Institute, Toronto, ON, Canada

Doug Steel
School of Business, University of Houston-Clear Lake, Houston, TX, USA

Stephen Strother
Baycrest Center, Rotman Research Institute, Toronto, ON, Canada

Valeriya Tafintseva
IMT (CIGENE), Norwegian University of Life Sciences, Ås, Norway

Arthur Tenenhaus
Supélec, Gif-sur-Yvette, France

Jason B. Thatcher
College of Business and Behavioral Science, Clemson University, Clemson, SC, USA

The Alzheimer's Disease Neuroimaging Initiative (ADNI)
http://adni.loni.ucla.edu/wpcontent/uploads/how_to_apply/ADNI_Acknowledgement_List.pdf

Kristin Tøndel
IMT (CIGENE), Norwegian University of Life Sciences, Ås, Norway

Laura Trinchera
Rouen Business School, Rouen, France

Johan Trygg
Department of Chemistry, Computational Life Science Cluster (CLiC), Umeå University, Umeå, Sweden

Pierre Valette-Florence
IAE, Grenoble, France

Jon Olav Vik
IMT (CIGENE), Norwegian University of Life Sciences, Ås, Norway

Dennis Wei
Electrical Engineering and Computer Science Department, University of Michigan, Ann Arbor, MI, USA

Svante Wold
Umeå University, Umeå, Sweden

Ryan T. Wright
School of Management, University of Massachusetts, Amherst, MA, USA

Lin Yang
Division of Clinical Epidemiology, McGill University Health Centre, Montréal, Canada

Arthur Tenenhaus
Supélec, Gif-sur-Yvette, France

[illegible]

The Alzheimer's Disease Neuroimaging Initiative (ADNI)
[illegible]

Part I
Keynotes

PLS-Based Multivariate Metamodeling of Dynamic Systems

Harald Martens, Kristin Tøndel, Valeriya Tafintseva, Achim Kohler, Erik Plahte, Jon Olav Vik, Arne B. Gjuvsland, and Stig W. Omholt

Abstract In this paper, we discuss the use of bi-linear methods for assessing temporal dynamics, in particular with regard to the understanding of complex biological processes. We show how the dynamics in multivariate time series measurements can be summarized efficiently by principal component analysis. Then we demonstrate how the development and use of complex, high-dimensional nonlinear differential equation models can be facilitated by multivariate metamodeling using nonlinear PLS-based subspace data modeling. Different types of metamodels are outlined and illustrated. Finally, we discuss some cognitive topics characterizing different modeling cultures. In particular, we tabulate various metaphors deemed relevant for how the time domain is envisioned.

Key words: Metamodeling, Complex systems, Differential equations, Time domain metaphors, Nonlinear dynamics, Multivariate subspace modeling, PLS regression, Chemometrics

H. Martens (✉)
IMT (CIGENE), Norwegian University of Life Sciences, Ås, Norway
Nofima Ås, Ås, Norway
e-mail: Harald.Martens@Nofima.no

K. Tøndel • V. Tafintseva • E. Plahte • J.O. Vik • A.B. Gjuvsland • S.W. Omholt
IMT (CIGENE), Norwegian University of Life Sciences, Ås, Norway
e-mail: kristin.tondel@umb.no; valeta@umb.no; erik.plahte@umb.no; jonovik@gmail.com; arne.gjuvsland@umb.no; stig.omholt@umb.no

A. Kohler
Centre for Integrative Genetics, Norwegian University of Life Sciences, Ås, Norway
e-mail: achim.kohler@umb.no

H. Abdi et al. (eds.), *New Perspectives in Partial Least Squares and Related Methods*, Springer Proceedings in Mathematics & Statistics 56, DOI 10.1007/978-1-4614-8283-3_1,

1 Background

1.1 Modeling in Biology

Herman Wold, the inventor of the partial least squares (PLS) framework of pragmatic, robust data-modeling, probably could not have fully envisioned the current enormous increase in available data and systems knowledge. But his visionary overview from 1983 (see Fig. 1) concerning quantitative systems analysis and the broad scope of PLS soft modeling is still applicable [11]. It indicates that low-rank PLS modeling may be used for linking the life sciences to many other relevant fields of science, ranging from natural sciences like physics and chemistry to medicine, cognitive science and psychology. We here show that it can also be linked to nonlinear applied mathematics, and therefore can be used for example in detailed modeling of nonlinear spatiotemporal dynamics. Advances in instrumentation and computers have allowed scientific knowledge to be collected in previously unknown quantities. For instance, quantitative information in Biology is now being collected in large data bases or repositories, in three more or less disjoint domains:

- Actual measurements from lots of biological samples in various multichannel instruments, including "omics" data such as mRNA, proteins and metabolites;
- Databases summarizing biological knowledge such as gene ontology (http://www.geneontology.org) and metabolic networks (e.g., http://www.genome.jp/kegg/pathway.html).
- Various repositories of physiological and regulatory models from Biology, (see, e.g., http://www.cellml.org, http://sbml.org http://www.physiome.org.nz/xml_languages/fieldml)

The question is how to combine and utilize all this knowledge efficiently. For instance, mathematical modeling is expected [8] to play an increasing role in Biology and Medicine. Explicit modeling of biological mechanisms offers the most compact, quantitative representation of complex biological knowledge. However, to bring "hard" modeling concepts from physics and physical chemistry into "soft" fields of bio-science represents a clash of science cultures: On one hand, explicit mathematical modeling was traditionally used for describing relatively simple, low-dimensional, homogeneous, isolated physical systems: how can they be extended to describe the far more complex, high-dimensional, heterogeneous systems of Biology? On the other hand, while the bio-sciences today make extensive use of computational statistics, mathematics was never the favorite subject for main-stream biologists. So how can Biology and medicine best benefit from applied mathematics? And how can the development and use of mathematical models benefit from—and provide benefit to—the use of information from massive databases of biological measurements and networks derived from "-omic" data? This is a question of mathematical methodology, but also about scientific culture.

Multivariate top-down data modeling has the potential to provide bio-scientists with tools to overview each of these domains and to bridge between data from different sources and of different nature, be it from actual measurements, postulated net-

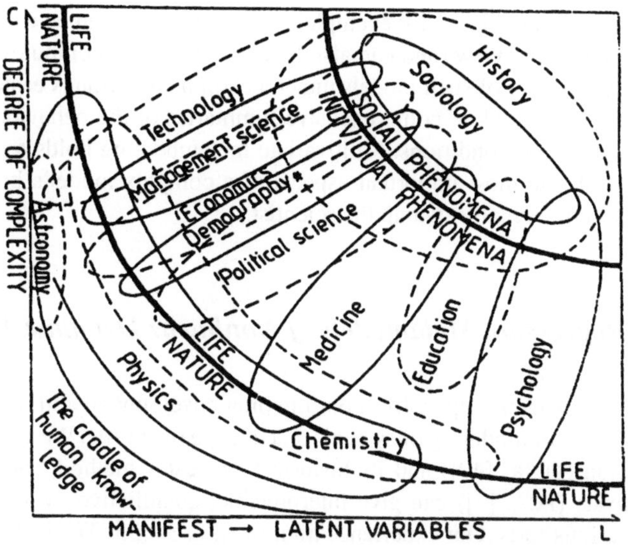

Fig. 1: The pedigree and broad scope of PLS soft modeling. The “father” of PLS, Herman Wold’s overview of the range of applications of PLS-based multivariate data modeling (From [38])

works or mathematical model simulations. In this paper we suggest ways in which multivariate data modeling based on the principle of partial least squares (PLS) can facilitate the use of mathematical modeling in Biology. In particular, based on our current experience with PLS-based metamodeling, we claim that the relevant subspace approximation of PLS regression can improve the understanding of the time domain, in the sense of enhancing the quantification and interpretation of complex temporal dynamics in living systems.

Clearly, three complex systems must be addressed simultaneously: (a) the biological system under scrutiny, (b) the perceptual and cognitive capacity of the scientist and (c) the computational capacity of the modeling hardware and software. The mathematical model must be complex enough to describe the biological system adequately for the given purpose. But the model development and its computational use should be under scientist’s cognitive control, without being limited to the scientist’s prior understanding. The numerical routines used for implementing the model in a computer must be robust and sufficiently accurate, and the computer implementation must offer solutions without unacceptable delays.

Predicting the behavior of a complex mathematical model just by looking at its set of equations is usually impossible. The number of chunks of information (variables, parameters, etc.) to monitor and combine is often beyond the capacity of human working memory. Moreover, our mind cannot logically or intuitively envision the consequences of nonlinear operations repeated several times in sequence. In addition, traditional methods of theoretical analysis in mathematics are difficult to apply to complex, nonlinear high-dimensional dynamic models. In particular, a large system of coupled nonlinear ODEs embedded in a large spatiotemporal grid, as needed in the modeling of, for example, the human heart, can be extremely difficult to assess and overview (and can also be cumbersome to compute). Looking at graphs of the behavior of such models under a certain condition is easier. It is even more informative to look at comprehensive summaries of the behavior of a model under many different conditions. Multivariate metamodeling holds a potential for summarizing all the most important aspects of a complex model's behavior, in a way that does not mentally swamp the scientist.

1.2 PLS Regression Metamodels of Nonlinear Dynamic Models

This paper is a progress report from our ongoing development and use of Chemometrics methods for modeling of systems that change over time. We shall show how PLS regression (PLSR, [40]) and its historical "ancestor method," principal component analysis (PCA, [7]) can give interpretable, quantitative dynamic subspace models of various sorts. Our experience till now ([22, 34, 35, 37]) is that PLS-based multivariate metamodeling offers several benefits to the developers of and to the users of large, mechanistic models in general. This is also confirmed by, for example, Sobie's use of PLS regression in sensitivity analysis and in constraining model parameters ([29, 30]). But since dynamic models of biological mechanisms are often highly nonlinear, we have found it advantageous to extend the PLS regression in various ways in order to handle strong nonlinearities. This will be illustrated below. A mathematical model $\mathscr{M}$ may be symbolized as

$$\text{Outputs} = \mathscr{M}(\text{Inputs}) \tag{1}$$

where the inputs represent model parameters and initial conditions, and the outputs represent simulated "phenotypes" or properties, often in terms of spatial and/or temporal data. Contemporary mathematical models $\mathscr{M}$ of, for example, the function of the heart are nonlinear and heterogeneous, and have complicated dynamics at several different spatiotemporal scales. Such high-dimensional models, with numerous nonlinear positive and negative feedback structures, are too complex for classical mathematical analysis, and their behavior is therefore difficult to predict theoretically. Hence, contemporary mechanistic models $\mathscr{M}$ are not only slow to compute, but also difficult to define, overview, control, compare, and improve. This is a typical arena where multivariate metamodeling is useful. What is a metamodel? It is a simple model of a complex model: a simplified statistical description of the behavior of a complicated mathematical model. It is sometimes also called a "surrogate model" [6]. For scientist developing or using a large, complicated mechanistic model $\mathscr{M}$ of, say, a complex biological system, a metamodel is an approximation model ($\mathscr{A}$) that summarizes the behavior of $\mathscr{M}$ in a way that relates different inputs to different outputs. In general, a complex model $\mathscr{M}$ will have a number of intermediate and output variables that represent necessary steps in the computation of $\mathscr{M}$, but that are of little interest for the modeler and/or little relevance for a given application of the model. In the metamodel $\mathscr{A}$ irrelevant variables are down-weighted and variables that co-vary are lumped together. Moreover, the original model $\mathscr{M}$ may contain computationally slow model elements, which may be replaced by computationally fast approximations in metamodel $\mathscr{A}$.

A multivariate metamodel is obtained in two steps:

- Extensive simulations with $\mathscr{M}$ to probe the desired ranges in the input parameters (all relevant input combinations) and record all relevant outputs (including intermediates). This results in large tables of input and output data.
- Analysis of the obtained input and output data from $\mathscr{M}$, more or less as if they were normal empirical data.

Multivariate metamodeling of a model $\mathscr{M}$ may be used for a wide range of purposes. Traditionally it was primarily used for sensitivity analysis and computational speed-up [4, 5, 12, 28]. But it can also be used to discover "hidden" patterns of co-variation in the model, to simplify models, to compare different models or to fit a model to empirical data. In each case a multivariate metamodel $\mathscr{A}$ summarizes and reveals what model $\mathscr{M}$ has really been doing, seen from a certain perspective and stated purposes, and limited to the conditions tested in the simulations. Different types of multivariate metamodels $\mathscr{A}$ may be developed to reveal different aspects of model $\mathscr{M}$. The main distinction is between:

$$\begin{array}{ll}\text{simple output metamodeling:} & \text{Outputs} = f(\text{Outputs})\\ \text{classical metamodeling:} & \text{Outputs} = f(\text{Inputs})\end{array}$$

and

$$\text{inverse metamodeling: Inputs} = f(\text{Outputs}).$$

But explicitly dynamic metamodels are also informative, for example:

$$\text{autoregressive metamodeling: Outputs}_{t\,=\,\text{now}}{=}f(\text{Outputs}_{t\,=\,\text{past}})$$

and

$$\text{ODE metamodeling: Output rates}_t = f(\text{Output states}_t).$$

The different metamodel types, alone or in combination, give insight into how the original model $\mathscr{M}$ behaves in practice. That can be quite different from what is apparent when simply looking at the mathematical equations that constitute $\mathscr{M}$. Thus, through multivariate metamodeling, theory-driven models ($\mathscr{M}$), built deductively and bottom-up, may for many purposes be illuminated—and sometimes even replaced by—one or more data-driven models ($\mathscr{A}$), built inductively and top-down.

Many different statistical methods for supervised and unsupervised learning may be used successfully for multivariate metamodeling. We have found that various versions of PLSR (see, e.g., [18]) are particularly well suited, due to their simplicity and versatility. When optimized with cross-validation/jackknifing ([19, 31]) and displayed in extensive graphics, multivariate metamodeling provides interpretable linear subspace models with excellent predictive validity. These PLS-based approximation models can improve insight and overview as well as predictive precision and computational speed-up. Our metamodeling development till now has focused on relatively simple models from bio-spectroscopy [13] and physiology (see below). But the PLS-based multivariate metamodeling techniques are generic and we expect them to be useful also for more complex models and in other application fields.

Today, we have a range of tools for cost-effective designs for computer experiments ([23, 34, 41]) and a versatile family of PLS-based regression methods for handling high-dimensional, rank deficient, and sometimes N-way simulation data, often with highly nonlinear input/output relationships ([20, 34, 36, 37]). In the following, some aspects of PCA and PLS-based metamodeling will be illustrated through examples. We shall here outline the general development of PLS-based metamodeling, and focus on various ways to use it for time-dependent modeling. Then we shall outline some cognitive differences between mathematical modeling of mechanisms ($\mathscr{M}$) and statistical approximation modeling ($\mathscr{A}$), and list some relevant metaphors used in dealing scientifically with the concept of time.

1.3 PCA and PLSR Similar to Taylor Expansions of Model $\mathscr{M}$?

First, we shall discuss how a bilinear PCA and PLS-based metamodel $\mathscr{A}$ may be regarded as a generalized series-expansion of its target model $\mathscr{M}$. This series expansion is not obtained by traditional Taylor-expansion of $\mathscr{M}$ (deriving a cumbersome sequence of derivatives of $\mathscr{M}$), but by a much simpler approach: structured computer simulations, followed by multivariate self-modeling ($\mathscr{A}$) that is similar to a truncated Taylor-expansion of $\mathscr{M}$, based on the simulation results.

Prior to the development of PLSR, in Chemometrics—like in many other fields—the main tool for quantitative approximation of large data matrices was PCA. PLSR and PCA share many properties: they both rely on so-called bilinear subspace modeling. PCA is still a very useful tool for exploratory data modeling, as an example from biotechnological process dynamics in this section will show. However, PCA and PLSR are often used for more or less assumption-free self-modeling of data tables. Thus they represent a data-driven paradigm very different from that of theory-driven mechanistic modeling. How can the two paradigms be reconciled? Let us start with PCA, the simplest bilinear method.

Usually, measured variables from real-world systems are more or less intercorrelated: they share common patterns in how they vary from sample to sample. This makes it easier to distinguish interesting signals from uninteresting noise in data, which we may expect to be random and uncorrelated. Valid signal patterns—the variations that we are usually interested in—have often been caused by shared causal structures, affecting two or more variables. The causal structure behind any two-way data matrix $\mathbf{D}$ (of order $n \times p$) may generally be written:

$$\mathbf{D} = f(g_i, h_k) + \varepsilon \tag{2}$$

where g_i and h_k are functions of two types of causal phenomena, affecting the rows (objects) with index $i = 1, 2, \ldots, n$ and columns (variables) with index $k = 1, 2, \ldots, p$, respectively, and where ε represents the uninteresting, presumably random, error. There may be several such sets of underlying (latent) functional causes:

$$\mathbf{D} = f\left(g_{i,1}, h_{k,1}, g_{i,2}, h_{k,2}, \ldots\right) + \varepsilon \tag{3}$$

Usually, the nature of some, or all, of these causes $g_{i,a}, h_{k,a}$, $a = 1, 2, \ldots$ are unknown to us. But provided that sufficiently many informative variables have been measured in sufficiently many informative objects or generated by simulation under sufficiently many conditions, we can discover a lot about these unknown phenomena by PCA.

Prior to the analysis, the input data **D** are usually preprocessed to improve linearity and remove known interferences. To balance the different types of variables in the data, they are then scaled to comparable units to generate a data table **X**.

In PCA, the matrix **X** is mean-centered and decomposed by the so-called singular value decomposition to identify its singular values and orthonormal left-hand and right-hand singular vectors (i.e., the eigenvectors of mean-centered $\mathbf{X}\mathbf{X}^{\mathsf{T}}$ and $\mathbf{X}^{\mathsf{T}}\mathbf{X}$, respectively, with ^{T} denoting the transpose operation). Based on, for example, cross-validation or common sense, the number of statistically valid principal components, A, is determined, and the bilinear approximation model of PCA is set up as:

$$\mathbf{X} = \mathbf{x}_0 + \sum_{a=1}^{A} \left(\mathbf{t}_a \mathbf{p}_a^{\mathsf{T}}\right) + \mathbf{E}_A = \mathbf{x}_0 + \mathbf{T}_A \mathbf{P}_A^{\mathsf{T}} + \mathbf{E}_A \tag{4}$$

where $\mathbf{x}_0$ represents a center vector (i.e., a mean vector or centroid) around which the bilinear model is developed. The so-called "scores" $\mathbf{T}_A = [\mathbf{t}_1, \mathbf{t}_2, \ldots, \mathbf{t}_A]$ represent the left-hand singular vectors, scaled by their relative importance (their singular values) that define the main patterns of co-variation in the objects (i.e., the rows of **X**). The "loadings" $\mathbf{P}_A = [\mathbf{p}_1, \mathbf{p}_2, \ldots, \mathbf{p}_A]$ represent the corresponding right-hand singular vectors, and show how these patterns manifest themselves in the different variables (columns). The matrix $\mathbf{E}_A$ represents the unmodeled residual, presumably containing small, unsystematic and uninteresting errors etc. PCA, which may be considered as the "mother of all multivariate data modeling methods," has for many years been the first choice for overviewing a single table of empirical data **X**. However, PCA is equally useful for summarizing variation patterns in tables of simulation data from computer experiments. Two, or three, dimensional plots of the first few principal components are usually very informative and easy to understand. But is this pragmatic data-approximation model from Eq. 4 a proper scientific model? Our answer is "yes," based on our own practical experience as well as on the following theoretical basis.

Svante Wold, a highly productive Nestor in Chemometrics, showed that a PCA solution, consisting of the sum of the first few principal components, may be seen as a truncated Taylor expansion of whatever multivariate causal structures (known or unknown) have given rise to **X** which is the two-way data table at hand ([39]). Model-based approximation is an essential activity in science. It is interesting to note that Robert Rosen wrote that the standard practice of truncation "is a good example of modeling within mathematics," and showed that this applies to Taylor series expansion ([27], pp. 78–9). Hence, this must apply for PCA as well.

Moreover, Rosen discussed the special case when one of the ways of the available data represents time. He showed that Taylor's theorem for truncated series expansion of time series data has some fundamental properties in linking diachrony (what happens over an extended series of instances) to synchrony (what happens at a single instance), and also to something very much like recursivity (a concept central to scientific thinking). Hence, one may expect Rosen's positive evaluation of truncated Taylor expansions to apply equally well to PCA of multivariate time series. Two-block PLSR provides models of the type $\mathbf{Y} \approx f(\mathbf{X})$. This is slightly more complicated than the one-block PCA, in the sense that PLSR defines the bilinear model of $\mathbf{X}$ from a sequence of one-component eigen-analyses of previously unmodeled $\mathbf{X}^{\mathsf{T}}\mathbf{Y}\mathbf{Y}^{\mathsf{T}}\mathbf{X}$, instead of one joint eigen-analysis of $\mathbf{X}^{\mathsf{T}}\mathbf{X}$. A number of different algorithmic formulations of this PLSR modeling process exists. While not yet proven, we expect it to be possible to show that PLSR represents some sort of multivariate Taylor expansion of a function of $\mathbf{X}$ and $\mathbf{Y}$, in a way that reflects their underlying, unknown causal relation. Let that be a challenge of the next PLS generation!

1.4 PCA Modeling of a Multivariate Dynamic System

An introductory example will now be given. It is analogous to simple output metamodeling, but based on real process measurements. The purpose is to show how a complex dynamic system with highly nonlinear behavior can be inspected and quantified in terms of its underlying systematic structure. The main "factors" ("variation phenomena") are summarized as abstract "components" in a bilinear model by PCA. In this case the data come from a real-world industrial process, and consist of high-dimensional spectral measurements. Figure 2a (see, [21]) shows the multichannel infrared spectra of a biotechnological batch fermentation process, read at more or less regular intervals over a 26 h period. These data were submitted to PCA analysis.

Figure 2b shows the scores for the first three PCA components. It is obvious that the process passes from time = 0 till time = 26 h through several distinct phases, leading from initial state profile S_1 via state profiles near "intermediate end state profiles" S_2, S_3 and S_4 to its final state S_5. Each of the reaction phases provides a gradual transition from one state to the next. The nature of these five states is unknown, but was assumed to reflect a sequential depletion of various carbon sources in the growth medium. The spacing of regular observation points along the trajectory shows that the speed of the process varies considerably.

Quantitative information about these five process state profiles S_1–S_5 was gained by post-processing of the orthogonal PCA solution from Fig. 2b. The rotational ambiguity of the PCA solution was here overcome by "simplex intersect" between trajectory extrapolations (see [16]) after 2, 19, and 21.5 h, as shown in Fig. 2b. The characteristic infrared spectra $S = [S_1\ S_2\ S_3\ S_4\ S_5]$ of the initial, intermediate and final end states were thus estimated. From these, the process dynamics was quantified in terms of its initial condition profile S_1 and a sequence of linear directions ($S_2 - S_1$, $S_3 - S_2$, $S_4 - S_3$ and $S_5 - S_4$) in which the process moves in the state space (Fig. 2c). By regressing the observed spectra in $\mathbf{X}$ on the five estimated state profiles S, the "concentration" levels $\mathbf{C} = [\mathbf{c}_1, \mathbf{c}_2, \mathbf{c}_3, \mathbf{c}_4, \mathbf{c}_5]$ (i.e., the "constituent concentrations")

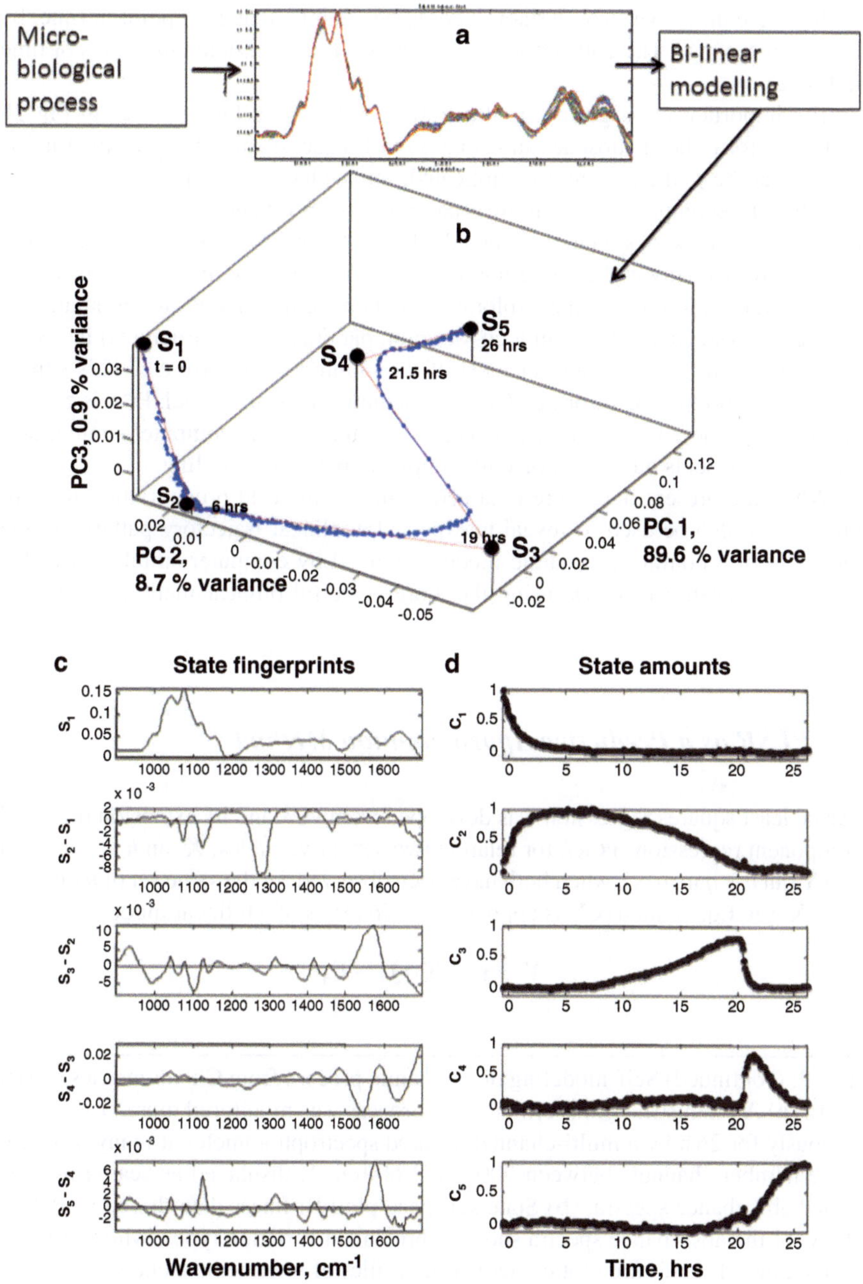

Fig. 2: (continued)

of these five unknown state variables, could be quantified at each point in time, by linear "unmixing." The concentrations are shown as conventional functions of time in Fig. 2d.

The important message in this illustration is that the bilinear scores in Fig. 2b and the reformulated subspace information in Fig. 2c, d show how detailed information can be gained about a complicated, multi-phase dynamic process, simply by bilinear decomposition of multivariate process observations, without any prior theory. The time series data represented a lot of "snapshots," but time as such was not treated explicitly in the modeling. Instead, the clear covariance of the measured spectral variables allowed the exploratory metamodeling process to automatically isolate the systematic, interesting co-variation patterns (eigen-structures) in a simple subspace model, leaving out most of the uninteresting noise etc. This simple subspace could then be reformulated into a tentative mixture model. PCA is equally applicable to the visualization and checking of outputs from computer simulations, and thus represents our first choice for simple output metamodeling.

While the present data were obtained from a real-world process, the same approach can also be used to reveal unexpected nonlinear trajectory patterns in the output from a nonlinear dynamic model, obtained by computer simulations. This will be demonstrated below, using the more powerful bilinear method of PLS regression.

1.5 PLSR as a Predictive Approximation Method

Partial least squares regression was developed from PCA and its extension principal component regression (PCR), for relating two sets of variables, $\mathbf{X}$ (an $n \times p$ matrix) and $\mathbf{Y}$ (an $n \times q$ matrix), when both have been observed for the same set of n objects. Like $\mathbf{X}$ (see Eq. 4), matrix $\mathbf{Y}$ is approximated in terms of a bilinear model:

$$\mathbf{Y} = \mathbf{y}_0 + \mathbf{T}_A \mathbf{Q}_A^\mathsf{T} + \mathbf{F}_A \tag{5}$$

Fig. 2: (continued) Self-modeling of a dynamic process from Chemometrics (From [21]). (**a**) An industrial milk fermentation process was monitored more or less continuously for 26 h by a multi-channel infrared spectrophotometer at many different wave-number channels between 900 and 1,600 cm-1, displayed as scattering corrected absorbance spectra. (**b**) State subspace plot: Bi-linear data-driven modeling (PCA) of the absorbance spectra showed three main variation types, whose orthogonal temporal scores are plotted here. The profiles of intermediate end states S_2, S_3 and S_4 were estimated by extrapolation according to the simplex intersect method (see [16]). (**c**) Multivariate description of the main types of process dynamics: The initial state profile S_1 and the four subsequent average rate profiles $S_2 - S_1$, $S_2 - S_3$, $S_3 - S_4$, $S_4 - S_5$. (**d**) Time series of state variables corresponding to S_1–S_5, estimated by linear unmixing (see [18]) where $\mathbf{y}_0$ represents the model center (i.e., the mean), $\mathbf{Q}_A$ is the coupling between the input variables in $\mathbf{Y}$ and the A components $\mathbf{T}_A$, and $\mathbf{F}_A$ contains the unmodeled residuals. But it should be noted that $\mathbf{T}_A$—the sequence of the A orthogonal PLS components (PCs)—is defined as a function of $\mathbf{X}$, not of $\mathbf{Y}$:

$$\mathbf{T}_A = (\mathbf{X} - \mathbf{x}_0)\,\mathbf{V}_A \tag{6}$$

where $\mathbf{V}_A$ (a $p \times A$ matrix) represents the estimated weights. Therefore the model of $\mathbf{Y}$ may equivalently be written

$$\mathbf{Y} = \mathbf{b}_{0A} + \mathbf{X}\mathbf{B}_A + \mathbf{F}_A \tag{7}$$

with $\mathbf{B}_A = \mathbf{V}_A\mathbf{Q}_A^{\mathrm{T}}$ and $\mathbf{b}_{0A} = \mathbf{y}_0 - \mathbf{x}_0\mathbf{B}_A$. In other words: with this model, $\mathbf{Y}$ can be predicted directly from $\mathbf{X}$ but not vice versa.

The difference between PCA/PCR and PLSR is simply that, in PCA and PCR, the weights $\mathbf{V}_A$ are defined to explain maximal covariance within $\mathbf{X}$, while in PLSR, the weights $\mathbf{V}_A$ are defined to explain maximal covariance between $\mathbf{X}$ and $\mathbf{Y}$.

PLSR was originally published ([38]) for multivariate calibration, (i.e., to facilitate the conversion of low-cost, non-selective input variables $\mathbf{X}$ into selective output predictions of high-cost variables $\mathbf{Y}$, see [18]). Very soon, however, it was employed also for a wide range of other purposes, from classical analysis of variance and inverse discriminant analysis, via classical mixture modeling and inverse multivariate calibration to time series analysis. A number of extensions of the basic PLSR have since been published. For instance, various PLS extensions are now used for relating several types data tables, such as multi-block and multi-matrix PLS regressions and several different sets of variables and/or objects have been measured.

We have found the N-way extension of PLSR [1, 2] particularly useful in metamodeling of models that give 3-way outputs (conditions $\times$ times $\times$ "phenotype" properties), along with various nonlinear PLSR extensions. In the following, we outline some of the metamodeling published with the use of PLS-regression and extensions thereof.

2 Dynamic Multivariate Metamodeling

2.1 *Metamodeling of Dynamic Processes*

A multivariate time series of the state variables in a process requires that the mathematical model is first defined in terms of its algebraic structure, parameter values and initial states. Then the model is run in a recursion algorithm or in numerical integration, for sufficient time to generate simulated time series of the internal state variables x_t. The full vector of state variables x_t may contain both properties ("how?") and physical coordinates ("where?"), such as chemical concentrations and 3D spatial positions, at time t. These simulated vectors of states x_t and their rates $\dot{x}_t = \frac{dx_t}{d_t}$ may then be studied as functions of time, to understand the system's behavior.

Often, the internal states of a complex system, represented by the state variables in the dynamic model $\mathscr{M}$, cannot be measured directly (because of measurement errors and selectivity problems). But with a suitable transfer function, the state variables x_t can be transformed into predictions of what can be measured empirically, y_t. These predictions can then be compared to actual measurements of y_t, and a lack-of-fit criterion defined. To find the right model for $\mathcal{M}$ and to estimate the optimal set of parameter values for a given system, this lack-of-fit criterion is minimized by repeating the simulation and it has to be repeated again and again. Traditionally, this has been a very tedious and difficult process.

Apparently, multivariate metamodeling can reduce this problem, since bilinear predictive metamodels, once established by simulation/regression, are much faster to run than large nonlinear ODE/PDE-based models, and a much larger number of combinations of parameter values can therefore be explored. Moreover, the metamodels produce effective measures of the sensitivity of the original differential equation model to the different model inputs and give maps of the correlation patterns between the different variables of the systems that are easy to overview and interpret. Metamodels may also be used directly to predict parameter values from experimental data. The following examples will show how PLSR extensions can be applied in various ways to the time domain. The first one is an illustration of autoregressive metamodeling.

2.2 PLS-Based Analysis of a Near-Chaotic Recursive System

The end of the eighties was perceived by many scientists as a post-modern revolution against what was considered as an overly reductive modeling tradition, idealizing classical physics, which had developed along with the modernism in western culture at large over the last century. Suddenly, a number of books on fractals, sensitivity to initial condition, positive feedback, cooperative processes, chaos and self-organization emerged. They demonstrated that apparently random processes—some of astonishing beauty—could be generated by deterministic sources in even very simple dynamic systems. This gave many "soft scientists" the courage to pursue "hard" sciences while maintaining respect for the complexity of reality.

This renewed scientific pragmatism, humility, and optimism corresponded well to the ethos already established in Chemometrics over the previous two decades, so many of us embraced it gladly. High computational capacity became available to anyone who wanted it and research funding was plentiful. This called for playful experimentation with models. On the other hand, the danger of fluffy philosophy and uncritical use of scientific concepts became clear: scientific focus on simplicity, interpretability, and reproducibility was needed now more than ever. And today's problems in communicating nonlinear mathematics to non-mathematical biologists were even greater then.

Per definition, the behavior of a chaotic process cannot be predicted very far into the future. But high-dimensional dynamic systems may have both chaotic and non-chaotic dimensions. Moreover, chaotic processes may sometimes be successfully forecast for a short time. From observations of a given system, how well can such short-term forecasting be expected to be?

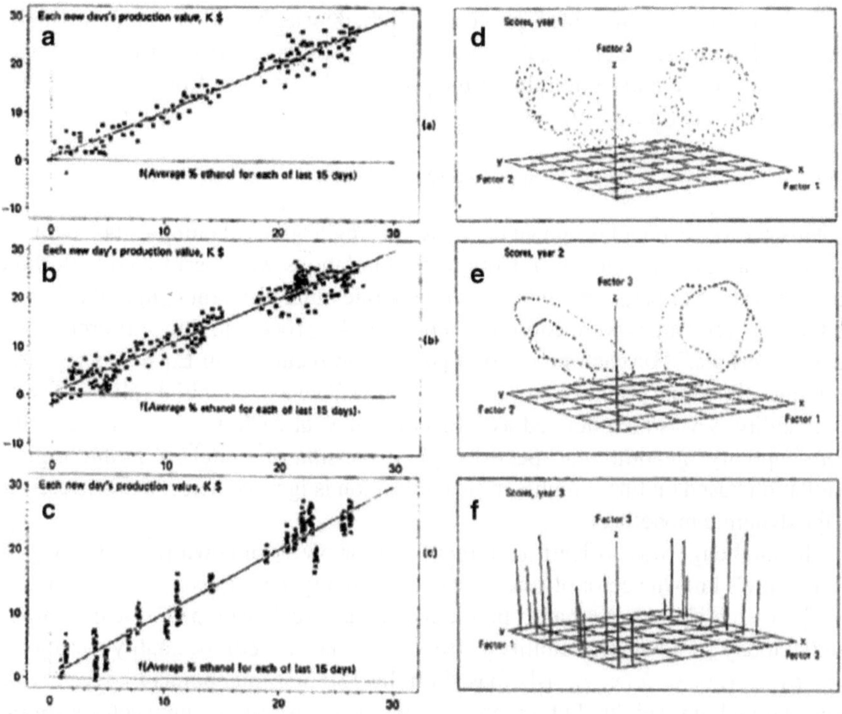

Fig. 3: An early foray into "soft" PLS-based metamodeling of a "hard" mathematical model: autoregressive PLSR forecasting in a near-chaotic system The simulated time series outputs from a highly nonlinear process model (a production process affected by a recursive Verhulst dynamics process) run under three different conditions (parameter a in Eq. 8 was equal to 2.625, 2.615, and 2.605, respectively). (**a**–**c**) Future value y_{t+1} forecast from present and past measurements $x_t + d$, $d = 0, 1, 2, \ldots, 14$ for the three conditions. (**d**–**f**) The scores of the three first PLS PCs, showing the trajectory of the process for the three conditions: (**a**) a near chaotic process. (**b**) a complicated limit cycle. (**c**) a fixed set of states repeatedly visited (vertically connected, for visual clarity) (From [17])

Figure 3 shows our first, feeble PLS-based analysis of the behavior of a highly nonlinear dynamic model, to test the possibility of forecasting future behavior of a complex system from present and past measurements. The following example was developed for an audience of applied process monitoring scientists: A hypothetical process was created so as to display three different degrees of complexity in its behavior. The purely computational example was designed to illustrate a hypothetical industrial whose profitability y_t was affected by variations in the quality of its product, x_t, over time $t = 0, 1, 2, \ldots$, according to some "unknown," underlying model $\mathscr{M}$. The question was: Can we forecast tomorrow's profitability y_{t+1} from measurements of today's and yesterday's quality $x_t, x_{t-1}, x_{t-2}, \ldots$?

Technically, the "unknown" process model $\mathscr{M}$ was based on simple recursive Verhulst dynamics, generating the so-called "logistic map:" For a given state variable x, the state at a discrete future time point $t+1$ is defined from its value at the previous discrete time point, t as

$$x_{t+1} = a \cdot x_t \cdot (1 - x_t). \tag{8}$$

This recursive model is known to create varying degrees of complex behavior for different values of parameter a. Computer simulations were performed with three different values of parameter a known to generate quite different temporal behavior patterns. These values were chosen to represent the process in three different "production periods." For each production period, the recursion in Eq. 8 was repeated 300 times (300 time points or "production days") from a fixed initial value of $x_t = 0$. Profitability y_t was then defined as a function of x_t at each day $t = 0, 1, 2, \ldots, 300$ (for simplicity, an additional spectroscopic expansion of model $\mathscr{M}$ in the original simulation based on linear multivariate calibration is ignored here, as it is irrelevant to the dynamic modeling).

The challenge was to learn to forecast tomorrow's unknown profitability y_{t+1} from today's known vector of present and past quality variations $[x_t, x_{t-1}, x_{t-2}, \ldots, x_{t-K}]$. A time-shifted PLS-based prediction model was developed to learn how to predict today's known profitability y_t from yesterday's vector of quality variations: $y_t = f\left([x_{t-1}, x_{t-2}, \ldots, x_{t-(K+1)}]\right)$, based on the 150 first time points $t = 1, \ldots, 150$. Conventional, un-weighted PLS regression was used, for regressing vector $\mathbf{y}$ (of order 150×1) on the matrix of the quality assessments $\mathbf{X}$ (of order 150×15) at the $K = 15$ previous days. Full cross-validation showed that three major and four minor PLS components were required to attain minimal error when forecasting $\mathbf{y}$ from $\mathbf{X}$.

Once the prediction model had been established, it was used to forecast tomorrow's profitability from today's vector of known quality variations: $y_{t+1} = f\left([x_t, x_{t-1}, x_{t-2}, \ldots, x_{t-15}]\right)$, for each of the next time points $t = 151, 152, \ldots, 300$. This was done for each of the three production periods. The resulting forecasting ability of the three production periods is shown in Fig. 3a–c. While the forecasts are not perfect, the results are quite good, given the erratically-looking raw time series from the Verhulst dynamics (not shown here).

However, the forecasts showed different structures for the three production periods. To shed more light on the nature of these differences, the scores for the first three PLS components were plotted for all 300 production days, to reveal the trajectories of the process. These are displayed in Fig. 3a–c for each of the three production periods, respectively. In all three cases, the process state is seen to wobble between two quite distinct regions. Within each of these two regions, the degree of complexity depended on the value of parameter a. In Fig. 3a it was very complex (two "strange attractors?"). At the intermediate value of a, the process settled in a complicated limit cycle (see Fig. 3b). The third value of the model parameter made the process repeatedly visit a limited number of discrete states (i.e., like a periodic attractor, see Fig. 3c). Hence, the complex Verhulst process generated some more or less "chaotic" behaviors in model $\mathscr{M}$. But in each case, the purely data-driven metamodels $\mathscr{A}$—obtained by reduced-rank PLS-based autoregressive modeling—revealed clear trajectory patterns.

This continuous-process simulation example in Fig. 3 showed us that—as expected—time-shift PLS regression of multichannel data from an unknown, highly nonlinear process may reveal complex, but systematic trajectory patterns, even in the absence of any causal mathematical theory. The score plots of the most relevant dynamic behavior of the system (Fig. 3d–f) as seen by autoregressive PLSR might have been difficult with traditional autoregressive time series analysis.

2.3 Data-Driven Development of the Essential Dynamical Model Kernel

The next example is an illustration of ODE metamodeling: To what extent would it be possible to formulate an explicit mathematical model from time series data? Can a PLS-based model be found that captures the essence of the systematic dynamics of the system, and would that obtained model be meaningful and not misleading?

Figure 4 shows a time-series oriented PLS-based metamodeling, presented at the PLS2009 meeting in Beijing [20]. The topic here was to see if it were possible to develop a meaningful dynamic model of the mechanisms controlling an unknown process, based on PLS regression of a limited amount of empirical time series data. In other words, using a "secret" nonlinear mechanistic models $\mathscr{M}$ to generate time series data by simulation, is it possible to use only these time series data to develop a metamodel $\mathscr{A}$ that in turn can reconstruct $\mathscr{M}$, or at least a nonlinear dynamic model that catches the essence of $\mathscr{M}$ and has similar behavior as $\mathscr{M}$ under the conditions simulated?

In this case the "unknown" model $\mathscr{M}$ was defined in terms of a "secret" nonlinear dynamic model relating three state variables x_1, x_2, x_3 to each other. When integrated numerically over n points in time, the model generated an "observed" time series data table of size $(n \times 3)$. The model $\mathscr{M}$ was a simple ODE, in which each of the three variables' rates $\frac{dx_k}{dt}$, $k = 1,2,3$, could be related to all three state variables x_1, x_2, x_3. To complicate matters, the ODEs in $\mathscr{M}$ were defined in a way that mimics biological complexity: the state-to-rate mechanisms were not constant: they depended on the process state itself (i.e., on its position $[x_1, x_2, x_3]$ in the three-dimensional state space).

A fractional factorial 2^{3-1} design was used for defining four different sets of the initial state vector $\mathbf{x}_0 = (x_{01}, x_{02}, x_{03})$, from which the model $\mathscr{M}$ was integrated numerically. Both the structure and parameter values of the model $\mathscr{M}$ were kept "secret" from the first author, who only did the metamodeling based on the time series data only.

From these data, a PLS-based regression model of rates $= f(\text{states})$ was developed, relating rates

$$\dot{\mathbf{x}}_t = \left[\frac{dx_{1,t}}{dt}, \frac{dx_{2,t}}{dt} \frac{dx_{3,t}}{dt}\right] \tag{9}$$

to states

$$\mathbf{x}_t = [x_{1,t}, x_{2,t}, x_{3,t}], \qquad t = 0, 1, \ldots, 100. \tag{10}$$

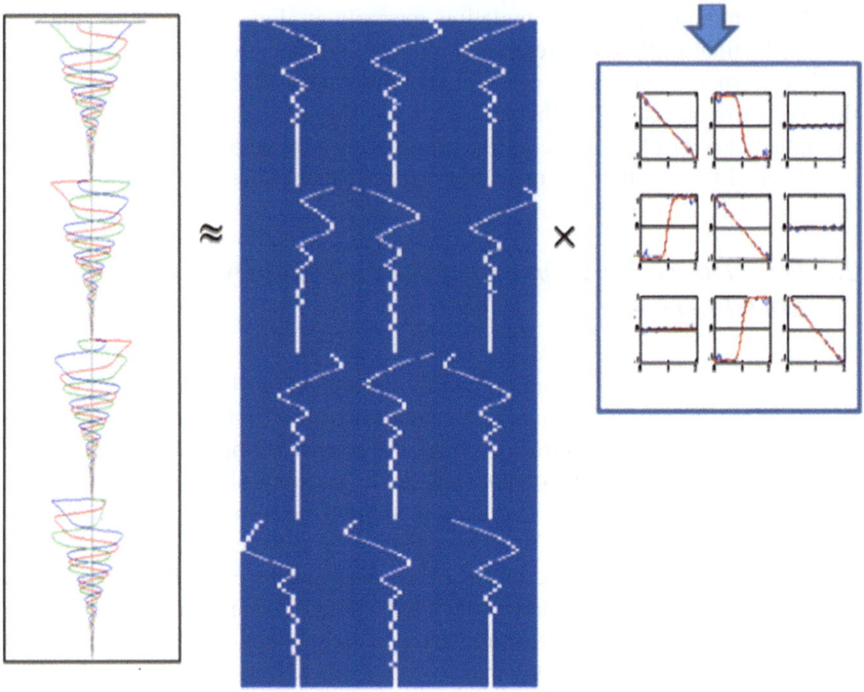

Fig. 4: Data-driven generation of nonlinear differential equation (ODE) system. Example of nonlinear (nominal-level) PLS regression. A complex system is here to be characterized in terms of a nonlinear dynamic mathematical model, linking three state variables. The input data consisted of four sets of time series, each containing them time series of the three different state variables obtained by numerical integration for a new set of initial states. Each of the 3 state variables were split into 20 category variables (white = 1, dark = 0) and the set of these 60 nominal variables were together used as **X**-variables. The three state variables were also differentiated with respect to time, and used as three **Y**-variables. Conventional PLS regression was employed based on the linear model of rates = $f(\text{states})$, (i.e., $\mathbf{Y} \approx \mathbf{XB}$). Cross-validation showed that four PCs gave optimal prediction of rates **Y** from state categories **X**. The nominal-level regression coefficients **B** at optimal PLS regression rank was finally split to show how different levels of each of the tree states affected each of the three rates (From [20])

If the underlying model had been known to be of the constant, linear type, the true, unknown $\dot{\mathbf{x}}_t = f(\mathbf{x}_t)$ would have been the same, irrespective of the values in $\mathbf{x}_t$:

$$\dot{\mathbf{x}}_t = \mathbf{xJ} \tag{11}$$

where the elements of the so-called Jacobian (i.e., the 3×3 matrix **J**) control the dynamic behavior of the system. This constant Jacobian could then have been estimated by conventional linear regression, namely by defining

$$\mathbf{Y} = \left[\frac{dx_{1,t}}{dt}, \frac{dx_{2,t}}{dt}, \frac{dx_{3,t}}{dt}\right], \qquad t = 0, 2, \ldots, n \tag{12}$$

and

$$\mathbf{X} = [x_{1,t}, x_{2,t}, x_{3,t}], \qquad t = 0, 1, \ldots, n. \tag{13}$$

The regression coefficients in a full rank linear regression (and hence a PLSR model with $A = 3$ components in Eq. 7) would yield estimates of the Jacobian matrix:

$$\hat{\mathbf{J}} = \hat{\mathbf{B}}_{A=3}. \tag{14}$$

In addition, estimates of the uncertainty standard deviations of the regression coefficients could have been used for estimating confidence intervals of the elements in the Jacobian.

However, in order to handle possibly nonlinear individual $\text{rate}_j = f(\text{state}_k)$ relationships, where the rates are not fixed for the system but instead change with the levels of the state variables, a nonlinear version of PLSR was needed. A second-order polynomial extension of **X** was tried, but did not give satisfactory results: obviously the nonlinearity at play here was not of the simple second-degree type.

So a more versatile non-linear PLSR version was developed instead (a balanced nominal-level PLSR): Instead of using the three original state variables directly as regressors, each of the three state variables was split into 20 category variables, and the resulting $3 \times 20 = 60$ 0/1-variables were used as regressors **X** (see, e.g., Fig. 4). Cross-validation was used for determining the optimal model rank A, and the "local Jacobian" matrix was obtained by partitioning the resulting nominal-level PLS regression coefficient matrix $\mathbf{B}_A$ as illustrated in Fig. 4.

We expected the nominal-level PLSR to yield an ODE model with good predictive ability, but not necessarily the "true" model. However, as it turned out, the partitioned regression coefficient matrix $\hat{\mathbf{B}}_A$ showed—to our surprise—a structure closely resembling the correct, "unknown" model form. In other words, in this particular case, the unknown model $\mathcal{M}$ was very well identified from the time series data only, both in terms of sign and curvature, within the estimated confidence limits. The true "unknown" model $\mathcal{M}$ had been defined by the following rates of change for the three state variables:

$$\begin{aligned} y_1 &= \tfrac{dx_1}{dt} = -x_1 \quad -1 - S(x_2) \; +0 \\ y_2 &= \tfrac{dx_2}{dt} = S(x_1) \; -x_2 \qquad\quad +0 \\ y_3 &= \tfrac{dx_3}{dt} = 0 \qquad +S(x_2) \qquad -x_3 \end{aligned} \tag{15}$$

where $S(.)$ defines a positive sigmoid (Hill) curve. The regression coefficients in Fig. 4 show that flat, near-zero curves were obtained for the non-existing rate-state $\frac{y_j}{x_k}$ relationships $\frac{y_1}{x_3}$, $\frac{y_2}{x_3}$, and $\frac{y_1}{x_1}$ (marked by 0 in Eq. 15). Positive sigmoids were found for relationships $\frac{y_2}{x_1}$, $\frac{y_3}{x_2}$, and a negative sigmoid for $\frac{y_1}{x_2}$, as specified in Eq. 15. Constant negative relationships were found for the self-degradation terms $\frac{y_1}{x_1}$, $\frac{y_2}{x_2}$, and $\frac{y_3}{x_3}$ again as expected.

To what extent was this unambiguous result caused by luck, or by the fact that the time series data were error free and highly informative? This experiment was repeated for five other "unknown" models, with the same result. Apparently, in these cases the time series data had sufficient information to constrain the system more or less completely. Using time series from more than just four initial conditions gave the same conclusion, with even smaller jackknife estimated parameter uncertainties.

In general we believe that one cannot expect a purely data-driven nominal-level PLSR-based time series analysis to be able to identify nonlinear dynamic systems completely. In very "sloppy," or over-parameterized, models one must expect ambiguity in empirical metamodeling solutions. But we do expect it to be possible to identify the relevant dynamic "essence" of an unknown nonlinear dynamic system, based on sufficiently informative, multivariate time-series data. This might give useful hints for how to model complex real-world systems (e.g., patients), from the analysis of time series data obtained by extensive monitoring with multichannel instrumentation. It might also be useful for reducing a large, overly detailed mathematical model $\mathscr{M}$, by determining which elements in $\mathscr{M}$ are essential and which may be safely ignored. Work is now in progress [33] to elucidate the ambiguity in fitting a given nonlinear dynamic model $\mathscr{M}$ to empirical time series data. Our preliminary results show that if a given non-linear dynamic model is over-parameterized relative to the information content in the available time series data, highly ambiguous parameter estimates can be obtained. For instance, for a set of error-free time series data, we were able to find a wide range of different parameter combinations that gave perfect fit. In addition, it seems that such sets of equivalent parameter combinations are highly structured.

That corresponds well with our previous studies of metamodeling of models generating curved temporal developments [10]. There, the "opposite" metamodeling process was employed. About 40 different mathematical models of widely different types, ranging from trigonometric functions, cumulative statistical distributions, growth curves, kinetic models and ODEs, each capable of generating widely different line curvatures (arches, sigmoid, etc.), were defined. Each of them was submitted to extensive computer simulations. When their thousands of output curves were combined in one very big data matrix and approximated by a PCA-based metamodel, the results showed that the behavioral repertoire of all 40 models could be fitted into one joint, simple "kernel" model. It appears that the behavior of this class of nonlinear mathematical models is far simpler than the diversity of mathematical forms within the class.

2.4 Dynamic Metamodeling of 3-Way Output Structures

For mathematical models whose output phenotypes have spatiotemporal character, the obtained output data come in an N-way array format (e.g., p state variable phenotypes $\times$ q points in space $\times$ m points in time $\times$ n input conditions). Can the good performance of bilinear approximation methods like PCA and PLSR analysis of two-way data tables be carried to data arrays with more than two ways?

The last example illustrates the use of $N-$ PLS [1–3] in inverse multivariate metamodeling. We have recently extended this method to handle highly nonlinear input-output relationships, as an analogy to the so-called Hierarchical Cluster-based PLSR (HC-PLSR, [35, 37]). The purpose of this two-step extension is to obtain improved predictive ability and increased insight into the input-output relationships for models with too complicated input-output relations to be approximated by normal PLS regression. In the first step in HC-PLSR, all the observations are used together, to develop one joint, "global" PLSR model. To pick up simple, smooth nonlinearities, this global PLSR step may be of the "polynomial" type, (i.e., it may optionally employ squares and cross-products of the original **X**-variables as extra **X**-variables).

To pick up more abrupt, drastic nonlinearities and other heterogeneities in the model's input-output relationship, the PLSR for the metamodeling is extended in a second step as follows: Based on a clustering on the scores from the first, global PLSR model, the objects are separated into local groups with more homogeneous, linear structures. For each such cluster of objects, a local PLSR submodel is then developed. To apply this hierarchical combination model to new observations, the new objects are first classified in **X** with respect to the submodels, and then **Y** is predicted from **X** via the submodels, either based on only the closest local submodel, or by a weighted sum of the predictions from all the relevant local submodels, with the cluster membership probabilities as weights. Thereby even abrupt nonlinearities can be successfully modeled.

The same type of two-step hierarchical clustering extension will now be illustrated for $N-$ PLS regression. Figure 5 illustrates the use of Hierarchical Cluster-based $N-$ PLS for inverse metamodeling of a complicated, state-of-the-art dynamic model $\mathscr{M}$ of the mammalian circadian clock (i.e., our biological day/night clock, [36]). The model contains a number of nonlinear ODE elements, coupled together in a complicated feedback structure. It was found that model $\mathscr{M}$ with different input parameters gave a wide range of output patterns. A preliminary, global inverse $N-$ PLS regression $\mathscr{A}_{\text{global}}$ gave the score plot in Fig. 5a. Here, each point actually represents a whole table multivariate time series (16 output phenotypes given at different timesteps in the simulation). This is an illustration of how multivariate subspace analysis can compress temporal data.

However, the preliminary, global metamodel $\mathscr{A}_{\text{global}}$ did not fit the data well enough and so it appears that the input-output relationship in model $\mathscr{M}$ did not follow a simple additive structure. To obtain a simpler and better metamodeling, the score plot in Fig. 5a was used for identifying six local, more easily modeled, subsets of objects. In Fig. 5b the simulated time series outputs are color coded according to this clustering. For each of these six clusters $c = 1,2,\ldots,6$, a local metamodel $\mathscr{A}_c$ was developed, again by inverse $N-$ PLS regression. This gave an informative separation of the simulations according to the temporal response of the analyzed dynamic model $\mathscr{M}$, and increased the predictive ability of the metamodel. Recently, a polynomial version of the Hierarchical Cluster-based (2-way) PLS regression was used in the converse direction for classical metamodeling for global hierarchical sensitivity analysis [37]. This revealed complex parameter interaction patterns in a model of the mouse heart muscle cell.

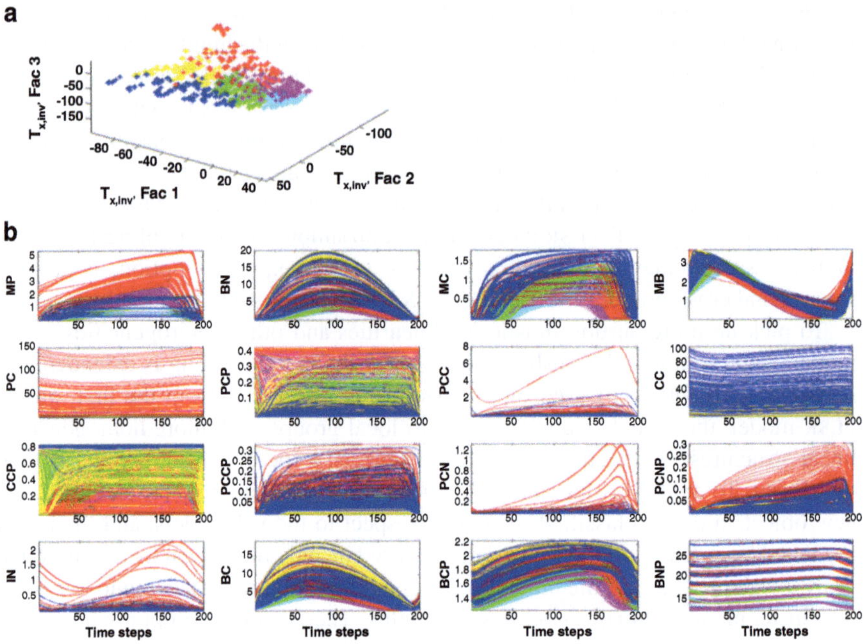

Fig. 5: Time-dimension in hierarchical PLSR: $N-$PLSR. Example of hierarchical N-way PLS regression used in inverse multivariate metamodeling of a dynamic model. The results from a fuzzy clustering on the **X**-scores from an $N-$PLS-based metamodel of a dynamic model of the mammalian circadian clock with six clusters are shown (from [36]). (**a**) Plot of the **X**-scores for the first three factors (Factors 1–3) from the global inverse metamodeling (where **X** is the 3-way state variable time series array and **Y** is the parameter data). The clustering was done on the first 19 score factors. The observations are colored according to their cluster memberships. (**b**) Circadian clock state variable time series for the observations belonging to each cluster, colored according to the cluster memberships. All state variables are given in *nM* units

Work is in progress to generalize and speed up the HC-based $N-$PLS regression modeling, based on N-way limitations pointed out by Wold ([39]) and on suggestions in [18]. But already, Bro's nonlinear N-way regression approach and its N-way PCA-extension ("PARAFAC," see [3]), appear to be versatile tools for metamodeling of nonlinear dynamic models. When implemented in the HC-based setting (Fig. 5) and combined with nominal-level modeling (Fig. 4) in a sparse setting (see, e.g., [9, 32]), we expect it to yield particularly simple metamodels, with reduced risk of false discovery. Applied to dynamic time series data as in Figs. 2, 3, or 5b, this approach can provide powerful "soft" multivariate metamodeling of "hard" mathematical models with high-dimensional spatiotemporal outputs. Hopefully, this will contribute to the realization of Herman Wold's vision for the integrative function of PLS methods (Fig. 1).

3 Discussion

Bilinear approximation: A proper science model? The low-rank structure models obtained by bilinear data-modeling are often referred to as "mathematical models" by PLS and PCA-practitioners. But for scientists trained in main-stream mechanistic modeling, our pragmatic, unassuming use of the term "model" may be perceived as alien. To what extent can bilinear approximation models be regarded as valid scientific models?

Throughout the last century, Physics was more or less implicitly taken as the role model by many other sciences. However, the traditional reductionist focus such as in Physics for example, preferring simplicity, homogeneity and generality, has created a wide-spread frustration among scientists working in more complex systems such as living cells. A search is now on for new scientific paradigms that retain science's critical, quantitative ambitions but which allows more rational handling of real-world systems. Robert Rosen ([27]) went as far as claiming that the life sciences should from now on be the ideal, from which scientists—including future physicists—should take inspiration. We share his vision. Herman Wold's original overview (see Fig. 1) indicates how the PLS principle could be used for interdisciplinary integration, particularly in the "holistic," integrative scientific setting outlined by Munck ([25]). But at the same time we argue for the use of powerful data analytic methodologies that can detect and quantify the mixture of known and unknown phenomena characterizing complex systems, while guarding against wishful thinking. And we recommend an increased use of mathematical modeling in the life sciences, with respect to formalizing our understanding of how Biology works (i.e., the dynamics of life itself). To simplify the mathematical modeling of complex biological processes, metamodeling may be used, primarily because it gives the scientists better control of the modeling process.

3.1 Cognitive Aspects of Temporal Modeling

The differences in models controlling scientists' thought models, perceptions, and practical use of mathematical modeling such as, for example, in Biology, are interesting from a cognitive science perspective [24]. One major perceptual and cognitive distinction appears to be whether a model is thought to describe the system from the "outside" or from the "inside." The difference in mental models in the two modeling cultures of "external" and "internal" system representation is enormous.

The "external," data-driven modeling in computational Statistics and Chemometrics is based inductively on empirical data: many properties described for many objects or situations. It can give a valuable overview of complex systems in a certain context, without the need for explicit causal theory or hypotheses. But the scope of the modeling is limited to the range and reliability of the empirical data, and the risk of false discovery can be high. Moreover, it reveals how the system appears externally, but gives little or no feel for how the system really works internally.

On the other hand, the "internal," theory-driven approach builds on a compact thought model of how the mechanisms in a system work. It requires much less new empirical data, since it builds primarily on theories or hypotheses derived from previous, already digested experimental data. Of course, if the employed theory is misleading then the mathematical model will also be misleading. Models are defined for different purposes, ranging from small, strongly simplified models of specific aspects of a system, to large models intended to give comprehensive, more or less realistic representations of the system.

When it comes to the modeling of time-varying systems, "external" statistical and Chemometrics analyses of observational data rely on more or less static models to assess its dynamics: A system is characterized by a set of "snapshot" data describing sets of objects, individuals or situations separated in time and space, and collected as time series data. These are summarized in terms of a statistical model, capable of describing the system top-down with respect to its observed dynamic behavior: "This is what we saw." The time series data may be described (modeled and/or plotted) in terms of their main intercorrelation structure (see Figs. 2b and 3) or related to time itself (see Figs. 2d and 5b). But little or no attempt is made to represent the system from the "inside"—formulating the causal mechanisms that explain how it actually works—"what influences what and in which way?"

On the other hand, "internal" modeling approaches such as, for example, in computational Biology, have the ambition of giving a deeper understanding of how nature works. For this purpose, one employs explicit mathematical modeling of spatiotemporal dynamics; the causality of the biological system is—tentatively—described, from the "inside," by a model $\mathscr{M}$. Because time is not considered a cause in itself, it is usually not parameterized explicitly; instead time is involved as index or argument t in the solutions describing the behavior of functions of t. These "bottom up" models are intended to describe—more or less precisely—how material, energy and information is obtained, utilized, and lost by the system. But this approach requires extensive computer simulations and "external" data modeling to see what the model is actually doing. Simple nonlinear dynamic models were employed to generate the data behind Figs. 3 and 4, while a full state-of-the-art model generated the data for Fig. 5.

3.2 Metaphors for Time

The second law of thermodynamics renders biological processes more or less irreversible. What happens at a given time in a given point in space will have consequences at several other points at several later times. And these points may, in turn, affect the initially given point again, forming feedback mechanisms with different time delays. Thus "everything may be related to everything" though a complex, time-varying web. To disentangle such a web of causalities and to distinguish causality from mere correlation is difficult [26] and may require intervention and a strict temporal control. However, as the previous illustrations indicate, the cacophony in data from a complex multivariate time series may be made more accessible by compact graphical representations of its underlying rhythms and harmonies. These may be identified by PLS based data modeling along the arrow of time, which can reveal the important inter-correlations and time-delay patterns. The following introduction to Fig. 3 was given (in 2008) by Martens and Martens [24]:

> In the Norwegian language we use the word "tidsrom," translating directly to "time-space," for the English word "time-span." This time-space concept of course has nothing to do with the four-dimensional time-space concepts of physics, consisting of three spatial and one temporal dimension. "Tidsrom" is a pure time concept, but the word "rom" (space, room) indicates that time somehow has more than one dimension in itself.
>
> This corresponds well with concepts from modern nonlinear dynamics as well as with our everyday experience. An object or phenomenon which on one time scale seems soft, pliable, variable will on another timescale appear hard, unyielding, constant. You can swim in the water, even dive into it, but you are knocked flat if you fall into it from an airplane.
>
> Therefore objects or phenomena (holons) with long time constants act as the solid, fixed framework of other holons with shorter time constants. Conversely, through the law of mass action, myriads of holons with short time constants form an apparently solid ("average") basis for the holons with longer time constants. It appears that this multivariate time structure is an important factor in the grand self-organization of our existence.

In Biology, processes take place on widely different time scales, from the near-instant passing of electrical fields between cells via the beating of the heart to the evolution of the heart over eons. In computational Biology, separation of the time scales is done by mathematical transformations from time to frequency, by spatiotemporal averaging and differentiation etc. This means that biological models may handle time in a variety of ways. One may argue that the difference between envisioning/seeing a dynamic system from the "outside" or "inside" is in particular determined by how prior theory is balanced against new empirical data. But distinctions may also be observed in how time is conceived.

How humans think about time is a matter of much interest in cognitive science, and discrepancies in temporal thought models may cause interdisciplinary conflict. The famous cognitive linguist George Lakoff and coworkers [15] have pointed out how mathematics in general is based on bodily metaphors (i.e., "how the embodied mind brings mathematics into being"). In particular, with respect to the mathematical handling of dynamics, Lakoff [14] once suggested that time can be understood in three different metaphors: (1) Time coming towards you (as seen from the locomotive of a train); (2) Time leaving behind you (as seen from the last wagon of the train); (3) Time passing in front of you (as if a train is seen moving across a flat landscape in the distance). His point was that each metaphor is useful, but mixing two or more metaphors uncritically may create confusion.

Remaining in this domain of train-spotting, we tentatively add some more metaphors: (4) Time moving along a trajectory, as a roller-coaster cart seen moving along its 3D track with t as a parameter (as opposed to the more linear metaphor 3, where time is thought of as an axis or variable); and (5) Time frozen as chronicles (as looking at the train's time table); (6) Time encapsulated in a train's blueprint (a model of the train's engine and wheels). These six time-metaphors are summarized in Table 1, with references to illustrations in this paper.

The choice of both the mathematical model form and graphical display mode reflects how a process is thought of scientifically. These choices will, in turn, affect the way results are perceived, interpreted, and remembered.

Generally speaking, time metaphors #1 and 2 are necessary steps in any dynamic study. But our perceptual and cognitive capacity is limited and so the other metaphors are needed to identify and reveal the essentials of the system. We believe that a conscious use of several of these metaphors can give access to a dynamical process from different angles. This can reveal unexpected patterns of system behavior inaccessible if using only one traditional time metaphors. Thereby it becomes easier to use the fruitful combination of mathematical, theory-driven modeling and statistical, data-driven modeling to its full advantage.

For instance, in model formulation, time may be used explicitly as a variable under metaphor #3, symbolized by the letter "t" in mathematical models of the type

$$y_t = f(t). \tag{16}$$

The model form and parameter values of $f(.)$ may be estimated from data or chosen from theory. Once established, this function may be used in forecasting

$$y_{t,\,\text{Future}} = f(t,\ \text{Future})\ . \tag{17}$$

Alternatively, the model formulation may, under metaphor #6, uses time only as an index identifying observed variables y_t or assumed state variables x_t. The interrelationships between these variables may then be quantified in various ways in different scientific cultures: by purely empirical statistical time series analysis (e.g., ARMA) semi-empirical cybernetic process approximations (e.g., Kalman Filter) or theory-driven causal mechanistic modeling (e.g., ODEs).

These three modeling traditions within the blueprint of metaphor #6 differ greatly in scope and ambition. For instance, of the three, the mechanistic modeling may be the more difficult, but has the highest ambition of description of the system deeply from "inside." Based on our experience till now, we believe that PLS-based methodology can contribute insight, stability and computational compaction in all three cultures, by identifying a dynamic system's relevant subspace dimensions and by using these for statistically stable and graphically accessible system descriptions.

How time is used in graphical displays also affects what kind of information can be gleaned from data (be it raw data or statistically obtained results). For instance,

Table 1: Some metaphors of time in train-spotting and in science

Metaphor #	Train view	Science view	Illustration
1	Look forward from inside locomotive	Compute a recursive/ ODE model	Data for Figs. 3–5
2	Look backward from inside last wagon	Record time series data	Figure 2a
3	See train cross a plain, from a hill-side	Plot or model measurements y_t vs. time	Figures 2d and 5b
4	See roller-coaster spiral, from hill-side	Plot trajectory in state subspace	Figures 2b and 3a–c
5	See train's time table	Just looking at numbers	Data tables
6	Plan the train's engine & wheels	Develop a recursive/ODE model	Equations 11, 14, and 15

the way the axes are chosen and data points are represented in a bivariate plot can affect the way the process is understood. Plotting time series variables with time itself as "abscissa" (horizontal axis) represents metaphor # 3: a bird's eye view of a temporally continuous process, viewed from far away (Figs. 2d, 5b). This graphical use of time makes the plots easy to read: using time as "x-axis" gives the mind a solid "floor" to stand on. But this wastes 50 % of the 2D graphical dimension-capacity on something that is already known: time. Plotting instead the time series data in a state (sub-) space (metaphor # 4) allows the mind to see complex multivariate nonlinear behaviors in the process trajectory and this can give a very different types of insight (Figs. 2b, 3a–c).

4 Conclusions

Quantitative scientific knowledge is presently being accumulated in terms of large repositories of measurements, ontologies and models. Mechanistic mathematical modeling is increasingly employed to encapsulate scientific knowledge about complex biological systems. Multivariate metamodeling, based on data from large, well-designed computer simulations, can facilitate this modeling process, by making it easier to overview what nonlinear dynamic models actually do, to compare alternative models, to reduce the computational load and to fit models to large amounts of data from measurements and ontologies. With proper extensions for handling strong non-linearities, PLS-regression and extensions thereof provide one useful alternative for such metamodeling.

To facilitate communication between different science communities, one should be aware of differences in how prior theory is employed, and in the metaphors used for envisioning time.

References

[1] Andersson CA and Bro R (2000). The N-way Toolbox for MATLAB. *Chemometrics and Intelligent Laboratory Systems*, **52**, 1–4.
[2] Bro R (1996): Multiway calibration. Multilinear PLS. *Journal of Chemometrics,* **10**, 47–61.
[3] Bro R and Kiers HAL (2003) A new efficient method for determining the number of components in PARAFAC models. *Journal of Chemometrics,* **17**, 274–286.
[4] Cacuci DG (2003). *Sensitivity and Uncertainty Analysis: Theory Vol 1.* New York, Chapman and Hall / CRC.
[5] Cacuci DG, Ionescu-Bujor M, Navon IM (2005). *Sensitivity and Uncertainty Analysis: Applications to Large-scale Systems Vol 2.* New York, Chapman and Hall / CRC.
[6] Gorissen D, Crombecq K, Couckuyt I, and Dehaene T (2009). Automatic Approximation of Expensive Functions with Active Learning (url). In: *Foundations of Computational Intelligence Volume 1: Learning and Approximation: Theoretical Foundations and Applications, Part I: Function Approximation* (A-E. Hassanien, A. Abraham, A.V. Vasilakos, and W. Pedrycz,eds). Berlin, Springer Verlag, (pp. 35–62). See also http://www.sumo.intec.ugent.be/?q=sumo_toolbox.

[7] Hotelling H (1933). Analysis of a complex of statistical variables into principal components. *Journal of Educational Psychology*, **24**, 417
[8] Hunter P, Coveney PV, de Bono B, Diaz V, Fenner J, Frangi AF, Harris P, Hose R, Kohl P, Lawford P, McCormack K, Mendes M, Omholt S, Quarteroni A, Skår J, Tegner J, Randall Thomas S, Tollis I, Tsamardinos I, van Beek JHGM, and Viceconti M (2010). A vision and strategy for the virtual physiological human in 2010 and beyond. *Philosophical Transactions of the Royal Society A: Mathematical, Physical and Engineering Sciences*, **368**, 2595–2614.
[9] Indahl U (2005). A twist to partial least squares regression. *Journal of Chemometrics*, **19**, 32–44.
[10] Isaeva J, Martens M, Sæbø S, Wyller, JA, and Martens, H (2012). The modelome of line curvature: Many nonlinear models approximated by a single bi-linear metamodel with verbal profiling. *Physica D: Nonlinear Phenomena*, **241**, 877–889.
[11] Jøreskog, K. and Wold, H., (eds.) (1982). *Systems under Indirect Observation. Causality, Structure. Prediction.* Amsterdam, North-Holland.
[12] Kleijnen JPC (2007). *Design and Analysis of Simulation Experiments.* New York, USA, Springer.
[13] Kohler, A., Sulé-Suso, J., Sockalingum, G.D., Tobin, M., Bahrami, F., Yang, Y., Pijanka, J., Dumas, P., Cotte, M., van Pittius, D.G., Parkes, G., and Martens, H. (2008). Estimating and correcting Mie scattering in synchrotron-based microscopic FTIR spectra by extended multiplicative signal correction (EMSC). *Applied Spectroscopy*, **62**, 259–266.
[14] Lakoff G. (1989) Personal communication to H. Martens
[15] Lakoff G and Nunez RE (2004). *Where mathematics comes from: How the embodied mind brings mathematics into being*. New York, Basic Books.
[16] Martens, H. (1979). Factor analysis of chemical mixtures. Non negative factor solutions for spectra of cereal amino acids. *Annals Chemica Acta,* **112**, 423–442.
[17] Martens H and Martens M (1993) NIR spectroscopy: applied philosophy. Introductory chapter. In K.I.Hildrum,, T. Isaksson, T.Naes and A.Tandberg, (Eds.) *Near Infra-Red Spectroscopy. Bridging the Gap between Data Analysis and NIR Applications*. Chichester, UK, Ellis Horwood. (pp 1–10).
[18] Martens H and Næs T (1989). *Multivariate Calibration*. Chichester (UK): Wiley.
[19] Martens H and Martens M (2001). *Multivariate Analysis of Quality. An Introduction.* Chichester, UK, Wiley.
[20] Martens H (2009). Non-linear multivariate dynamics modeled by PLSR. In V.E.Vinzi, M.Tenenhaus and R.Guan, (eds.) *Proceedings of the 6th International Conference on Partial Least Squares and Related Methods,* Beijing 4–7, 2009. Publishing House of Electronics Industry, http://www.phei.com.cn, pp. 139–144.
[21] Martens and Kohler, A (2009). Mathematics and Measurements for High-throughput Quantitative Biology. *Biological Theory*, **4**, 29–43.
[22] Martens H, Veflingstad SR, Plahte E, Martens M, Bertrand D, and Omholt SW (2009). The genotype-phenotype relationship in multicellular pattern-generating models: The neglected role of pattern descriptors. *BMC Systems Biology*, **3**, 87. doi:10.1186/1752-0509-3-87.
[23] Martens H, Måge I, Tøndel K, Isaeva J, Høy M, and Sæbø S (2010). Multi-level Binary Replacement (MBR) design for computer experiments in high-dimensional nonlinear systems. *Journal of Chemometrics*, **24**, 748–756.
[24] Martens M and Martens H (2008). The senses linking mind and matter. *Mind and Matter*, **6**, 51–86.
[25] Munck L (2007). A new holistic exploratory approach to systems biology by Near Infrared Spectroscopy evaluated by Chemometrics and data inspection. *Journal of Chemometrics*, **21**, 406–426.
[26] Pear L (2009). *Causality: Models, Reasoning and Inference*. Cmabridge: Cambridge University Press.
[27] Rosen R (1991). *Life Itself. A Comprehensive Inquiry into the Nature, Origin and Fabrication of Life*. Columbia, Columbia University Press.
[28] Saltelli A, Chan K, and Scott (2000). *EM: Sensitivity Analysis*. New York, NY, Wiley.

[29] Sarkar AX and Sobie EA (2010). Regression analysis for constraining free parameters in electrophysiological models of cardiac cells. *PLoS Computational Biology*, **6**.
[30] Sobie EA (2009). Parameter sensitivity analysis in electrophysiological models using multivariable regression. *Biophysical Journal,* **96**, 1264–1274.
[31] Stone M (1974). Cross-validatory choice and assessment of statistical predictions. *Journal of the Royal Statistical Society B*, **36**, 111–147.
[32] Sæbø S, Almøy T, Aarøe J and Aastveit AH (2008). ST-PLS: a multi-directional nearest shrunken centroid type classifier via PLS. *Journal of Chemometrics,* **22**,54–62.
[33] Tafintseva, V., Tøndel K, Ponosov A, and Martens H (in preparation).
[34] Tøndel K, Gjuvsland A B, Måge I and Martens H (2010). Screening design for computer experiments: Metamodeling of a deterministic mathematical model of the mammalian circadian clock. *Journal of Chemometrics*, **24**, 738–747.
[35] Tøndel K, Indahl UG, Gjuvsland AB, Vik JO, Hunter P, Omholt SW, and Martens H (2011). Hierarchical Cluster-based Partial Least Squares Regression is an efficient tool for metamodelling of nonlinear dynamic models. *BMC Systems Biology,* **5**, 90.
[36] Tøndel K, Indahl UG, Gjuvsland AB, Omholt SW, and Martens H (2012). Multi-way metamodelling facilitates insight into the complex input-output maps of nonlinear dynamic models. *BMC Systems Biology*, **6**, 88.
[37] Tøndel K, Vik JO, Martens H, Indahl UG, Smith N, and Omholt SW (2013). Hierarchical multivariate regression-based sensitivity analysis: a n effective tool for revealing complex parameter interaction patterns in dynamic models. *Chemometrics and Intelligent Laboratory Systems*, **120**, 25–41.
[38] Wold H (1983). Quantitative systems analysis: The pedigree and broad scope of PLS soft modeling. In H. Martens and H. Russwurm, (eds.) *Food research and data analysis*. London, Applied Science Publisher LTD, p. 409.
[39] Wold, S (1974). A theoretical foundation of extrathermodynamic relationships (linear free energy relationships). *Chemica Scripta*, **5**, 97–106.
[40] Wold S, Martens H and Wold H (1983). The multivariate calibration problem in chemistry solved by the PLS method. In A. Ruhe and B. Kågström, (Eds.) *Proceedings of the Conference on Matrix Pencils*. Heidelberg, Springer Verlag. (pp 286–293).
[41] Ye KQ (1998). Orthogonal column Latin hypercubes and their application in computer experiments. *Journal of the American Statistical Association*, **93**, 1430–1439.

You Write, but Others Read: Common Methodological Misunderstandings in PLS and Related Methods

George A. Marcoulides and Wynne W. Chin

Abstract PLS and related methods are currently enjoying widespread popularity in part due to the availability of easy to use computer programs that require very little technical knowledge. Most of these methods focus on examining a fit function with respect to a set of free or constrained parameters for a given collection of data under certain assumptions. Although much has been written about the assumptions underpinning these methods, many misconceptions are prevalent among users and sometimes even appear in premier scholarly journals. In this chapter, we discuss a variety of methodological misunderstandings that warrant careful consideration before indiscriminately applying these methods.

Key words: Structural equation models, Path models, Confirmatory factor analysis, Multiple regression, Path analysis, Covariance structure analysis, Latent class, Mixture analysis, Equivalent models, Power, Model identification, Formative indicators, Reflective indicators, Mode A, Mode B, Scale invariance

1 Introduction

We begin this section with a short story, which led to the central theme and focus of the chapter. Not long ago, the first author received an action letter from a journal editor concerning a manuscript he had submitted for publication consideration.

G.A. Marcoulides (✉)
Research Methods and Statistics, Graduate School of Education
and Interdepartmental Graduate Program in Management, A. Gary Anderson Graduate School of Management, University of California, Riverside, CA 925121, USA
e-mail: georgem@ucr.edu

W.W. Chin
Department of Decision and Information Systems, C. T. Bauer College of Business, University of Houston, Houston, TX 77204-6021, USA
e-mail: wchin@uh.edu

H. Abdi et al. (eds.), *New Perspectives in Partial Least Squares and Related Methods*, Springer Proceedings in Mathematics & Statistics 56, DOI 10.1007/978-1-4614-8283-3_2,

The editor was essentially asking for assistance in tackling a comment provided by one of the manuscript reviewers. The reviewer's comment specifically stated that "...you did not do this analysis correctly you need to follow the procedures outline in Marcoulides [1]...." You can imagine the dismay, especially since the procedures were not outlined in the article in the manner referenced by the reviewer. When the editor was contacted and told that the original writings had been misunderstood by the reviewer, he responded with instructions that "...perhaps you should write more clearly next time..."

Interestingly enough, this experience mirrors one of our favorite passages provided by Galton [2] in his book *Natural Inheritance*:

> ...some people hate the name statistics, but I find them full of beauty and interest. Whenever they are not brutalized, but delicately handled by the higher methods, and are warily interpreted, their power dealing with complicated phenomena is extraordinary (p. 62).

Galton's words are as *à propos* today as when he wrote them. Whenever issues examined are complex, both theoretical and procedural, people with limited information and/or limited knowledge will likely develop misunderstandings. Like a rumor that contains half-truths, conceptualizations and insights often contain partially correct information. Unfortunately as characterizations of what constitutes good research, they can often lead people away from important understandings (e.g., see the detailed commentary offered by Marcoulides, Chin, and Saunders [3] on comparisons between various modeling techniques).

PLS, and related methods currently enjoy widespread popularity in the behavioral, information, social, and educational sciences. The number of contributions to the literature in terms of books, book chapters, and journal articles applying or attempting to develop extensions to these methods are appearing at an incredible rate. A major reason for their appeal is that these methods allow researchers to examine models of complex multivariate relationships among all types of variables (observed, weighted component, or latent) whereby the magnitude of both direct and indirect effects can be evaluated. Another reason is due to the availability of easy to use computer programs that often require very little technical knowledge about the techniques. For example, programs such as AMOS [4], EQS, [5], LISREL, [6], LVPLS, [7], Mplus, [8], Mx, [9], PLS-Graph, [10, 11], PLS-GUI, [12, 13], RAMONA, [14], SAS PROC CALIS, [15], SEPATH, [16], Smart-PLS, [17], VisualPLS, [18, 19], and XLStat PLSPM, [20], are all broadly available for the analyses of models. Using these programs a variety of complex models can be examined and include confirmatory factor analysis, multiple regression, path analysis, models for time-dependent data, recursive and non-recursive models for cross-sectional, longitudinal data and multilevel data, covariance structure analysis, and latent class or mixture analysis to name a few. A frequent assumption made when using these models is that the relationships among the considered variables are linear (although modeling nonlinear relationships is also becoming increasingly popularity, for details see, [21]).

Unfortunately, once advanced modeling methods become widely available in easy to use software packages, they also tend to be quickly abused. Although these methods are based on a number of very specific assumptions and much has been written about adhering to the assumptions underpinning these techniques, many misconceptions are prevalent among users and sometimes even appear in premier scholarly journals. Applied researchers conducting analyses based on these modeling methods generally proceed in three stages: (i) a theoretical model is hypothesized, (ii) the overall fit of data to the model is determined, and (iii) the specific parameters of the proposed model are evaluated. A variety of problems might be encountered at each of these stages, especially if assumptions are not kept in mind and examined. Indeed as indicated by Marcoulides et al. [3], applied researchers often do not pay sufficient attention to the stochastic assumptions underlying particular statistical models. This lack of attention to espoused statistical theory can also divert focus from precise statistical statements, analyses, and applications, by purporting to do what cannot be legitimately done with the particular recommended approach and even encouraging others to engage in similar activities. An example that immediately comes to mind is the investigation that attempted to make comparisons between modeling methods provided in Goodhue, Lewis, and Thompson (GLT, [22]). In this investigation, the authors ignored basic issues that most statisticians would consider essential preliminaries to any attempt to apply these methods in practice. So does this suggest that applied investigators (and for that matter, those acting as reviewers, since this manuscript made it through a full review cycle) do not pay much attention to the fundamental assumptions behind the modeling approaches? Is it because they find it difficult to follow the original sources or have misunderstood the writings of the original developers leading to attempts to legitimize bad practices simply because they lack a deep understanding of the mathematical details? In either case, when Marcoulides et al. [3] objected to the inappropriateness of the GLT comparisons, describing them as attempts to compare apples with oranges, the associate editor handling the original manuscript submission stated that "simply pointing out the flaws in GLT is not an effective critique without the inclusion of a direction towards the solution (January 30, 2012, AE Report)." It seems that this associate editor and the overseeing senior editor most likely had not understood the mathematical details of the methodology. Furthermore, they did not realize that in this situation no exact solution was possible and only an approximation was possible (for approximation details see [23]).

The purpose of this chapter, therefore, is to discuss a variety of methodological misunderstandings that warrant careful consideration before applying these methods indiscriminately and obtaining inappropriate interpretations. Naturally many of these issues have been addressed before, in one form or another, either by us or by a number of other researchers. Thus, although in general this paper may be viewed as containing nothing that has not been said before, our approach to the topics is intended to be more informative and didactic. For additional details and more abstruse discussion, we refer readers to the original sources and other essential outlets.

The issues to be addressed here are: (i) modeling perspectives for conducting analyses, (ii) equivalent models, (iii) sample size, (iv) identification issues, (v) myths about coefficient α, (vi) the use of correlation and covariance matrices, and, finally, (vii) comparisons among PLS and related methods. Extensive listing to resources that are readily available in the literature outlining all the issues in detail will be avoided including how one can legitimately use these methods. Also barring inclusion of statements such as "do not try this at home" or "use at your own risk" on all commercially available software programs, in what follows we attempt to summarize main concerns and provide guidelines towards using PLS and related methods. While many of our comments will occasionally be quite critical and may even come across as inappropriate and derogatory, but, as indicated by Cudeck [24], we believe that "it is good for one's character, not bad for it, to acknowledge past errors and clearly be capable of learning (p. 317)." Steiger [25] compared entry into the practice of using these modeling methods as akin to trying to merge onto a busy superhighway filled with large trucks and buses driving fast in reverse. Without doubt, the knowledge base required for understanding such analyses is continually expanding, but it is essential if one is to avoid professional embarrassment.

2 Overview of Modeling Perspectives for Conducting Analyses

The fitting and testing of any theoretical model can be considered from three general modeling perspectives or approaches [26]. The first is the so-called strictly confirmatory approach in which a single initially proposed theoretical model is tested against obtained empirical data and is either accepted or rejected. The second situation is one in which a finite number of competing or alternative theoretical models are considered. All proposed models are assessed and the best is selected based upon which model fits the observed data best using any number of currently available fit criteria. The third situation is the so-called model generating approach in which an initially proposed theoretical model is repeatedly modified until some acceptable level of fit is obtained. Of course, we strongly believe that the decision regarding which approach to follow should generally be based on the initial theory. A researcher who is firmly rooted in his or her theory will elect a different approach than one who is quite tentative about the various relationships being modeled or one who acknowledges that such information is what actually needs to be explored and determined. Nevertheless, once a researcher has determined that an initially proposed model is to be abandoned, the modeling approach is no longer confirmatory. Under such circumstance, the modeling approach has clearly entered an exploratory mode in which revisions to the model occur, either by simply adding and/or removing parameters in the model or even completely changing or modifying the initially proposed model both in terms of latent variables, observed variables, and/or their path connections and correlations.

The notion of changing aspects of a PLS model fits quite well with the original ideologies of its founder Wold [27] who indicated that "... PLS is primarily designed for research contexts that are simultaneously data-rich and theory skeletal (p. 26), ... it is an evolutionary process," one in which "... at the outset the arrow scheme ... is more or less tentative ...." Indeed, Wold [27] saw absolutely nothing wrong with "... getting indications for modifications and improvement, and gradually consolidating the design ..." until a final model is selected. For example, if the values of the loadings for a latent variable show high correlations with an observed variable that has not yet been considered for inclusion as one of its indicators, this might be subsequently deemed worthy of inclusion among its indicators. In a simple confirmatory factor analytic model $\Sigma_{xx} = \Lambda\Phi\Lambda' + \Theta$, where Σ_{xx} is the covariance/correlation matrix of the observed **x** variables, Λ the factor loading matrix, Φ the factor correlation matrix, and Θ is the error matrix, which is commonly set up by a priori imposing a number of restrictions on Λ and Φ, such an approach would entail changing (fixing or freeing) other additional aspects of these particular matrices (for further details on model restrictions in factor analysis, see [28]).

When considering competing theoretical models, the number of possibilities to compare are feasible for small sets of variables. For example, with only two observed variables, there are only four possible models to examine. For 3 variables, there are now 64 total possible models to examine. However, with more variables in play, the number of possible model combinations can become prohibitively large. For example, even with just 6 observed variables there are $1,073,741,824$ possible model combinations to examine. One way to think about the total number of models among P investigated variables is to consider the number of possible ways each pair can be connected, to the power of the number of pairs of variables and is determined by $4^{[P(P-1)/2]}$ [29]. Nonetheless, when examining all possible models becomes impractical, various heuristic optimization or automated search algorithms can be used [30]. Although heuristic search algorithms are specifically designed to determine the best possible model based upon some objective function solution, they do not guarantee that the optimal solution is found—though their performance using empirical testing or worst cases analysis indicates that in many situations they seem to be the only way forward to produce concrete results [31]. So as the models become more complicated, automated procedures can at least make "chaotic situation(s) somewhat more manageable by narrow(ing) attention to models on a recommendation list" ([32], p. 266). Heuristic model search procedures have recently made their way into the general modeling literature. Examples of such numerical heuristic model search procedures include: ant colony optimization ([33–36]), genetic algorithms [37], ruin-and-recreate [38], simulated annealing [39], and Tabu search [30, 40]—and over the years a great variety of modifications have been proposed to these procedures (e.g., [41, 42]). As indicated, all of these methods focus on the evaluation of an objective function, which is usually based upon some aspect of model fit (e.g., the Lagrange multiplier or the Stone-Geisser Criterion; for additional details see [43]).

Model searches and modifications are extremely difficult especially whenever the number of possible variables and potential models are high. Thus, automated algorithms have the potential to be quite helpful for examining models, particularly where all available information has been included in a specified model and when this information is not sufficient to obtain an acceptable model fit. Nevertheless—despite the fact that such searches can usually determine the best models according to a given fit criteria—all final generated models must be cross-validated with new data before any real validity to the final models can be claimed. This is quite important as specification searches are completely "data-driven exploratory model fitting" and, as such, can capitalize on chance [44]. For example, in cases where equivalent models are encountered (see next section), such searches will only lead one to a list of feasible models and then it becomes the responsibility of the researcher to decide which model to accept as the best model. To date, no automated search can make such a decision for a researcher. As noted by Marcoulides et al. [30], as long as researchers keep in mind that the best use of model searches is to narrow attention to a reduced list (a sort of top-ten list), the algorithms will not be abused in empirical applications. More research on these algorithms is clearly needed to establish which one works best with a variety of models. For now, we believe that the Tabu search is one of the best available automated specification search procedures to provide valuable assistance in modeling applications. Unfortunately, to date no commercially available program offers this automated search option, although many programs do provide researchers with some rudimentary options to conduct specification searches and improve model to data fit.

3 Equivalent Models

While equivalent models has also received considerable methodological attention over the past couple of decades, it does not seem to be well understood or even considered by applied researchers using modeling techniques (e.g., [45–58]). Equivalent models are a set of models that yield identical (a) implied covariance, correlation and other observed variable moment matrices when analyzing the same data, (b) residuals and fitted moment matrices, and (c) fit function and related goodness-of-fit indices (e.g., chi-square values and p-values). Distinguishing between equivalent models cannot be achieved simply by using any currently available fit indices. Model equivalence can only be realistically managed via substantive considerations and/or considerations pertaining to design and data collection features (apart from the case of multiple-population versions of single-group equivalent models, where statistical distinction becomes possible with appropriate group constraints, if substantively correct [54]).

Two hypothesized models (denoted simply as M_1 and M_2), would be considered equivalent if the model implied covariance or correlation matrices are identical (which can be written as $\hat{\Sigma}_{M_1} = \hat{\Sigma}_{M_2}$). Let us consider this notion in the following equation:

$$\Sigma(\underline{\theta}) = \Lambda\Phi\Lambda' + \Theta = \Lambda(I_q - B)^{-1}\Psi(I_q - B')^{-1}\Lambda' + \Theta \tag{1}$$

where $\Sigma(\underline{\theta})$ is the model implied matrix (i.e., either $\hat{\Sigma}_{M_1}$ or $\hat{\Sigma}_{M_2}$), Λ the factor loading matrix, Λ' its transpose, Φ the factor correlation matrix, Θ is the error matrix, B is the matrix of structural regression coefficients relating the latent variables between themselves, Ψ is the covariance matrix of the structural regression residuals, and I_q is the $q \times q$ identity matrix (where q is the number of latent variables in the model, with the usual assumption the matrix $I_q - B$ is full rank). This equation implies that different matrices appearing in its right-hand side may lead to identically reproduced covariance/correlation matrices in its left-hand side. This is because from the sums and products of the matrices one cannot uniquely deduce the individual matrices on the right-hand side of the equation.

This statement highlights the fact that model equivalence is not defined by the data, but rather by an algebraic equivalence between hypothesized model parameters. In turn, because of this model equivalence, the values of any considered statistical tests or goodness-of-fit indices of model fit will always be identical. Thus, even when a hypothesized model fits well according to the examined fit criteria, there can still be other equivalent models with identical fit—even if the theoretical implications or substantive interpretations of those models are radically different. In fact, as presented by Raykov and Marcoulides [55], there may even potentially be an infinite series of equivalent models to an initially hypothesized one. Identifying equivalent models can be a very difficult and time-consuming task. But there is clearly a compelling reason for undergoing such a difficult activity. Unfortunately, many researchers conducting various modeling activities do not seem to realize that alternative models might exist and that these others need to be considered.

A number of researchers have proposed a taxonomy that can be used to distinguish among several different types of equivalent models: namely, *observationally equivalent* and *covariance equivalent* (see, e.g., [49, 53, 57, 59] to name but a few). Two models are considered *observationally equivalent* only if one model can generate every probability distribution that the other model can generate. Observational equivalence is model equivalence in the broadest sense, and can be shown using data of any type. In contrast, models are considered *covariance equivalent* if every covariance (correlation) matrix generated by one model can be generated by the other. Thus, observational equivalence encompasses covariance (model) equivalence; that is, observational equivalence requires the identity of individual data values, whereas covariance equivalence requires the identity of summary statistics such as covariances and variances. We note that observationally equivalent models are always going to be covariance equivalent, whereas covariance equivalent models might not necessarily be observationally equivalent. Additional distinctions made include the mathematical notions of global and local equivalence, thereby signifying *globally equivalent* models and *locally equivalent* models. For two models to be globally equivalent, *a function* must exist that translates *every* parameter of one model into the parameters of another model. If only a *subset* of one model's parameter set is translatable into the parameter set of another model, the models are then considered *locally equivalent*. Local equivalence does not guarantee that the implied covariance or correlation matrices of the two models will be the same.

Categorizations of strategies for approaching the problem of equivalent models that have been considered in the extant literature include: (i) those that occur either before data collection, and (ii) those after data collection. The strategies consist of the four rules developed by Stelzl [60] and the more general rule by Lee and Hershberger [47], those based on graph theory that translate the model relationships into statistical relations (see [29, 61, 62]), the rank matrix approach that uses the rank of the matrix of correlations among the parameters of the proposed model [63], the data mining type automated heuristic searches [42], the information complexity criterion (ICOMP, [58]) with the one providing the lowest value representing the least complex of models, those that use computational problems associated with model misspecification as a way to distinguish among equivalent models (e.g., [64]), the comparison of the R^2 values among models [6], and the examination of extended individual case residuals (EICR, [65]). Of course, as expected, each of these strategies has their proponents and opponents. Regardless of which approach is used, we believe that the consideration of equivalent models must become a standard part of the process of defining, testing, and modifying models. Unfortunately, and despite nearly decades of reports arguing convincingly for the importance of model equivalence in the model-fitting enterprise, to date many researchers do not take the extra effort to even consider the potential presence of equivalent models. We strongly believe that researchers must take the initiative and effort required to thoroughly examine the potential presence of equivalent models. We also strongly believe that replication and cross-validation of models are additional essential activities when utilizing such advanced modeling techniques. Perhaps the best affirmation of this ideology was provided by Scherr, who declared that

> ... the glorious endeavor that we know today as science has grown out of the murk of sorcery, religious ritual, and cooking. But while witches, priests, and chefs were developing taller hats, scientists worked out a method for determining the validity of their results: they learned to ask: Are they reproducible ([66], p. *ix*)?

4 Sample Size Issues

The issue of sample size and model identification are very different and separate issues. We emphatically declare that even a study using a large fraction from a population of interest (e.g., $N = 10{,}000$) may still posit a model for which parameters cannot be determined. This is because the issue of model estimation is closely tied to the issue of model identification and not sample size. Sample size is tied to power and stability of estimates whereas model identification is tied to existence and uniqueness of a solution (see details provided in the next section). This appears to be an issue that many researchers regrettably confuse as equivalent when it is in fact not at all comparable.

The recent PLS and related modeling literature is replete with examinations and discussions (some bad, some good) concerning the performance of PLS analyses with various sample sizes (e.g., [67–75]). Indeed, and despite popular belief, the evidence is quite clear that PLS like any other statistical technique is in no way immune to the distributional assumption concerning the need for an adequate sample size. This goes back to Hui and Wold [72] who determined that PLS estimates improved and their average absolute error rates diminished as sample sizes increased. Similarly, Chin and Newsted [70] determined that small sample sizes (e.g., $N = 20$) do not permit a researcher to detect low valued structural path coefficients (e.g., 0.20) until much larger sample sizes (i.e., between $N = 150$ and $N = 200$) are reached. Small sample sizes could only be used with higher valued structural path coefficients (e.g., 0.80), and even then "... with reasonably large standard errors ..." ([70], p. 333). Similarly, Marcoulides and Saunders [73], Chin and Dibbern [76] and Chin [77] all noted the deleterious impact of non-normal data on PLS estimates and the need for markedly large sample sizes. Ultimately, a researcher needs to consider the distributional characteristics of the data, the potential presence of missing data, the psychometric properties of the variables included in the model, and the magnitude of the relationships considered before definitely deciding on an appropriate sample size to use.

These results and recommendations corroborate Wold's earlier writings and theorems in which he indicated that PLS estimates

> ... are asymptotically correct in the joint sense of consistency (large number of cases) and consistency at large (large number of indicators for each latent variable ... [78], p. 266),

implying in the statistical sense that estimation error decreases as N increases (i.e., as $N \rightarrow \infty$, the estimation error tends to 0), or simply that any estimated PLS coefficients will converge on the parameters of the model as both sample size and number of indicators in the model become infinite (see also Falk and Miller [79]; McDonald [80]). This same statistical interpretation and recommendation is provided by Hui and Wold [72] who indicate that PLS "estimates will in the limit tend to the true values as the sample size N increases indefinitely, while at the same time the block sizes increase indefinitely but remain small relative to N (p. 123)."

Lu [81] and Lu, Thomas, and Zumbo [82] have also warned researchers about the bias that arises from a failure to use large number of indicators for each latent variable (i.e., consistency at large) and labeled it "finite item bias." Dijkstra [83] and Schneeweiss [84] provided some discussion about the magnitude of standard errors for PLS estimators resulting from not using enough observations (consistency) and indicators for each latent variable (consistency at large). Schneeweiss [84] also provided closed form equations that can be used to determine the magnitude of finite item bias relative to the number of indicators used in a model. Using these equations, Schneeweiss ([84], p. 310) indicated that item bias is generally small when many indicators, "each with a sizeable loading and an error which is small and uncorrelated (or only slightly correlated) with other error variables" are used to measure each factor in the model. These warnings clearly echo well established concerns that a determination of the appropriate sample size (which depends on many factors) is also an essential aspect of the whole modeling process.

Although sample size plays an important role in almost every statistical technique applied in practice and there is universal agreement among researchers that the larger the sample the more stable the parameter estimates, there is no agreement as to what constitutes large. This topic has received much attention in the broad statistical literature, but no easily applicable and clear-cut criteria have been determined, only some general rules of thumb have been proposed. For example, some researchers cautiously suggested the general rule of thumb that the sample size should always be more than 10 times the number of free model parameters [85, 86].

To complicate matters, due to the partial nature of the PLS algorithm, the total number of free model parameters should not be the basis for sample size requirements. Being a components based approach, sample size requirements may differ in terms of obtaining stable component weights, measurement paths, and structural model paths. Chin [10] suggested that a researcher using the PLS path weighting scheme should examine the largest of two possibilities: (a) the block with the largest number of formative indicators (i.e., the largest so-called mode *B* measurement equation) or (b) the dependent variable with the largest number of independent variables impacting it (i.e., the largest so-called structural equation). Chin [10] then concluded by saying

> *If one were to use a regression heuristic* of 10 cases per predictor, the sample size requirement would be 10 times either (a) or (b), whichever is the greater (p. 311, emphasis added).

Here Chin used the example heuristic rule of 10 in conjunction with the path weighting scheme. But the main focus was on considering how to determine the largest regression analysis during the PLS iterative algorithm for estimating required sample size for obtaining stable estimates for either (1) weights for PLS components or (2) model paths (i.e., measurement and structural estimates).

Many researchers seem unaware that the equations for a PLS analysis can change depending on the choice of the inner weighting scheme and that the weight estimates for the PLS components are not necessarily affected by the structural model. Chin [10] noted that

> If one is *not* using a path-weighting scheme for inside approximation, then only the measurement model with formative indicators are considered for the first stage of estimation. At the extreme, we see that a factor- or centroid-weighting scheme with all reflective (mode A) measures will involve only a series of simple regressions. Under this condition, it may be possible to obtain stable estimates for the weights and loadings of each component independent of the final estimates for the structural model (p. 311).

Unfortunately, many applied researchers without adequate statistical understanding of the PLS algorithm have unreflectively applied the example rule of 10 that Chin [10] provided. Beyond identifying the constraining regression equation in a PLS analysis, Marcoulides et al. [74] also noted that it seems there is a "reification of the 10 case per indicator rule of thumb (p. 174)" by most PLS researchers ignoring Chin and Newsted's [70] statement that

> for a more accurate assessment, you would specify the effect size for each regression analysis and look up the power tables provided by Cohen '[87] or Green's '[88] 'approximation to these tables' (p. 327).

Clearly, many other researchers (e.g., [3, 70, 73, 74, 89–93]) have indicated that no rule of thumb can be applied indiscriminately to all situations. This is because the appropriate size of a sample depends on the many other factors noted earlier. When these issues are carefully considered, samples of varying magnitude may be needed to obtain reasonable parameter estimates.

In spite of these cautiously proposed rules of thumb available in the PLS literature, there continue to be sweeping claims made by some researchers that PLS modeling can be or should be used (and often, instead of the covariance-based approach) because it makes no sample size assumptions or because "... sample size is less important in the overall model... ([79], p. 93)." Unfortunately, some of these studies even appear in top-tiered journals and frequently report results based on ridiculously low sample sizes, despite the overall inferential intentions of the studies and the actually magnitude of the parent populations of interest. To make things worse, they also try to legitimize these actions by making references to the original developers of the PLS approach. Even a cursory preview of articles using PLS over the past decade reveal a plethora of problematic comments concerning sample size. Recently Ringle, Sarstedt, and Straub [94] documented a sizable number of articles published in MISQ (one of the top-tiered information systems journals) that reported using PLS due to small sample size. Included in some of those articles, were comments such as: (i) "the PLS approach does not impose sample size restrictions ... for the underlying data ... ([95], p. 237)," (ii) "... PLS, a component based approach that is suitable with smaller data sets ... ([96], p. 685)," and (iii) "... PLS ... provides the ability to model latent constructs even under conditions of non-normality and small- to medium-size samples ... ([97], p. 49)." To be fair to these authors, similar troubling comments concerning sample size in PLS modeling abound in almost every other substantive area we examined, consequently the issue is not unique to the information systems field.

All three of the above mentioned articles reported on results from studies in which they had examined and fulfilled the general 10 cases per indicator rule of thumb mentioned above. Specifically, in the Bhattacherjee and Premkumar's study [95], the largest number of indicators per construct in the confirmatory factor analysis (CFA) conducted was 4 and the authors reported using samples sizes between 54 and 77 (depending on the specific construct examined, see p. 237). The Bassellier and Benbasat's study [96] also conducted a CFA using 109 observations and 3–4 item scales. Finally, the Subramani's study [97] used 131 observations in a CFA with 3–4 item scales and the largest number of paths to any construct was 6. Thus, all three above mentioned studies followed the general rule of thumb guidelines regarding sample size.

Nevertheless, as discussed earlier, the generic rule of thumb of 10 cases per indicator does not always ensure accurate and sufficiently stable estimates. So is it the case that many PLS users simply ignore essential preliminaries with regards to sample issues when using these methods in practice? Why is it that many do not seem to carefully examine model parameters along with indexes of their stability across repeated sampling from the studied population? These indexes—the parameter standard errors—also play an instrumental role in constructing confidence intervals for particular population parameters of interest (e.g., [98–101]). Is it not obvious to them that models estimated using questionable sample sizes with extremely unstable estimates and wielding huge standards errors and confidence intervals should be sufficient evidence for an investigator to question the generalizability of results and validity of conclusions drawn? Questionable sample sizes can also cause standard errors to be either overestimated or underestimated. Overestimated standard errors can also result in significant effects being missed, while underestimated standard errors may result in overstating the importance of effects ([73, 93]).

In order to determine the precision of estimation and find standard errors, two approaches can be considered: (a) using analytic approaches (such as the delta or Taylor series expansion method, see, e.g., [98, 101, 102]) or (b) using computer-intensive re-sampling methods (see, e.g., [10, 27]). Unfortunately, finding formulas for standard errors of PLS estimates using the delta or Taylor series expansion method is not a trivial task [83], but recent work [98, 101, 102] has proved promising. As a consequence, Monte Carlo simulation re-sampling methods continue to dominate the field. The principle behind a Monte Carlo simulation is that the behavior of a parameter estimate in random samples can be assessed by the empirical process of drawing many random samples and observing this behavior.

There are, actually, two kinds of Monte Carlo re-sampling strategies that can be used to examine parameter estimates and related sample size issues. The first strategy can be considered a "reactive" Monte Carlo analysis (such as the popular Jackknife or Bootstrap approaches [10, 27, 73, 98]) in which the performance of an estimator of interest is judged by studying its parameter and standard error bias relative to repeated random samples drawn with replacement from the original observed sample data. This type of Monte Carlo analysis is currently quite popular (particularly the Bootstrap approach), despite the fact that it may often give "an unduly optimistic impression of accuracy or stability" of the estimates ([83], p. 86) and there are no generally applicable results as yet of how good the underlying approximation of sampling by pertinent re-sampling distributions is within the framework of latent variable modeling [102].

The second, a less commonly known strategy, can be considered a "proactive" Monte Carlo simulation analysis [25, 73, 103, 104]. In a proactive Monte Carlo analysis, data are generated from a population with hypothesized parameter values and repeated random samples are drawn to provide parameter estimates and standard errors. The approach can also be used in a reactive manner to judge obtained estimates of parameter values and determine the magnitude of standard errors based upon the sample size actually used in a study. Thus, a proactive Monte Carlo analysis (sometimes also referred to as a power analysis) can be used to both examine parameter estimate precision and the necessary sample size needed to ensure the precision of parameter estimates [73, 93].[1] To date, only a few IS articles have conducted Monte Carlo based PLS power analyses (e.g., [105, 106]).

It is surprising to note that some researchers believe that such power analyses are not useful after a study has been completed. For example, Walden [107] proclaimed that "... no one should ever ask for an after the fact power analysis on a sample that shows results." He believes this because a

> power analysis asks ... how big does a sample need to be to detect an effect of a certain size with some probability .... This question does not make sense after the sample has been collected and the null hypothesis rejected, for several reasons ... there is no probability to be evaluated ... if an effect is observed, the sample is clearly large enough to observe an effect ... if you detected an effect, you had the power you needed. (March 5, 2012, AIS World).

Walden [107] seems to have forgotten that researchers always conduct statistical hypothesis testing under conditions of uncertainty. In other words, researchers

---

[1] These generation of Monte Carlo data can easily be done using the statistical analysis program M*plus* [8]. M*plus* has a fairly easy-to-use interface and offers researchers a flexible tool to analyze data using all kinds of model choices. Detailed illustrations for using the M*plus* Monte Carlo simulation options can also be found in [93], in the M*plus User's Guide* [8], and at the product Web site www.statmodel.com

makes a decision about the "true state of affairs" in a studied population, based on information only from part of it (the sample) which typically is a fairly small fraction of the population of interest. Because one functions in this situation of uncertainty, the decision may be incorrect. This is the reason one may commit one of two types of errors—a Type I or a Type II error (one cannot commit both types of errors as they are mutually exclusive possibilities [108]). Hence, even when there may appear to be overwhelming evidence supporting a null hypothesis, as long as the sample is not identical to the population one can never claim to have definitively proved the validity of the null hypothesis. Such a scenario can only occur when the entire population of interest is exhaustively studied. Any time sample data are used there is no guarantee that a null hypothesis that is rejected based on the observed data, is actually true in the population. Thus, one always runs the risk of committing an error. By at least determining the magnitude of the power of a test of the null hypothesis [i.e., determining $(1-\beta)$, which is the complement to the probability of making a Type II error] one can gain some probabilistic insight into any decision about the true state of affairs.

Looking at a number of simple power analyses in studies that employed PLS modeling techniques, one can quickly deduce the importance of power analyses and the fallacy of Walden's [107] argument. To illustrate this point, let us first consider a confirmatory factor analysis (CFA) model in which two correlated factors (ϕ_{21}), each of which has three continuous factor indicators and the following factor loading $\Lambda = [\lambda_{11}, \lambda_{21}, \lambda_{31}, \lambda_{42}, \lambda_{52}, \lambda_{62}]$ and error variance $\Theta = [\theta_{11}, \theta_{22}, \theta_{33}, \theta_{44}, \theta_{55}, \theta_{66}]$ matrix structures. Assume that the data are generated with varying values for the factor inter-correlations (ϕ_{21} between 0.1 and 0.9), each factor loading (λ between 0.4 and 0.9), for the error variances (θ between 0.19 and 0.84) and, consequently, for the indicator reliabilities (between 0.16 and 0.81) and examined with both normal and non-normal distributions. The non-normal data are generated under conditions of moderate non-normality (i.e., skewness set to range between 1 and 1.5, and kurtosis set to range between 1 and 1.5). For ease of presentation, no missing data patterns are considered. To ensure stability of results, the number of sample replications is set at 5,000. To simplify matters further, we focus only on the factor correlation parameter (ϕ_{21}), although any other model parameter could be similarly examined (for additional details see [73]).

Table 1 presents the results of a Monte Carlo simulation based on a pre-selected $N = 100$ sample size. The boldfaced column values correspond to the various factor loadings considered (i.e., the values of λ), while the boldfaced row values correspond to the considered factor inter-correlations (i.e., the values of ϕ_{21}). The entries provided in Table 1 correspond to the computed value of the power of the study to reject the hypothesis that the factor correlation in the population is zero (i.e., the probability of rejecting the null hypothesis when it is actually false). As can be seen by examining the entries provided in Table 1, power remains relatively high when indicators with sizeable factor loadings (and thereby more reliable indicators) are used to measure factors. For example, a power value of 0.97 is achieved when indicators with factor loadings equal to 0.90 are used to examine a 0.50 valued correlation between the two factors. Even so, power estimates deteriorate when examining low valued factor correlations, especially when using poor quality indicators. For exam-

Table 1: Power values determined for normally distributed data with no missing values ($N = 100$)

	λ					
ϕ_{21}	**0.9**	**0.8**	**0.7**	**0.6**	**0.5**	**0.4**
0.1	0.13	0.12	0.13	0.13	0.10	0.06
0.2	0.46	0.31	0.26	0.24	0.18	0.10
0.3	0.85	0.82	0.71	0.48	0.36	0.21
0.4	0.97	0.96	0.90	0.81	0.57	0.27
0.5	0.97	0.96	0.96	0.92	0.77	0.41
0.6	1.00	1.00	1.00	0.96	0.91	0.53
0.7	1.00	1.00	1.00	0.99	0.93	0.66
0.8	1.00	1.00	1.00	1.00	0.97	0.77
0.9	1.00	1.00	1.00	1.00	1.00	0.82

ple, a power value of only 0.10 is achieved when indicators with factor loadings equal to 0.50 are used to examine a 0.10 valued factor inter-correlation.

Table 2: Power values determined for normally distributed data with no missing values ($N = 50$)

	λ					
ϕ_{21}	**0.9**	**0.8**	**0.7**	**0.6**	**0.5**	**0.4**
0.1	0.11	0.11	0.12	0.13	0.15	0.09
0.2	0.29	0.27	0.24	0.22	0.21	0.13
0.3	0.59	0.52	0.45	0.37	0.30	0.22
0.4	0.87	0.78	0.70	0.62	0.46	0.24
0.5	0.97	0.93	0.87	0.72	0.54	0.30
0.6	1.00	0.99	0.94	0.88	0.70	0.46
0.7	1.00	1.00	1.00	0.93	0.74	0.50
0.8	1.00	1.00	1.00	0.97	0.83	0.57
0.9	1.00	1.00	1.00	1.00	0.87	0.58

Table 2 presents the results in which a much smaller pre-selected $N = 50$ sample size is used. As can be seen from these results, power again tends to remain relatively high when psychometrically sound indicators measure factors, particularly when examining very high valued factor correlations. Tables 3 and 4 present the results of Monte Carlo simulations for the same two sample sizes ($N = 50$ and $N = 100$) but instead under conditions of non-normality. Unfortunately, when the power values are examined under such conditions of non-normality, their deterioration is evident and quite disconcerting. In fact, none of the values provided in Tables 3 and 4 are above 0.40, indicating that using these sample sizes a researcher would just not be able to reject false null hypotheses concerning the factor inter-correlation.

So what samples sizes would be needed to achieve a sufficient level of power, say equal to 0.80 (considered by most researchers as acceptable power)? The results of such a Monte Carlo analysis under conditions of normality and non-normality are

Table 3: Power values determined for non-normally distributed data with no missing values ($N = 50$)

	λ					
ϕ_{21}	**0.9**	**0.8**	**0.7**	**0.6**	**0.5**	**0.4**
0.1	0.11	0.09	0.11	0.12	0.09	0.04
0.2	0.12	0.12	0.11	0.12	0.10	0.05
0.3	0.14	0.14	0.13	0.15	0.11	0.06
0.4	0.17	0.15	0.15	0.16	0.12	0.08
0.5	0.19	0.19	0.19	0.19	0.15	0.08
0.6	0.20	0.22	0.21	0.22	0.16	0.11
0.7	0.23	0.23	0.23	0.24	0.20	0.12
0.8	0.25	0.26	0.26	0.27	0.23	0.12
0.9	0.30	0.31	0.30	0.32	0.26	0.18

Table 4: Power values determined for non-normally distributed data with no missing values ($N = 100$)

	λ					
ϕ_{21}	**0.9**	**0.8**	**0.7**	**0.6**	**0.5**	**0.4**
0.1	0.08	0.07	0.08	0.11	0.10	0.04
0.2	0.09	0.07	0.09	0.11	0.11	0.06
0.3	0.14	0.11	0.11	0.12	0.12	0.07
0.4	0.17	0.15	0.12	0.12	0.12	0.09
0.5	0.20	0.21	0.16	0.17	0.15	0.10
0.6	0.25	0.25	0.24	0.19	0.16	0.12
0.7	0.28	0.28	0.26	0.27	0.21	0.14
0.8	0.33	0.34	0.33	0.29	0.24	0.18
0.9	0.40	0.38	0.36	0.36	0.27	0.21

provided in Tables 5 and 6. As can be seen by examining the entries in Table 5, relatively small sample sizes can often be used when psychometrically sound indicators are available to examine high valued factor inter-correlations. However, when trying to examine low valued factor inter-correlations using poor quality indicators, much larger sample sizes are needed. It is important to note that the results presented in Table 5 corroborate those presented by Hui and Wold [72], Chin and Newsted [70], and Schneeweiss [84] that small sample sizes do not permit a researcher to detect low valued model coefficients until much larger sample sizes are reached. However, the problem can be much more disconcerting than these researchers originally reported, as can be seen by examining the entries in Table 6. When moderately non-normal data are considered, the sample sizes needed sometimes become astronomical, despite the inclusion of highly reliable indicators in the model. These results are evidence that determining the appropriate sample size even for a simplistic CFA depends on many model characteristics, including the psychometric properties of the indicators, the strength of the relationships among the factors, and the distributional characteristics of the data. It is also important to note that for only a limited number of normally distributed data conditions would the PLS rule of thumb of 10 cases per indicator really suffice in these example models considered. As indicated previously, a researcher must consider the distributional characteristics of the data, potential missing data, the psychometric properties of the variables examined, and the magnitude of the relationships considered before deciding on an appropriate sample size to use or to ensure that a sufficient sample size is actually available to study the phenomena of interest.

Table 5: Sample sizes needed to achieve power = 0.80 with normally distributed data and no missing values

	λ					
ϕ_{21}	**0.9**	**0.8**	**0.7**	**0.6**	**0.5**	**0.4**
0.1	916	1,053	1,261	1,806	2,588	4,927
0.2	256	292	371	457	764	1,282
0.3	96	99	147	223	317	672
0.4	46	57	71	98	186	343
0.5	25	34	43	66	111	220
0.6	16	20	23	44	78	175
0.7	15	15	17	33	61	134
0.8	15	15	17	25	46	109
0.9	15	15	17	25	42	99

Table 6: Sample sizes needed to achieve power = 0.80 with non-normally distributed data and no missing values

	λ		
ϕ_{21}	**0.9**	**0.8**	**0.7**
0.1	15,646	24,574	31,381
0.2	4,922	5,766	6,251
0.3	2,357	2,623	2,817
0.4	1,331	1,536	1,715
0.5	931	1,018	1,203
0.6	653	707	864
0.7	467	545	639
0.8	386	433	486
0.9	345	351	407

It is also quite useful to examine in detail the previously mentioned studies and determine the extent to which they exhibited a sufficient level of power to support the results and validity of conclusions drawn. We note that all of these studies indicated they had examined and fulfilled the 10 cases per indicator rule of thumb. Table 7 presents a power analysis of the CFA model from the study by Bhattacherjee and Premkumar [95]. Two values are provided in each cell of Table 7, the inter-construct correlation reported in the published study (for full details see [95], Table 2, p. 239), and the determined level of power for each examined inter-correlation.

Because no specific details were provided in the published study about the distributional characteristics of the data or any apparent missing data patterns, the power analyses were conducted by assuming normally distributed data with no missing data patterns. The factor loadings reported for the scaled items in the study were all in the 0.80–0.96 range and the sample sizes were between 54 and 77 observations, depending on the construct examined (see Table 1, p. 238). As can be seen by examining the entries in Table 7, and assuming normally distributed data with no missing data patterns, power would be considered quite high for all the inter-construct correlations examined in the Bhattacherjee and Premkumar's study [95].

Table 8 presents a power analysis of the CFA from the study by Bassellier and Benbasat [96]. Once again, two values are provided in Table 8. The inter-correlations among the constructs reported in the published study (see [96], Table

Table 7: CFA inter-construct correlations and power values for Bhattacherjee and Premkumar [95] ($N = 54$)

CBT study (time t2 – t3)	U2	A2	D3	S3	U3	A3
Attitude (A2)	0.74[a] (0.94)[b]					
Disconfirmation (D3)	0.45 (1.00)	0.44 (0.94)				
Satisfaction (S3)	0.59 (1.00)	0.58 (0.100)	0.46 (1.00)			
Usefulness (U3)	0.65 (1.00)	0.61 (1.00)	0.60 (1.00)	0.64 (1.00)		
Attitude (A3)	0.63 (1.00)	0.69 (1.00)	0.54 (0.99)	0.80 (1.00)	0.71 (1.00)	
Intention (I3)	0.63 (1.00)	0.58 (1.00)	0.52 (0.97)	0.53 (0.98)	0.79 (1.00)	0.61 (1.00)

[a]Inter-construct correlation
[b]Power

Table 8: CFA inter-construct correlations and power values for Bassellier and Benbasat [96] ($N = 109$)

CBT study	1	2	3	4	5	6	7
1. Intentions for partnerships							
2. Organizational overview	0.406[a] (0.97)[b]						
3. Organizational unit	0.352 (0.89)	0.809 (1.00)					
4. Organizational responsibility	0.504 (1.00)	0.601 (1.00)	0.676 (1.00)				
5. IT-business integration	0.516 (1.00)	0.628 (1.00)	0.589 (1.00)	0.595 (1.00)			
6. Knowledge networking	0.283 (0.73)	0.453 (0.99)	0.405 (0.96)	0.308 (0.81)	0.341 (0.87)		
7. Interpersonal communication skills	0.381 (0.94)	0.485 (0.99)	0.386 (0.95)	0.345 (0.88)	0.478 (0.98)	0.474 (0.98)	
8. Leadership skills	0.427 (0.97)	0.589 (1.00)	0.600 (1.00)	0.516 (1.00)	0.652 (1.00)	0.517 (1.00)	0.582 (1.00)

[a]Inter-construct correlation
[b]Power

6, p. 688), and the appropriately determined level of power for each examined inter-correlation. Because this study also did not provide any specific details about the distributional characteristics of the data or any apparent missing data patterns, the analyses were conducted under the assumption of normality and no missing data patterns. The factor loadings reported for scaled items in the study were all in the 0.71–0.89 range and the sample size was 109 observations (see Table 5, p. 687). As can be seen from the entries in Table 8, and assuming normally distributed data with no missing data patterns, power would also be considered quite high for almost all the inter-construct correlations examined in the Bassellier and Benbasat's study [96]. The only power value that is lower (0.73) than the commonly accepted cutoff point of 0.80 is for the correlation between the constructs of "knowledge networking (#6)" and "intentions for partnerships (#1)."

Finally, Table 9 presents a power analysis of the CFA from the study by Subramani [97]. The same two values are provided in the table; the inter-construct correlations reported in the published study (see [97]; Table 2, p. 61) and the determined level of power for each examined correlation. As with the previously examined studies, this study also did not provide any specific details about the distributional characteristics of the data or missing data patterns. As such, the analyses were again conducted under the assumption of normality and no missing data patterns. The factor loadings were not specifically reported for scaled items in the study, but were apparently "uniformly high (p. 59)" and between 0.71 to "above 0.80 (p. 59)," using a sample with 131 observations. As can be seen by examining the entries in Table 9, and assuming normally distributed data with no missing data patterns, power would be considered quite low (and in some cases well below the commonly accepted cutoff point of 0.80), even for many of the reported statistically significant inter-construct relationships. For example, although the relationship between the Operational Benefits (#5) and IT USE for Exploitation (#1) was reported as being statistically significant (0.179, $p < 0.05$), its power value on the basis of the proactive Monte Carlo simulation analysis is determined to be equal to 0.40 (assuming of course normally distributed data with no missing values—if in fact, the data were not normally distributed a much lower power value would be expected). In other words, the computed value of the power of the study to reject the hypothesis that the factor inter-correlation in the population is zero was determined to be quite low. It is particularly important to note that, despite the fact that Subramani's study [97] utilized a larger sample size than either of the other two previously examined studies (which as we saw exhibited sufficiently high levels of power), the low valued factor correlations examined in Subramani's study [97] actually deteriorated the power of the statistical tests conducted. And, although Subramani's study [97] clearly fulfilled the frequently used rule of thumb of using 10 cases per indicator, it appears that the generalizability of some of the results and the validity of the conclusions drawn from this study may be questionable.

As indicated by Marcoulides and Saunders [73] the selection of an appropriate sample size that will ensure an adequate level of power clearly depends on many factors. These include the psychometric properties of the variables considered, the strength of the relationship among the variables, the complexity and size of the

Table 9: CFA inter-construct correlations and power values for Subramani [97] ($N = 131$)

CBT study	1	2	3	4	5	6	7	8	9	10
1. IT use for exploitation										
2. IT use for expiration	0.188[a] (0.41)[b]									
3. Business process specificity	0.321* (0.89)	0.039 (0.09)								
4. Domain-knowledge specificity	0.329* (0.89)	0.468* (1.00)	0.200 (0.49)							
5. Operational benefits	0.179* 0.40	0.343* (0.96)	0.163 (0.32)	0.550* (1.00)						
6. Strategic benefits	0.258* (0.70)	0.352* (0.96)	0.257* (0.69)	0.410* (0.99)	0.489* (1.00)					
7. Competitive performance	0.086 (0.16)	0.005 (0.07)	0.013 (0.07)	0.158 (0.30)	0.173 (0.36)	0.274* (0.77)				
8. Uncertainty	0.049 (0.09)	0.198 (0.14)	0.077 (0.14)	0.197 (0.46)	−0.028 (0.08)	−0.044 (0.08)	0.131 (0.23)			
9. Retailer replaceability	0.028 (0.09)	0.070 (0.14)	−0.100 (0.22)	−0.111 (0.23)	−0.230** (0.72)	−0.310** (0.90)	−0.670** (1.00)	0.130 (0.23)		
10. Size	0.132 (0.24)	0.165 (0.32)	−0.152 (0.26)	0.205* (0.52)	−0.019 (0.07)	−0.015 (0.07)	−0.01* (0.07)	0.328** (0.89)	0.190** (0.42)	
11. Years of association	0.137 (0.24)	0.117 (0.23)	−0.187* (0.56)	0.097 (0.21)	−0.002 (0.08)	0.055 (0.10)	0.007 (0.09)	0.133 (0.25)	−0.069 (0.14)	0.320** (0.89)

Shaded cells are correlations reported significant (but below the 0.80 power threshold)

$*p < 0.05, **p < 0.01$

[a]Inter-construct correlation

[b]Power

model, the amount of missing data, and the distributional characteristics of the variables. Examining all these issues using a proactive Monte Carlo simulation analysis will at least provide researchers with some insight concerning the stability and power of the parameter estimates that would be obtained across repeated sampling from the studied population. Ignoring these issues could lead to important effects being completely missed in a study or lead to overstating the importance of effects in a study.

5 Model Identification Issues

The examination of issues related to model identification began in the early part of the last century with the work of Albert [109, 110], Koopmans and Reiersol [111], and Ledermann [112]. Identification basically consists of two specific aspects: existence and uniqueness (e.g., [63, 113, 114]). For example, in the context of a factor analysis model, *existence* and *uniqueness* would imply the following: (1) *Existence*: Does a factor decomposition (e.g., $\Sigma = \Lambda\Lambda' + \Psi$) exist in the population (for a given number of factors m), where Λ is a $p \times m$ factor loading matrix with rank m, and Ψ is a unique variance diagonal (error) matrix with positive elements? (2) *Uniqueness*: Assuming the existence of a factor decomposition, is it the *only* decomposition possible? In other words, are there no other matrices (e.g., some other matrices such as Λ_2 and Ψ_2 different from Λ and Ψ) that can give the same matrix Σ (i.e., $\Sigma = \Lambda_2\Lambda_2' + \Psi_2$, with rank of Λ_2 not greater than m)?

Model identification can also be categorized in one of two ways: (1) *global identification*, in which *all* of a model's parameters are identified, or in contrast, (2) *local identification*, in which at least one—but not all—of a model's parameters is identified. Globally identified models are locally identified, but locally identified models may or may not be globally identified. Global identification is a prerequisite for drawing inferences about an entire model. When a model is not globally identified, local identification of some of its parameters permits inferential testing in only that section of the model. Some researchers have referred to this as *partial identification* [28]. Kano [115] has suggested that we focus more attention on examining what he referred to as *empirical identification*, rather than the more commonly considered notion of *mathematical identification*. At least in theory (although somewhat debatable in practice) parameters that are not identified do not influence the values of ones that are identified.

In addition to categorizing models as either globally or locally identified, they can also be classified as (1) *under-identified*, (2) *just-identified*, or (3) *over-identified*. Consider for example a model involving four ($P = 4$) observed variables. Such a model would be determined as having a correlation or covariance data matrix with altogether $\frac{1}{2}P(P+1) = \frac{4\times5}{2} = 10$ nonredundant elements. Now let us consider the case of an under-identified model. Under-identification occurs when not enough relevant data are available to obtain unique parameter estimates. Using the notion of the degrees of freedom of any hypothesized model as the difference between the number of nonredundant elements in the data matrix and the number of parameters in the model, an under-identified model will have negative degrees of freedom. We note that when the degrees of freedom of a model are negative, at least one of its parameters is under-identified. Having positive degrees of freedom with any proposed model is a necessary but not a sufficient condition for identification. That is because having positive degrees of freedom does not guarantee that every parameter is identified. There can in fact be situations in which the degrees of freedom for a model are quite high (the so-called over-identified case) and yet some of its parameters remain under-identified [100]. Conversely, having negative degrees of freedom is a sufficient but not a necessary criterion for showing that a model is globally under-identified.

Two additional and frequently interchangeably used concepts are those of a "saturated model" and of a "just identified" model. A just identified model can be defined as an identified model that has zero degree of freedom, while a saturated model can be defined as a model that has zero degree of freedom [116, 117]. Nevertheless, as noted by Raykov et al. [117], the distinction between these two models is quite important since using them interchangeably can lead to consequential theoretical and empirical confusion, with potentially misleading substantive conclusions. Because a saturated model need not be (just) identified, the two concepts must be kept separate. Raykov et al. [117] proposed that (a) the notion of "the saturated model" be reserved for a particular saturated model (the one with unconstrained variable variances and covariances), and (b) that the reference "a saturated model" be used when the pertinent statement would be correct for any saturated model for that set of observed variables.

If theory testing is the main objective, the most desirable identification status of a model is *over-identification*, where the number of available data elements is more than those needed to obtain a unique solution. Although as indicated above, having positive degrees of freedom does not guarantee that every parameter in the model is identified. An over-identified model thereby implies that, for at least one parameter, there is more than one equation the estimate of a parameter must satisfy; only under these circumstances—the presence of multiple solutions—are models provided with the opportunity to be rejected by the data.

Although identification issues have major implications with respect to model fitting, they are frequently ignored due to their challenging technical intricacies [28]. To simplify matters, some researchers often make a specific assumption about existence and focus mainly on uniqueness aspects. For example, existence in factor analysis implies that factor decomposition exists for a given number of factors, whereas uniqueness assumes it is the only decomposition possible—in other words, it is commonly assumed that a factor decomposition does exist in the population of interest. The topic of model uniqueness has generally followed two early lines of research: one originating in Albert [109, 110] and Anderson and Rubin [118] and the other based on the work of Ledermann [112]—for a detailed overview see [28] and the references therein.

Anderson and Rubin [118] specifically proposed the following theorem (the so-called *Theorem 5.1*) in factor analysis for a *sufficient* condition of *uniqueness*:

Theorem 5.1. If any single row of a factor loading matrix Λ is deleted, there still remain two disjoint (i.e., non-overlapping) submatrices of rank m. Then the FA decomposition is *unique* (for a detailed proof of this theorem see [119]).

To illustrate, consider for example the following matrix:

$$\Sigma = \begin{pmatrix} 1 & 0.340 & 0.310 & 0.100 & 0.095 \\ 0.340 & 1 & 0.285 & 0.095 & 0.090 \\ 0.310 & 0.285 & 1 & 0.090 & 0.085 \\ 0.100 & 0.095 & 0.090 & 1 & 0.150 \\ 0.095 & 0.090 & 0.085 & 0.150 & 1 \end{pmatrix}$$

and a FA decomposition leading to a factor loading (Λ) matrix with rank $m = 2$:

$$\Lambda = \begin{pmatrix} 0.60 & 0.10 \\ 0.55 & 0.10 \\ 0.50 & 0.10 \\ 0.10 & 0.40 \\ 0.10 & 0.35 \end{pmatrix}$$

and a unique variance (Ψ) matrix equal to:

$$\Psi = \begin{pmatrix} 0.63 & 0 & 0 & 0 & 0 \\ 0 & 0.69 & 0 & 0 & 0 \\ 0 & 0 & 0.74 & 0 & 0 \\ 0 & 0 & 0 & 0.83 & 0 \\ 0 & 0 & 0 & 0 & 0.87 \end{pmatrix}.$$

If the first row in the factor loading matrix Λ were deleted to provide

$$\Lambda = \begin{pmatrix} & \\ 0.55 & 0.10 \\ 0.50 & 0.10 \\ 0.10 & 0.40 \\ 0.10 & 0.35 \end{pmatrix},$$

then the two possible disjoint submatrices

$$\begin{pmatrix} & \\ 0.55 & 0.10 \\ 0.50 & 0.10 \\ & \\ & \end{pmatrix} \text{ and } \begin{pmatrix} & \\ & \\ & \\ 0.10 & 0.40 \\ 0.10 & 0.35 \end{pmatrix}$$

would provide nonzero determinants, thereby signifying that there are two disjoint submatrices whose rank is $m = 2$ (we note that similar rank results would be obtained for the remaining possible submatrices; for complete details and a step by step analysis for conducting such examination, including a SAS PROC IML subroutine, see Table 1 in [28]). Consequently, the factor analysis decomposition $\Sigma = \Lambda\Lambda' + \Psi$ is unique. In other words, based upon these results there is no alternative factor decomposition (such as $\Sigma = \Lambda_2\Lambda_2' + \Psi_2$) of the matrix Σ.

The above Anderson and Rubin [118] theorem essentially requires that the relationship between the number of observed variables (p) and number of factors (m) be satisfied as $p \geq 2m + 1$ (i.e., the number of observed variables p has to be greater than twice the number of factors m [115]). In other words, what *Theorem 5.1* implies is that if the number of factors (m) selected is greater than $(p-1)/2$ observed variables, it will be difficult for the solution to be identified [120]. For example, if four factors were selected in a study with only eight observed variables, it will be difficult to get the solution to be identified. A number of other researchers (e.g., [121, 122]) have also provided alternative conditions to those proposed by Anderson and Rubin [118] and much research continues to date on this topic (e.g.,[114, 120]).

Anderson and Rubin [118] also proposed other important theorems for *necessary* conditions of *uniqueness*, which they called *Theorems 5.5–5.7*. A particular theorem that has very important practical implications to the practice of factor analysis and related models is *Theorem 5.6*. This theorem states: If any rotated factor loading matrix (rotated by a nonsingular matrix) has a column with *at most two non-zero elements*, then the FA decomposition is *not unique* (and therefore is *not* identified).

In other words, if a researcher extracts a factor whose factor loading estimates are quite small and do not differ significantly from zero except for at most two elements, then it may be reasonable to suspect that the factor is not *uniquely* identified. The example, the population factor loading matrix Λ and its estimate $\hat{\Lambda}$ would be illustrative of such a not *uniquely* identified third factor (see [115], p. 143):

$$\Lambda = \begin{pmatrix} 0.57 & 0.13 & 0.00 \\ 0.57 & 0.33 & 0.00 \\ 0.21 & 0.37 & 0.54 \\ 0.75 & 0.07 & 0.00 \\ 0.73 & 0.07 & 0.00 \\ 0.29 & 0.25 & 0.54 \\ 0.19 & 0.45 & 0.00 \\ 0.18 & 0.33 & 0.00 \\ 0.10 & 0.71 & 0.00 \\ 0.31 & 0.43 & 0.00 \\ 0.02 & 0.42 & 0.00 \\ 0.03 & 0.53 & 0.00 \end{pmatrix} \qquad \hat{\Lambda} = \begin{pmatrix} 0.58 & 0.13 & 0.02 \\ 0.59 & 0.33 & 0.03 \\ 0.24 & 0.35 & 0.05 \\ 0.74 & 0.61 & 0.01 \\ 0.72 & 0.07 & 0.01 \\ 0.31 & 0.23 & 0.55 \\ 0.20 & 0.45 & 0.02 \\ 0.17 & 0.33 & 0.02 \\ 0.12 & 0.71 & 0.03 \\ 0.30 & 0.40 & 0.01 \\ 0.01 & 0.43 & 0.02 \\ 0.03 & 0.45 & 0.02 \end{pmatrix}.$$

Consequently, a researcher should be very cautious whenever an estimated factor loading matrix looks anything like the one displayed on the right-hand side above.

Fortunately, most general statistics programs provide options to output the standard errors for rotated factor loadings and, consequently, a researcher can at least select to conduct hypothesis tests to determine whether the rotated factor loadings in the population significantly differ from zero (although once again the issue of power and sample size considered in the above section would again come into play). For simultaneous hypothesis testing, it may be wise to employ Bonferroni adjustments to control the overall Type I error however, it is a difficult decision because such smaller overall alpha levels often result in lowering statistical power. Alternatively, Kano [115] proposed that the Lagrangian multiplier test be used (this test is also quite commonly provided in some commercially available modeling programs; see, e.g., EQS [85]) to investigate such cases.

Despite the fact that the issues of model identification have been well documented, these do not appear to be well known or commonly considered by applied researchers using modeling methodologies. We strongly believe that researchers must become cognizant of the potential consequences of ignoring these issues and at least understand some of the basics involved in accordance to the model being tested.

6 Myths About the Coefficient α

Coefficient alpha is frequently used in empirical research as an index that informs about measurement instrument reliability. Most measurement instruments (e.g., inventories, questionnaires, self-reports or tests), are typically developed to provide an overall assessment (an overall score) of an underlying latent dimension by accumulating information about various aspects of the latent dimension across their components (e.g., questions or items). Coefficient alpha is applicable when the components of a given measurement instrument are dichotomous or polytomous and capitalizes on the interrelationships among the instrument components (specifically their covariance) to provide a reliability estimation index. The estimate is readily available in most statistical packages and can be easily obtained with them for use in any empirical research setting. Unfortunately, and despite its availability and widespread use, a number of troubling myths about coefficient alpha appear prevalent among researchers [123]. For example, many researchers incorrectly use it as an index of dimensionality, often declaring that a set of items can be judged to be unidimensional when $\alpha > 0.70$. Coefficient alpha assumes unidimensionality, but it is not a test of it. As we highlight below, however, such interpretations and reliance on alpha can be quite problematic and misleading (for complete details, see [54, 123–134]). We address here two specific myths held about the coefficient and clarify some inaccuracies and inconsistencies commonly encountered in the literature (for more details see [123]):

1. Alpha is only an index of internal consistency. In other words, it is an index of the degree to which a set of instrument components are interrelated (in terms of inter-item covariance). The higher the magnitude of this covariance, the higher the value of coefficient alpha. Such evidence, however, does not imply unidimensionality of the set of components of a considered instrument (see also [135], for an insightful discussion and counter-examples, as well as [124]). The coefficient

alpha merely assumes unidimensionality, it does not test for it. If a researcher is interested in assessing unidimensionality, alpha cannot provide such information. In order to examine the unidimensionality hypothesis itself, one should prefer the use of an exploratory or a confirmatory factor analysis. With a confirmatory factor analysis one can essentially statistically test this hypothesis and evaluate the extent to which it may be viewed as supported for a measurement instrument in a given data set.

2. Alpha is not in general a lower bound of reliability. Alpha is a lower bound of reliability only under certain specific measurement circumstances. For example, Raykov and Marcoulides [123] stressed that this property only holds with uncorrelated errors among a set of components (for further details see also [54, 134]). With correlated errors, the underestimation feature of alpha does not generally hold. If they are correlated, alpha may or may not be a lower bound of composite reliability, regardless of the number of underlying dimensionality of the measuring instrument under consideration.

7 The Use of Correlation and Covariance Matrices

Although much literature has addressed the issue of the potential differences that can occur when analyzing correlation matrices as covariance matrices (and vice versa), it does not seem to be well understood that applying a covariance structure to a correlation matrix can produce some combination of incorrect standard errors, parameter estimates, or test statistics, and may even alter the studied model and results (see also [24] and references therein). Researchers applying PLS and related methods often appear to arbitrarily analyze the matrix of choice (or perhaps even convenience) without realizing that it is possible and probably most likely that incorrect conclusions may be drawn because of this choice. Such choices are especially important when flawed attempts are made to compare the various methods and their performance under supposedly varying distributional characteristics (see, e.g., [22] in which PLS was compared to other commonly used modeling techniques—see detailed discussion in next section).

The message is quite simple. The model must be scale invariant [24]. In other words, a scale invariant covariance matrix is one that can be transformed into the associated correlation matrix by rescaling the model parameters by functions of standards deviations. Simply standardizing the covariance matrix may or may not affect the analysis, but it really depends on the model being considered. For example, using both the correlation and covariance matrices computed for a set of eight variables ($n = 72$) collected in a clinical setting originally reported in Jolliffe ([136]; see p. 40 for the observed data matrix), Marcoulides et al. [74] showed that the weighted composite $w_1 = 0.2, 0.4, 0.4, 0.4, -0.4, -0.4, -0.2, -0.2$ (explaining 35% of the total variation in the variables) would be obtained when the correlation matrix is used, compared to the obtained weighted composite $w_2 = 0.0, 1.0, 0.0, 0.0, 0.0, 0.0, 0.0, 0.0$ (explaining 99% of the total variation) when the covariance matrix is used. The main reason for this disparity is the considerable difference in the standard deviations caused by the differences in scale for each of the eight variables (which are 0.371, 41.253, 1.935, 0.077, 0.071, 4.037, 2.732, and 0.297, respectively). Cudeck [24] also presented three simple factor analytic models for the observed matrix $\Sigma = \Sigma(\gamma) = \Lambda\Phi\Lambda' + \Theta$, (where Λ is the factor loading matrix, Φ the factor correlation matrix, Θ is the error matrix, and γ is the model defined parameter vector), and showed that, unless the model under examination is indeed appropriate for scale changes, any rescaling that occurs modifies the model completely (in the example used two of the models were scale invariant and one was not). Thus, if a factor analysis model is invariant, it is always possible to obtain estimates of the parameters. The original model structure will not be modified, only the elements of the parameter vector γ.

Although an algebraic derivation is the best way to determine whether a model is scale invariant, unfortunately this is often quite cumbersome and particularly difficult with complex models. An easy to follow practical approach is to simply fit the proposed model structure twice: first to the observed covariance matrix and then again to the observed matrix of correlations. The model structure is likely invariant if at the minima the discrepancy function of each obtained model is equal. We emphasize that this equality is not in and of itself sufficient evidence [24]. Nevertheless, the structure is categorically not invariant if the two are not equal. Assuming that either a correlation or a covariance matrix may be interchangeably examined can prove to be tricky.

8 Comparisons Among Modeling Methods

Numerous researchers have attempted studies comparing the efficacy of PLS with that of other modeling approaches, often without ever addressing the issue of the legitimacy of these comparisons. For example, if the comparison is between multiple regression, PLS, or other related modeling technique, then it is trivial. This is because an analysis of the same data and model based on a single regression equation using multiple regression, PLS or other modeling approach will always result in identical estimates. Obtaining such identical estimates is due to the well known fact that a single regression equation is a just-identified model and fits the data in the exact same way irrespective of the minimized fit function. For instance, Goodhue et al. [22, 137] attempted such trivial comparisons and then reported on supposed differences in the methods without ever realizing that their comparisons were wrong (for complete details see [3]). In contrast, Hwang et al. [138] in their comparison study carefully stipulated the precise conditions of their analyses and fully acknowledge the limitations of their comparisons. They openly acknowledged the differences in the setup of the approaches in terms of model specification and parameter estimation ahead of any analyses conducted. They subsequently indicated that

> ...this leads to the specification of different sets of model parameters for latent variables (i.e., factor means and/or variances in covariance structure analysis versus compo-

nent weights in partial least squares)... The algebraic formulations underlying the three approaches seem to result in substantial difference in the procedures of parameter estimation.

They go on to point out again that the

... approaches estimate different sets of model parameters.... Thus, in this study we evaluate and report the recovery of the estimates of a common set of parameters... (p. 703).

They conclude by acknowledging their inability to provide correctly parameterized comparisons among the approaches and indicate that

... we generated simulated data on the basis of covariance structure analysis... we adopted the procedure because it was rather difficult to arrive at an impartial way of generating synthetic data for all three approaches... (p. 710).

An appropriate approach for correctly parameterized comparisons between PLS and other methods was recently proposed by Treiblmaier et al. [23]. This approach begins by distinguishing between models with observed variables ($\mathbf{x}$), composite variables ($\mathbf{F}$) and latent variables (F), and unambiguously implements an $\mathbf{F}$ that closely approximates an F for comparison purposes. Doing so, however, requires a two-step approach that splits the determinate part of the composite into two or more composites and then models them as latent variables. This method can be readily contrasted with other inappropriate comparisons that simply create substitute estimates of latent variables (as was done in [22]). As explained by Marcoulides et al. [3], specifying models in this manner does not eliminate the fact that they are differentially parameterized models (in other words, an $\mathbf{x} \rightarrow \mathbf{F}$ path is not the same as a $\mathbf{x} \rightarrow F$ path). Although substitution of estimates for $\mathbf{F}$ is routinely done when conducting such comparisons, there are well-known and clear consequences (see complete details provided in [23]), not the least of which that "... not all parameters will be estimated consistently" ([139], p. 37).

The above reasons summarize why Marcoulides et al. [3, 74] emphatically warned researchers that

... the comparison of PLS to other methods cannot and should not be applied indiscriminately

and referred to any inappropriate evaluations between methods as "comparing apples with oranges."

The central issue dictating the legitimacy of such comparisons revolves around the notion of differentially parameterized models. Ignoring the legitimacy of this concern can lead to incorrect conclusions or may lead to overstating the importance of observed results [74]. Goodhue, Lewis, and Thompson to some extent acknowledged this fact when they stated that

We owe you all an apology! We were so certain that we were right in the equivalence of the methods, but now we see that the issue is more complicated than we thought (You are probably not surprised, at least by this last phrase!). We were focusing on how the techniques were used in practice, and didn't see that how they are used in practice is, in fact, not equivalent (May 1, 2008, personal communication).

But, unfortunately, Goodhue et al. [22, 140] somehow ultimately disregarded this fact and attempted to report on results from incorrect comparisons. It is essential that researchers ensure that any observed differences encountered between methods are not merely a function of differentially parameterized models being analyzed. Ignoring this matter can and has repeatedly lead to the unfortunate incidence of overstating the importance of the outcomes observed as in the case of Goodhue et al. [22, 140].

References

[1] G.A. Marcoulides, Structural equation modeling for scientific research. *Journal of Business and Society*, **2**, pp. 130–138, 1989.

[2] F. Galton, *Natural inheritance*, New York: MacMillan, 1889.

[3] G.A. Marcoulides, W.W. Chin, and C. Saunders, When imprecise statements become problematic: A response to Goodhue, Lewis, and Thompson. *MIS Quarterly*, **36**, pp. 717–728, 2012.

[4] J.L. Arbuckle and W. Wothke, *Amos 4.0 User's guide*, Chicago, IL: SPSS, 1999.

[5] P.M. Bentler, *EQS structural equations program manual*. Encino, CA: Multivariate Software, 2004.

[6] K. G. Jöreskog, and D. Sörbom, *LISREL 8 user's reference guide*. Chicago, IL: Scientific Software International, Inc, 2005.

[7] J.B. Lohmoller, *Latent variable path modeling with partial least squares*, Heidelberg: Physica-Verlag, 1989.

[8] L.K. Muthén, and B.O. Muthén, *Mplus user's guide* (5th ed.). Los Angeles, CA: Muthén and Muthén, 2011.

[9] M.C. Neale, S.M. Boker, G. Xie, and H.H. Maes, *Mx: Statistical modeling* (5th ed.). Richmond, VA: Virginia Commonwealth University, 1999.

[10] W.W. Chin, The partial least squares approach for structural equation modeling. In G.A. Marcoulides (Ed.), *Modern methods for business research* (pp. 295–336), Mahwah, NJ: Lawrence Erlbaum, 1998.

[11] W.W. Chin, PLS *Graph User's Guide—Version 3.0*, Soft Modeling Inc, 1993–2003.

[12] Y. Li, *PLS-GUI: Graphic user interface for partial least squares (PLS-PC 1.8)–Version 2.0.1 beta*, University of South Carolina, Columbia, SC, 2005.

[13] N. Sellin, *PLSPATH—Version 3.01. Application manual*. Universität Hamburg, Hamburg, 1989.

[14] M.W. Browne, and G. Mels, Path analysis (RAMONA). In SPSS Inc. *SYSTAT 10 statistics II*, Chapter 7, (pp. 233–291), Chicago: Author, 2000.

[15] SAS Institute, *SAS PROC CALIS User's guide*, Cary, NC: Author, 1989.

[16] Statistica, *User's guide*, Tulsa, OK: Statistica Inc, 1998.

[17] C.M. Ringle, S. Wende and A. Will, *SmartPLS–Version 2.0*, Universität Hamburg, Hamburg, 2005.

[18] J.-R. Fu, *VisualPLS, Partial Least Square (PLS) Regression: An enhanced GUI for Lvpls (PLS 1.8 PC) Version 1.04*, National Kaohsiung University of Applied Sciences, Taiwan, ROC, 2006.

[19] J.-R. Fu, *VisualPLS: Partial least square (PLS) regression, an enhanced GUI for LVPLS (PLS 1.8 PC) Version 1.04*. http://www2.kuas.edu.tw/prof/fred/vpls/index.html, 2006.

[20] Addinsoft XLSTAT 2012, Data analysis and statistics software for Microsoft Excel, http://www.xlstat.com, Paris, France, 2012.

[21] R.E. Schumacker and G.A. Marcoulides, *Interaction and nonlinear effects in structural equation modeling*, Mahwah, NJ: Lawrence Erlbaum, 1998.

[22] D. Goodhue, W. Lewis, and R. Thompson, Comparing PLS to regression and LISREL: A response to Marcoulides, Chin, and Saunders. *MIS Quarterly*, **36(3)**, pp. 703–716, 2012.
[23] H. Treiblmaier, P.M. Bentler, and P. Mair, Formative constructs implemented via common factors. *Structural Equation Modeling*, **18**, pp. 1–17, 2010.
[24] R. Cudeck, Analysis of correlation matrices using covariance structure models. *Psychological Bulletin*, **105**, pp. 317–327, 1989.
[25] J.H. Steiger, Driving fast in reverse: The relationship between software development, theory, and education in structural equation modeling. *Journal of the American Statistical Association*, **96**, pp. 331–338, 2001.
[26] K.G. Jöreskog, Testing structural equation models. In K.A. Bollen and J.S. Long (Eds.), *Testing structural equation models* (pp. 294–316). Newbury Park, CA: Sage, 1993.
[27] H. Wold, Soft modeling: Intermediate between traditional model building and data analysis. *Mathematical statistics*, **6**, pp. 333–346, 1982.
[28] K. Hayashi, and G.A. Marcoulides, Examining identification issues in factor analysis. *Structural Equation Modeling*, **13**, pp. 631–645, 2006.
[29] C. Glymour, R. Scheines, R. Spirtes, and K. Kelly, *Discovering causal structure: Artificial intelligence, philosophy of science, and statistical modeling*, Orlando, FL: Academic Press, 1987.
[30] G.A. Marcoulides, Z.'Drezner, and R.E.'Schumacker, Model specification searches in structural equation modeling using Tabu search. *Structural Equation Modeling*, **5**, pp. 365–376, 1998.
[31] S.'Salhi, Heuristic search methods. In G.A.'Marcoulides (Ed.). *Modern methods for business research* (pp. 147–175). Mahwah, NJ: Lawrence Erlbaum, 1998.
[32] G.A. Marcoulides, and Z. Drezner, Specification searches in structural equation modeling with a genetic algorithm. In G.A. Marcoulides and R.E. Schumacker (Eds.), *New developments and techniques in structural equation modeling* (pp. 247–268). Mahwah, NJ: Lawrence Erlbaum, 2009.
[33] W.L. Leite, I.C. Huang, and G.A. Marcoulides, Item selection for the development of short-form of scaling using an ant colony optimization algorithm. *Multivariate Behavioral Research*, **43**, pp. 411–431, 2008.
[34] W.L. Leite, and G.A. Marcoulides, *Using the ant colony optimization algorithm for specification searches: A comparison of criteria*, Paper presented at the Annual Meeting of the American Education Research Association, San Diego: CA, 2009, April.
[35] G.A. Marcoulides, and W.L. Leite, Exploratory data mining algorithms for conducting searchers in structural equation modeling: A comparison of some fit criteria. In J.J. McArdle, and G. Ritschard (Eds.). *Contemporary Issues in Exploratory Data Mining in the Behavioral Sciences*, London: Taylor & Francis, 2013.
[36] G.A. Marcoulides, and Z. Drezner, Model specification searchers using ant colony optimization algorithms. *Structural Equation Modeling*, **10**, pp. 154–164, 2003.
[37] G.A. Marcoulides, and Z. Drezner, A model selection approach for the identification of quantitative trait loci in experimental crosses: Discussion on the paper by Broman and Speed. *Journal of the Royal Statistical Society, Series B*, **64**, pp. 754, 2002.
[38] G.A. Marcoulides, *Conducting specification searches in SEM using a ruin and recreate principle*, Paper presented at the Annual Meeting of the American Psychological Society, San Francisco, CA, 2009, May.
[39] G.A. Marcoulides, and Z. Drezner, Using simulated annealing for model selection in multiple regression analysis. *Multiple Regression Viewpoints*, **25**, pp. 1–4, 1999.
[40] G.A. Marcoulides, and Z. Drezner, Tabu search variable selection with resource constraints. *Communications in Statistics: Simulation & Computation*, **33**, pp. 355–362, 2004.
[41] Z. Drezner, and G.A. Marcoulides, A distance-based selection of parents in genetic algorithms. In M. Resenda and J.P. Sousa (eds.). *Metaheuristics: Computer decision-making* (pp. 257–278). Boston, MA: Kluwer Academic Publishers, 2003.
[42] G.A. Marcoulides, *Using heuristic algorithms for specification searches and optimization*, Paper presented at the Albert and Elaine Brochard Foundation International Colloquium, Missillac, France, 2010, July.

[43] G.A. Marcoulides, and M. Ing., Automated structural equation modeling strategies. In R. Hoyle (Ed.), *Handbook of structural equation modeling*, New York: Guilford Press, 2012.
[44] R. MacCallum, Specification searches in covariance structure modeling. *Psychological Bulletin*, **100**, pp. 107–120, 1986.
[45] S.J. Breckler, Applications of covariance structure modeling in Psychology: Cause for concern? *Psychological Bulletin*, **107**, pp. 260–273, 1990.
[46] S.L. Hershberger, The specification of equivalent models before the collection of data. In A. von Eye and C. Clogg (Eds.), *The analysis of latent variables in developmental research* (pp. 68–108). Beverly Hills, CA: Sage, 1994.
[47] S. Lee., and S.L. Hershberger, A simple rule for generating equivalent models in covariance structure modeling. *Multivariate Behavioral Research*, **25**, pp. 313–334, 1990.
[48] S.L. Hershberger, and G.A. Marcoulides, The problem of equivalent models. In G.R. Hancock and R.O. Mueller (Eds.), *Structural equation modeling: A second course* (2nd Ed.). G. R. Greenwich, CT: Information Age Publishing, 2012.
[49] T.C.W. Luijben, Equivalent models in covariance structure analysis. *Psychometrika*, **56**, pp. 653–665, 1991.
[50] R.C. MacCallum, D.T. Wegener, B.N. Uchino, and L R. Fabrigar, The problem of equivalent models in applications of covariance structure analysis. *Psychological Bulletin*, **114**, pp. 185–199, 1993.
[51] R. Levy, and G.R. Hancock, A framework of statistical tests for comparing mean and covariance structure models. *Multivariate Behavioral Research*, **42**, pp. 33–66, 2007.
[52] K.A. Marcus, Statistical equivalence, semantic equivalence, eliminative induction and the Raykov-Marcoulides proof of infinite equivalence. *Structural Equation Modeling*, **9**, pp. 503–522, 2002.
[53] R.P. McDonald, What can we learn from the path equations?: Identifiability, constraints, equivalence. *Psychometrika*, **67**, pp. 225–249, 2002.
[54] T. Raykov, Equivalent structural equation models and group equality constraints. *Multivariate Behavioral Research*, **32**, pp. 95–104, 1997.
[55] T. Raykov, and G.A. Marcoulides, Can there be infinitely many models equivalent to a given covariance structure model? *Structural Equation Modeling*, **8**, pp. 142–149, 2001.
[56] T. Raykov, and G.A. Marcoulides, Equivalent structural equation models: A challenge and responsibility. *Structural Equation Modeling*, **14**, pp. 527–532, 2007.
[57] T. Raykov, and S. Penev, On structural equation model equivalence. *Multivariate Behavioral Research*, **34**, pp. 199–244, 1999.
[58] L. J. Williams, H. Bozdogan, and L. Aiman-Smith, Inference problems with equivalent models. In G.A. Marcoulides and R.E. Schumacker (Eds.), *Advanced structural equation modeling: Issues and techniques* (pp. 279–314). Mahwah, NJ: Erlbaum, 1996.
[59] C. Hsiao, Identification. In Z. Griliches and M.D. Intriligator (Eds.), *Handbook of econometrics, Vol. 1* (pp. 224–283). Amsterdam: Elsevier Science, 1983.
[60] I. Stelzl, Changing a causal hypothesis without changing the fit: Some rules for generating equivalent path models. *Multivariate Behavioral Research*, **21**, pp. 309–331, 1986.
[61] J. Pearl, *Causality: Models, reasoning, and inference* (2nd ed.). New York: Cambridge University Press, 2009.
[62] B. Shipley, *Cause and correlation in biology*, New York: Cambridge University Press, 2000.
[63] P.A. Bekker, A. Merckens, and T.J. Wansbeek, *Identification, equivalent models, and computer algebra*, Boston: Academic Press, 1994.
[64] S.B. Green, M.S. Thompson, and J. Poirier, Exploratory analyses to improve model fit: Errors due to misspecification and a strategy to reduce their occurrence. *Structural Equation Modeling*, **6**, pp. 113–126, 1999.
[65] T. Raykov, and S. Penev, The problem of equivalent structural equation models: An individual residual perspective. In G.A. Marcoulides and R.E. Schumacker (Eds.). *New developments and techniques in structural equation modeling*, (pp. 297–321). Mahwah, NJ: Lawrence Erlbaum, 2001.
[66] G.H. Scherr, Irreproducible science: Editor's introduction. *The best of the Journal of Irreproducible Results*, New York: Workman Publishing, 1983.

[67] H. Apel, and H. Wold, *Simulation experiments on a case value basis with different sample lengths, different sample sizes, and different estimation models, including second dimension of latent variables*. Unpublished manuscript, Department of Statistics, University of Uppsala, Sweden, 1978.
[68] H. Apel, and H. Wold, Soft modeling with latent variables in two or more dimensions: PLS estimation and testing for predictive relevance. In K.G. Jöreskog and H. Wold (Eds.), *Systems under indirect observation, Part II*, (pp. 209–248). Amsterdam: North Holland, 1982.
[69] B.E. Areskoug, The first canonical correlation: Theoretical PLS analysis and simulation experiments. In K.G. Jöreskog and H. Wold (Eds.), *Systems under indirect observation, Part II* (pp. 95–118). Amsterdam: North Holland, 1982.
[70] W.W. Chin, and P R. Newsted, Structural equation modeling analysis with small samples using partial least squares. In R. H. Hoyle (Ed.), *Statistical strategies for small sample research* (pp. 307–341). Thousand Oaks, CA: Sage, 1999.
[71] B.S. Hui, *The partial least squares approach to path models of indirectly observed variables with multiple indicators*. Unpublished doctoral dissertation, University of Pennsylvania, Philadelphia, PA, 1978.
[72] B.S. Hui, and H. Wold, Consistency and consistency at large in partial least squares estimates. In K G. Jöreskog and H. Wold (Eds.), *Systems under indirect observation, Part II* (pp. 119–130). Amsterdam: North Holland, 1982.
[73] G.A. Marcoulides, and C. Saunders, PLS: A Silver Bullet? *MIS Quarterly*, **30**, pp. iv–viii, 2006.
[74] G.A. Marcoulides, W.W. Chin, and C. Saunders, A critical look at partial least squares modeling. *MIS Quarterly*, **33**, pp. 171–175, 2009.
[75] R. Noonan, and H. Wold, PLS path modeling with indirectly observed variables: A comparison of alternative estimates for the latent variable. In K. G. Jöreskog and H. Wold (Eds.), *Systems under indirect observation, Part II* (pp. 75–94). Amsterdam: North Holland, 1982.
[76] W.W. Chin, and J. Dibbern, A permutation based procedure for multi-group PLS analysis: Results of tests of differences on simulated data and a cross cultural analysis of the sourcing of information system services between Germany and the USA, In V.E. Vinzi, W.W. Chin, J. Henseler and H. Wang (Eds.), *Handbook of partial least squares concepts, methods and applications* (pp. 171–193), New York, NY: Springer Verlag, 2010.
[77] W.W. Chin, A permutation procedure for multi-group comparison of PLS models. Invited presentation, In M. Valares, M. Tenenhaus, P. Coelho, V.E. Vinzi, and A. Morineau (Eds.), *PLS and related methods, proceedings of the PLS-03 Iternational Symposium: "Focus on Customers,"* Lisbon, September 15th to 17th, pp. 33–43, 2003.
[78] K.G. Jöreskog, and H. Wold, *Systems under indirect observation, Part I & II*, North Holland: Amsterdam, 1982.
[79] R.F. Falk, and N.B. Miller, *A primer of soft modeling*. Akron, OH: The University of Akron Press, 1992.
[80] R.P. McDonald, Path analysis with composite variables. *Multivariate Behavioral Research*, **31**, pp. 239–270, 1996.
[81] I.R.R. Lu, *Latent variable modeling in business research: A comparison of regression based on IRT and CTT scores with structural equation models*, Doctoral dissertation, Carleton University, Canada, 2004.
[82] I.R.R. Lu, D.R. Thomas, and B.D. Zumbo, Embedding IRT in structural equation models: A comparison with regression based on IRT scores. *Structural Equation Modeling*, **12**, pp. 263–277, 2005.
[83] T. Dijkstra, Some comments on maximum likelihood and partial least squares methods. *Journal of Econometrics*, **22**, pp. 67–90, 1983.
[84] H. Schneeweiss, Consistency at large in models with latent variables. In K. Haagen, D.J. Bartholomew, and M. Deistler (Eds.). *Statistical modelling and latent variables*, Elsevier: Amsterdam, 1993.
[85] P.M. Bentler, *EQS structural equation program manual*, Encino, CA: Multivariate Software, Inc., 1995.

[86] L.-T. Hu, P.M. Bentler, and Y. Kano, Can test statistics in covariance structure analysis be trusted? *Psychological Bulletin*, **112**, pp. 351–362, 1992.
[87] J. Cohen, *Statistical power analysis for the behavioral sciences* (2nd ed.). Hillsdale, NJ: Lawrence Erlbaum, 1988.
[88] S.B. Green. How many subjects does it take to do a regression analysis? *Multivariate Behavioral Research*, **26**, pp. 499–510, 1991.
[89] A. Boomsma, The robustness of LISREL against small sample sizes in factor analysis models. In K.G. Jöreskog and H. Wold (Eds.), *Systems under indirect observation, Part I* (pp. 149–174). Amsterdam: North Holland, 1982.
[90] R. Cudeck, and S.J. Hensly, Model selection in covariance structure analysis and the "problem" of sample size: A clarification. *Psychological Bulletin*, **109**, pp. 512–519, 1991.
[91] D.L. Jackson, Revisiting sample size and the number of parameter estimates: Some support for the $N : q$ Hypothesis. *Structural Equation Modeling*, **10**, pp. 128–141, 2003.
[92] R.C. MacCallum, M.W. Browne, and H.M. Sugawara, Power analysis and determination of sample size for covariance structure modeling. *Psychological Methods*, **1**, pp. 130–149, 1996.
[93] L.K. Muthén, and B.O. Muthén, How to use a Monte Carlo study to decide on sample size and determine power. *Structural Equation Modeling*, **9**, pp. 599–620, 2002.
[94] C. Ringle, M. Sarstedt, and D. Straub, A critical look at the use of PLS-SEM in MIS Quarterly. *MIS Quarterly*, **36**, pp. iii–xiv, 2012.
[95] A. Bhattacherjee, and G. Premkumar, Understanding changes in belief and attitude toward information technology usage: A theoretical model and longitudinal test. *MIS Quarterly*, **28**, pp. 229–254, 2004.
[96] G. Bassellier, and I. Benbasat, Business competence of information technology professionals: Conceptual development and influence on IT-business partnerships. *MIS Quarterly*, **28**, pp. 673–694, 2004.
[97] M. Subramani, How do suppliers benefit from information technology use in supply chain relationships? *MIS Quarterly*, **28**, pp. 45–73, 2004.
[98] M.C. Denham, Prediction intervals in Partial Least Squares. *Journal of Chemometrics*, **11**, pp. 39–52, 1997.
[99] W. L. Hays, *Statistics*, Fort Worth, TX: Harcourt Brace Jovanovic, 1994.
[100] T. Raykov, and G.A. Marcoulides, *A first course in structural equation modeling* (2nd ed.). Mahwah, NJ: Lawrence Erlbaum, 2006.
[101] S. Serneels, P. Lemberge, and P.J. Van Espen, Calculation of PLS prediction intervals using efficient recursive relations for the Jacobian matrix. *Journal of Chemometrics*, **18**, pp. 76–80, 2004.
[102] T. Raykov, and G.A. Marcoulides, Using the delta method for approximate interval estimation of parameter functions in SEM. *Structural Equation Modeling*, **11**, pp. 621–637, 2004.
[103] G.A. Marcoulides, Evaluation of confirmatory factor analytic and structural equation models using goodness-of-fit indices. *Psychological Reports*, **67**, pp. 669–671, 1990.
[104] P. Paxton, P.J. Curran, K.A. Bollen, J. Kirby, and F. Chen, Monte Carlo experiments: Design and implementation. *Structural Equation Modeling*, **8**, pp. 287–312, 2001.
[105] A. Majchrak, C. Beath, R. Lim, and W.W. Chin, Managing client dialogues during information systems design to facilitate client learning. *MIS Quarterly*, **29**, pp. 653–672, 2005.
[106] J. Dibbern, W.W. Chin, A. Heinzl, Systemic determinants of the information systems outsourcing decision: A comparative study of German and United States firms. *Journal of the Association for Information Systems*, **13**, pp. 466–497, 2012.
[107] E. Walden, AIS World: Power analysis after the fact. Aisworld-bounces@listss.aisnet.org. [Monday, March 5, 2012 6:51pm], 2012.
[108] T. Raykov, and G.A. Marcoulides, *Basic statistics: An introduction with R*. London, UK: Rowman & Littlefield Publishers, Inc., 2012.
[109] A.A. Albert, The matrices of factor analysis. *Proceedings of the National Academy of Science*, **30**, pp. 90–95, USA, 1944.

[110] A.A. Albert, The minimum rank of a correlation matrix. *Proceedings of the National Academy of Science*, **30**, pp. 144–146, USA, 1944.
[111] T.C. Koopmans, and O. Reiersol, The identification of structural characteristics. *Annals of Mathematical Statistics*, **21**, pp. 165–181, 1950.
[112] W. Ledermann, On the rank of reduced correlation matrices in multiple factor analysis. *Psychometrika*, **2**, pp. 85–93, 1937.
[113] M. Sato, A study of identification problem and substitute use in principal component analysis in factor analysis. *Hiroshima Mathematical Journal*, **22**, pp. 479–524, 1992.
[114] M. Sato, On the identification problem of the factor analysis model: A review and an application for an estimation of air pollution source profiles and amounts. In *Factor analysis symposium at Osaka* (pp. 179–184). Osaka: University of Osaka, 2004.
[115] Y. Kano, Exploratory factor analysis with a common factor with two indicators. *Behaviormetrika*, **24**, pp. 129–145, 1997.
[116] K.A. Bollen, *Structural equations with latent variables*, New York, NY: Wiley, 1989.
[117] T. Raykov, G.A. Marcoulides, and T. Patelis, Saturated versus just identified models: A note on their distinction. *Educational and Psychological Measurement*, **73**, pp. 162–168, 2013.
[118] T.W. Anderson, and H. Rubin, Statistical inferences in factor analysis. In J. Neyman (Ed.), *Proceedings of the third Berkeley symposium on mathematical statistics and probability* (Vol. 5, pp. 111–150). Berkeley: University of California, 1956.
[119] M. Ihara, and Y. Kano, A new estimator of the uniqueness in factor analysis. *Psychometrika*, **51**, pp. 563–566, 1986.
[120] H. Yanai, K. Shigemasu, S. Maekawa, and M. Ichikawa, *Factor analysis: Its theory and methods*, Tokyo: Asakura-shoten (in Japanese), 1990.
[121] J.S. Williams, A note on the uniqueness of minimum rank solutions in factor analysis. *Psychometrika*, **46**, pp. 109–110, 1981.
[122] Y. Tumura, and M. Sato, On the identification in factor analysis. *TRU Mathematics*, **16**, pp. 121–131, 1980.
[123] T. Raykov, and G.A. Marcoulides, *Introduction to psychometric theory*. New York, NY: Routledge, 2011.
[124] R.P. McDonald, The dimensionality of tests and items. *British Journal of Mathematical and Statistical Psychology*, **34**, pp. 100–117, 1981.
[125] R.P. McDonald, *Test theory. A unified treatment*, Mahwah, NJ: Lawrence Erlbaum, 1999.
[126] P.M. Bentler, Alpha, dimension-free, and model-based internal consistency reliability. *Psychometrika*, **74**, pp. 137–144, 2009.
[127] L. Crocker, and J. Algina, *Introduction to classical and modern test theory*, Fort Worth, TX: Harcourt College Publishers, 1986.
[128] S.B. Green, and Y. Yang, Commentary on coefficient alpha: A cautionary tale. *Psychometrika*, **74**, pp. 121–136, 2009.
[129] S.B. Green, and Y. Yang, Reliability of summed item scores using structural equation modeling: An alternative to coefficient alpha. *Psychometrika*, **74**, pp. 155–167, 2009.
[130] W. Revelle, and R. Zinbarg, Coefficients alpha, beta, and the GLB: Comments on Sijtsma. *Psychometrika*, **74**, pp. 145–154, 2009.
[131] K. Sijtsma, On the use, the misuse, and the very limited usefulness of Cronbach's alpha. *Psychometrika*, **74**, pp. 107–120, 2009.
[132] K. Sijtsma, Reliability beyond theory and into practice. *Psychometrika*, **74**, pp. 169–174, 2009.
[133] T. Raykov, Cronbach's alpha and reliability of composite with interrelated nonhomogenous items. *Applied Psychological Measurement*, **22**, pp. 375–385, 1998.
[134] T. Raykov, Bias of coefficient alpha for congeneric measures with correlated errors. *Applied Psychological Measurement*, **25**, pp. 69–76, 2001.
[135] S.B. Green, R.W. Lissitz, and S.A. Mulaik, Limitations of coefficient alpha as an index of test unidimensionality. *Educational and Psychological Measurement*, **37**, pp. 827–838, 1977.
[136] I.T. Jolliffe, *Principal component analysis* (2nd Ed.). New York: Springer, 2002.

[137] D. Goodhue, W. Lewis, and R. Thompson, PLS, Small Sample Size, and Statistical Power in MIS Research. *HICSS '06 Proceedings of the 39th Annual Hawaii Conference on System Sciences*, pp. 202b, 2006.
[138] H. Hwang, N.K. Malhotra, Y. Kim, M.A. Tomiuk, and S. Hong, A comparative study of parameter recovery of the three approaches to structural equation modeling. *Journal of Marketing Research*, **67**, pp. 699–712, 2010.
[139] T. Dijkstra, Latent variables and indices: Herman Wold's basic design and partial least squares. In V.E. Vinzi, W.W. Chin, J. Henseler, and H. Wang (Eds.), *Handbook of partial least squares: Concepts, methods, and applications, computational statistics* (pp. 23–46). New York: Springer Verlag, 2010.
[140] D. Goodhue, W. Lewis, and R. Thompson, Does PLS have advantages for small sample size or non-normal data. *MIS Quarterly*, **36**, pp. 981–1001, 2012.

Correlated Component Regression: Re-thinking Regression in the Presence of Near Collinearity

Jay Magidson

Abstract We introduce a new regression method—called Correlated Component Regression (CCR)—which provides reliable predictions even with near multicollinear data. Near multicollinearity occurs when a large number of correlated predictors and relatively small sample size exists as well as situations involving a relatively small number of correlated predictors. Different variants of CCR are tailored to different types of regression (e.g. linear, logistic, Cox regression). We also present a step-down variable selection algorithm for eliminating irrelevant predictors. Unlike PLS-R and penalized regression approaches, CCR is scale invariant. CCR is illustrated in several examples involving real data and its performance is compared with other approaches using simulated data.[1] 1

Key words: Correlated component regression, Multicollinearity, High dimensional data, Big data, PLS regression, Variable selection, Suppressor variables, Scale invariance, Cross-validation

1 Background and Introduction

When correlation between predictor variables is moderate or high, coefficients estimated using traditional regression techniques become unstable or cannot be uniquely estimated due to multicollinearity (singularity of the covariance matrix). In the case of high dimensional data, where the number of predictor variables P approaches or exceeds the sample size N, such instability is often accompanied by perfect or near perfect predictions within the analysis sample. However, this seem-

[1] All data sets are available on the website statisticalinnovations.com

J. Magidson (✉)
Statistical Innovations Inc., Belmont, MA, USA
e-mail: jay@statisticalinnovations.com

H. Abdi et al. (eds.), *New Perspectives in Partial Least Squares and Related Methods*, Springer Proceedings in Mathematics & Statistics 56, DOI 10.1007/978-1-4614-8283-3_3,

ingly good predictive performance is usually associated with *overfitting*, and tends to deteriorate when applied to new cases outside the sample.

The primary "regularization" approaches that have been proposed for dealing with this problem are (1) penalized regression such as Ridge, Lasso and Elastic Net, and (2) dimension reduction methods such as Principle Component Regression, and PLS Regression (PLS-R). In this paper we describe a new method similar to PLS-R called Correlated Component Regression (CCR) and an associated step-down algorithm for reducing the number of predictors in the model to $P^* < P$. CCR has different variants depending upon the scale type of the dependent variable (e.g. CCR-linear regression for Y continuous, CCR-logistic regression for Y dichotomous, CCR-Cox regression for survival data). Unlike the other regularization approaches, the CCR algorithm shares with traditional maximum likelihood regression approaches the favorable property of scale invariance.

In this paper we introduce CCR, and describe its performance on various real and simulated data sets. The basic CCR algorithms are described in Sect. 2. CCR is contrasted with PLS-R in a linear regression key driver application with few predictors (Sect. 3) and in an application with Near Infrared (NIR) data involving many predictors (Sect. 4). We then describe the CCR extension to logistic regression, linear discriminant analysis (LDA) and survival analysis and discuss results from simulated data where suppressor variables are included among the predictors (Sect. 5). Results from our simulations suggest that CCR may be expected to outperform other sparse regularization approaches, especially when important suppressor variables are included among the predictors. We conclude with a discussion of a hybrid latent class CCR model extension (Sect. 6).

2 Correlated Component Regression

CCR utilizes $K < P$ correlated components, in place of the P predictors to predict an outcome variable. Each component S_k is an exact linear combination of the predictors, $X = (X_1, X_2, \ldots, X_P)$, the first component S_1 capturing the effects of those predictors that have direct effects on the outcome. The CCR-linear regression (CCR-LM) algorithm proceeds as follows:

Estimate the *loading* $\lambda_g^{(1)}$, on S_1, for each predictor $g = 1, 2, \ldots, P$, as the simple regression coefficient in the regression of Y on X_g , $\lambda_g^{(1)} = \frac{cov(Y,X_g)}{var(X_g)}$. Then S_1 is defined as a weighted average of all 1-predictor effects:

$$S_1 = \frac{1}{P} \sum_{g=1}^{P} \lambda_g^{(1)} X_g \tag{1}$$

The predictions for Y in the 1-component CCR model are obtained from the simple OLS regression of Y on S_1. Similarly, predictions for the 2-component CCR model are obtained from the simple OLS regression of Y on S_1 and S_2, where the second component S_2, captures the effects of suppressor variables that improve prediction by removing extraneous variation from one or more predictors that have direct effects. Component $S_{k'}$ for $k' > 1$, is defined as a weighted average of all 1-predictor partial effects, where the partial effect for predictor g is computed as the partial regression coefficient in the OLS regression of Y on X_g and all previously computed components $S_k, k = 1, \ldots, k'-1$. For example, for $K = 2$ we have:

$$Y = \alpha + \gamma_{1.g}^{(2)} S_1 + \lambda_g^{(2)} X_g + \varepsilon_g^{(2)} \tag{2}$$

and $S_2 = \frac{1}{P} \sum_{g=1}^{P} \lambda_g^{(2)} X_g$, or more simply[2] we can write $S_2 = \sum_{g=1}^{P} \lambda_g^{(2)} X_g$.

As mentioned earlier, predictions for Y in the K-component CCR model are obtained from the OLS regression of Y on $S_1, \ldots, S_K$. For example, for $K = 2$: $\hat{Y} = \alpha^{(2)} + b_1^{(2)} S_1 + b_2^{(2)} S_2$. In general, K^* components are computed, where the optimal value, K^*, is determined by M-fold cross-validation (CV). For $K = 1$, maximum regularization, no predictor correlation information is used in parameter estimation. As K is repeatedly incremented by 1, more and more information provided by the predictor correlations is utilized, and M-fold CV determines the value of K where near multicollinearity begins to deteriorate the predictive performance, the value for K^* being obtained accordingly. Deterioration occurs beginning at $K = 3$ for the example illustrated in Sect. 3, and thus $K^* = 2$.

Any K-component CCR model can be re-expressed to obtain regression coefficients for X by substituting for the components as follows:

$$\hat{Y} = \alpha^{(K)} + \sum_{k=1}^{K} b_k^{(K)} S_k = \alpha^{(K)} + \sum_{k=1}^{K} b_k^{(K)} \sum_{g=1}^{P} \lambda_g^{(k)} X_g = \alpha^{(K)} + \sum_{g=1}^{P} \beta_g X_g$$

Thus, the regression coefficient β_g for predictor X_g is simply the weighted sum of the loadings, where the weights are the regression coefficients for the components (component weights) in the K-component model: $\beta_g = \sum_{k=1}^{K} b_k^{(K)} \lambda_g^{(K)}$.

Simultaneous variable reduction is achieved using a step-down algorithm where at each step the least important predictor is removed, importance defined by the absolute value of the standardized coefficient $\beta_g^* = (\sigma_g / \sigma_Y) \beta_g$, where σ denotes the standard deviation. M-fold CV is used to determine the two tuning parameters: the number of components K and number of predictors P.

Consider an example with 6 predictors. For any given value for K, and say $M = 10$ folds, the basic CCR algorithm is applied 10 times, generating predictions for cases in each of the 10 folds[3] based on models with all 6 predictors, yielding a baseline (iteration = 0) CV-$R^2(K)$ for $P = 6$. In iteration 1, the variable reduction algorithm eliminates 1 predictor, which may not be the same predictor in all 10

---

[2] Going forward, the factor $1/P$ will be omitted which will not alter the predictions since multiplying S_k by P is offset by the OLS estimate for gamma (i.e., $\gamma_{k.g}^{(K)}$ becomes $\gamma_{k.g}^{(K)}/P$).

[3] The square of the correlation between these predictions and the observed Y yields CV-R^2.

subsamples, each resulting 5-predictor model being used to obtain new predictions for the associated omitted folds, yielding CV-$R^2(K)$ for $P = 5$. In iteration 2, the variable reduction process continues resulting in 10 4-predictor models, which yields CV-$R^2(K)$ for $P = 4$. Following the last iteration, $P^*(K)$ is determined as the value of P associated with the maximum CV-$R^2(K)$.

The basic idea is that by applying the proper amount of regularization through the tuning of K, we reduce any confounding effects due to high predictor correlation, thus obtaining more interpretable regression coefficients, and better, more reliable predictions. In addition, tuning P tends to eliminate irrelevant or otherwise extraneous predictors and further improve both prediction and interpretability.

Since the optimal P may depend on K, P should be tuned for each K, the optimal (P^*, K^*) yielding the global maximum for CV-R^2. Alternatively, as a matter of preference a final model may be based on a smaller value for P and/or K, such that the resulting CV-R^2 is within c standard errors of the global maximum, where $c \leq 1$.

Since K can never exceed P, for $P = K$, the model becomes saturated and is equivalent to the traditional regression model.[4] For pre-specified K, when P is reduced below K, we maintain the saturated model by also reducing K so $K = P$. For example, for $K = 4$, when we step down to 3 predictors, we reduce K so $K = 3$. Similarly, when we step down to 1 predictor, $K = 1$. This is similar to traditional stepwise regression with backwards elimination.

Prime predictors, those having direct effects, are identified as those having substantial loadings on S_1, and suppressor variables, as those having substantial loadings on one or more other components, and relatively small loadings on S_1. See Sect. 5 for further insight into suppressor variables.

Since CCR is scale invariant, it yields identical results regardless of whether predictions are based on unstandardized or standardized predictors (Z-scores). Other methods such as PLS-R and penalized regression (Ridge Regression, Lasso, Elastic Net) are not scale invariant and hence yield different results depending on the predictor scaling used.

3 A Simple Example with Six Correlated Predictors

Our first example makes use of data involving the prediction of car prices (Y) as a linear function of 6 predictors, each having a statistically significant positive correlation with Y (between 0.6 and 0.9).

- $N = 24$ car models
- Dependent variable: Y = PRICE (car price measured in francs)
- 6 Predictor Variables:
  - X_1 = CYLINDER (engine measured in cubic centimeters)
  - X_2 = POWER (horsepower)
  - X_3 = SPEED (top speed in kilometers/hour)

[4] See Appendix for proof of this equivalence.

- X_4 = WEIGHT (kilograms)
- X_5 = LENGTH (centimeters)
- X_6 = WIDTH (centimeters)

The OLS regression solution (Table 1a) imposes no regularization, maximizing R^2 in the training sample. This solution is equivalent to that obtained from a saturated ($K = P = 6$ components) CCR model. Since this solution is based on a relatively small sample and correlated predictors, it is likely to overfit the data and the R^2 is likely to be an overly optimistic estimate of the true population R^2. Table 1a shows only 1 statistically significant coefficient (0.05) and unrealistic (negative) coefficient estimates for 3 of the 6 predictors, which are problems that can be explained by model overfitting due to imposing no regularization.

Table 1: (a) (left) shows OLS Regression Coefficient results ($P = K = 6$) and (b) (right) shows R^2 and CV-R^2 for different numbers of components K and for the final CCR model ($P = 3, K = 2$)

	Unstandardized coefficients		Standardized coefficients		
	$\hat{\beta}$	Std. error	$\hat{\beta}^*$	t	Sig.
CYLINDER	**−1.9**	33.6	**−0.02**	−0.06	0.95
POWER	1,315.9	613.5	0.89	2.14	0.05
SPEED	**−472.5**	740.3	**−0.21**	−0.64	0.53
WEIGHT	45.9	100.0	0.18	0.46	0.65
LENGTH	209.6	504.2	0.15	0.42	0.68
WIDTH	**−505.4**	1,501.6	**−0.07**	−0.34	0.74
(Constant)	12,070.4	194,786.6		0.06	95

P	K	R^2	CV-R^2
6	1	0.7852	0.7457
6	2	0.8189	**0.7461**
6	3	0.8449	0.6732
6	4	0.8469	0.6455
6	5	0.8474	0.6371
6	6	0.8474	0.6342
3	2	0.8362	**0.7690**

To determine the value for K that provides the optimal amount of regularization, we choose the CCR model that maximizes the CV-R^2. For cross-validation we used 10 rounds of 6-folds, since 24 divides evenly into 6, each fold containing exactly 4 cars. Table 1b shows that $K = 2$ components provides the maximum CV-R^2 based on $P = 6$ predictors, and when the step-down algorithm is employed, CV-R^2 increases to 0.769 which occurs with $P^* = 3$ predictors.[5] While traditional OLS regression yields a higher R^2 in the analysis sample (0.847 vs. 0.836), the 2-component CCR model with 3 predictors yields a higher CV-R^2, suggesting that this CCR model will outperform[6] OLS regression when applied to new data.

Further evidence of improvement for the 2-component models over OLS regression is that the coefficients are more interpretable. Table 2 shows that the coefficients in the 2-component CCR models are all positive, which is what we would expect if we were to interpret them as measures of effect.[7]

[5] The analysis was conducted using the CORExpress® package (patent pending) [1].

[6] Since multiple rounds of 6-folds are performed, standard errors are available, which yield 95% confidence intervals for CV-R^2 of 0.746 ± 0.04 for the CCR model with 6 predictors and 0.769 ± 0.056 for the 3-predictor CCR model.

[7] Interestingly, each CCR model based on an insufficient amount of regularization ($K > 2$) provides uninterpretable coefficients, in each case exactly three coefficients turning out negative.

Table 2 Comparison of results from PLSR (a) (left) with unstandardized predictors, and (b) with standardized predictors, and CCR (c) without variable selection and (d) (right) with variable selection

	PLS with unstandardized predictors ($K^* = 3$)	PLS with standardized predictors ($K^* = 2$)	CCR ($K^* = 2$)	CCR with selection ($K^* = 2$)
Training R^2	0.83	0.81	0.82	0.84
CV-R^2	0.69	0.76	0.75	0.77
Predictors	$\hat{\beta}^*$	$\hat{\beta}^*$	$\hat{\beta}^*$	$\hat{\beta}^*$
CYLINDER	−0.02	0.19	0.19	0.00
POWER	0.43	0.31	0.37	0.45
SPEED	0.17	0.22	0.20	0.10
WEIGHT	0.48	0.18	0.17	0.44
LENGTH	−0.05	0.08	0.02	0.00
WIDTH	0.00	0.01	0.05	0.00

PLS-R with standardized predictors, the recommended PLS-R option when predictors are measured in different units, yields similar results to CCR here. When the predictors remain unstandardized, PLS-R yields more components ($K^* = 3$), two negative coefficients, and substantially worse predictions (CV-$R^2 = 0.69$), as the much larger variance for the predictor CYLINDER causes this predictor to dominate the first component, requiring two additional components to recover.

4 An Example with Near Infrared (NIR) Data

Next, we analyze high dimensional data involving $N = 72$ biscuits, each measured at each of $P = 700$ near infrared (NIR) wave-lengths corresponding to every other wavelength between the range 1,100–2,500 [2]. Since all 700 predictors are measured in comparable units in this popular PLS-R application, typically the 700 predictors are analyzed on an unstandardized basis, or standardized using Pareto scaling [3] where the scaling factor is the square root of the standard deviation. As shown above, results from PLS-R differ depending upon whether the predictors are standardized or not, while for the scale invariant CCR, no decision needs to be made regarding such standardization, predictions being identical in either case.

The goal of modeling here is to reduce costs of monitoring fat content by predicting the percent fat based on spectroscopic absorbance variables from the NIR frequencies. Following Kraemer and Boulesteix [4], we use $N = 40$ samples as the calibration (training) set to develop models based on the 700 wave lengths.

It is well known that for NIR data, a column plot of regression coefficients exhibit a sequence of oscillating patterns, the most important wavelength ranges being those with the highest peak-to-peak amplitude. For example, for these data, wavelengths in the 1,500–1,598 range yield a peak to peak amplitude of $0.109 - (-0.203) = 0.312$, based on a CCR model with $K = 9$ (see Fig. 1).

Table 3a compares the corresponding amplitudes obtained from CCR and both unstandardized and Pareto standardized PLS-R models, where the number of components is determined based on 10 rounds of 5-folds. As can be seen in Table 3a, all three models agree that absorbances from the 1,500–1,598 wavelengths tend to be among the most important (relatively large amplitude).

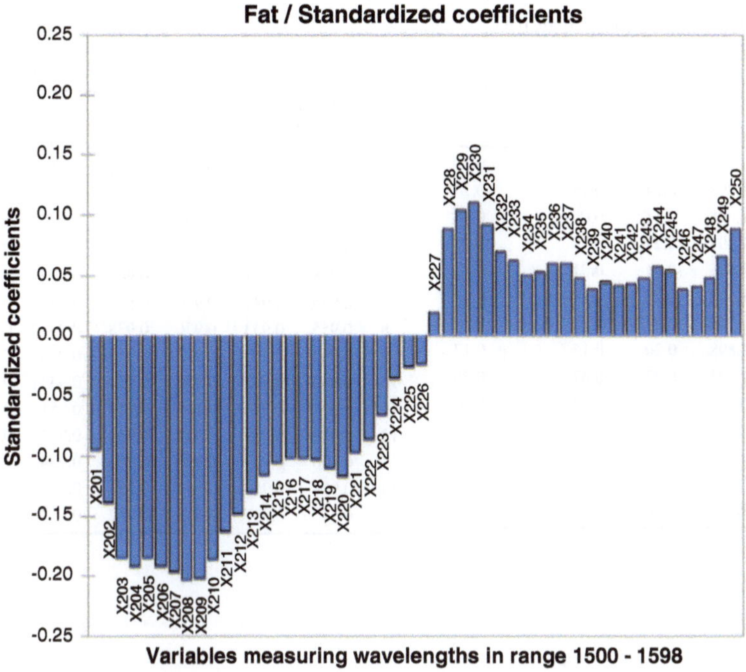

Fig. 1: Column plot of standardized coefficients output from XLSTAT-CCR

Previous analyses of these data excluded the highest 50 wavelengths since they were "...thought to contain little useful information" [5]. Table 3a shows that CCR identifies these wavelengths as least important (smallest amplitude), but the amplitude of 0.44 resulting from PLS-R suggests that these wavelengths are important.

Figure 2 shows the standardized coefficients for the 50 highest wavelengths for CCR and PLS-R models. As can be seen, the weights obtained from the CCR model are small and diminishing, the coefficients for the highest wavelengths being very close to 0. In contrast, PLS-R weights are quite high and show no sign of diminishing for the highest wavelengths (Fig. 2(right)), a similar pattern being observed for PLS-Pareto.

One possible reason that the conclusions from CCR and PLS-R differ regarding the importance of these high wavelengths is that its scale invariance property allows CCR to better determine that the high variability associated with these wavelengths is due to increased amounts of measurement error. In other words, the much higher amplitude obtained from PLS-R is likely due to the higher standard deviations of the absorbances in this range.

Table 3: (a) (left) Comparison of peak-to-peak amplitudes for various frequency ranges based on three models, with the most and least important ranges according to CCR in bold, and (b) (right) comparison of CV-R^2 (highest is bold) obtained from three models with ($P = 700$) and without ($P = 650$) the highest wavelengths included among the predictors

	Peak-to-peak amplitude based on standardized coefficients		
Wavelengths	CCR ($K = 9$)	PLSR ($K = 13$)	PLS-Pareto ($K = 13$)
1,100–1,198	0.16	0.12	0.19
1,200–1,298	0.24	0.15	0.24
1,300–1,398	0.11	0.08	0.13
1,400–1,498	0.27	0.25	0.21
1,500–1,598	**0.31**	**0.31**	**0.32**
1,600–1,698	0.23	0.14	0.15
1,700–1,798	0.27	0.24	0.22
1,800–1,898	0.20	0.15	0.17
1,900–1,998	0.07	0.47	0.36
2,000–2,098	0.22	0.37	0.30
2,100–2,198	0.16	0.17	0.15
2,200–2,298	0.18	0.30	0.29
2,300–2,398	0.18	0.55	0.47
2,400–2,498	**0.06**	**0.44**	0.25

	CCR		PLSR		PLS-Pareto	
K	$P = 700$	$P = 650$	$P = 700$	$P = 650$	$P = 700$	$P = 650$
1	0.237	0.232	0.260	0.257	0.247	0.245
2	0.506	0.589	0.345	0.461	0.412	0.477
3	0.759	0.860	0.736	0.725	0.721	0.736
4	0.914	0.932	0.906	0.835	0.922	0.882
5	0.948	0.946	0.916	0.928	0.933	0.917
6	0.948	0.951	0.919	0.947	0.927	0.949
7	0.945	0.947	0.930	0.942	0.936	0.946
8	0.955	0.953	0.936	0.938	0.944	0.948
9	**0.962**	0.960	0.932	0.952	0.946	0.952
10	0.960	**0.963**	0.939	0.958	0.946	0.961
11	0.957	0.959	0.942	**0.959**	0.951	**0.962**
12	0.958	0.959	0.949	0.958	0.952	0.961
13	0.958	0.959	**0.950**	0.956	**0.954**	0.9059
14	0.958	0.958	0.947	0.953	0.953	0.957
15	0.958	0.957	0.946	0.952	0.952	0.956

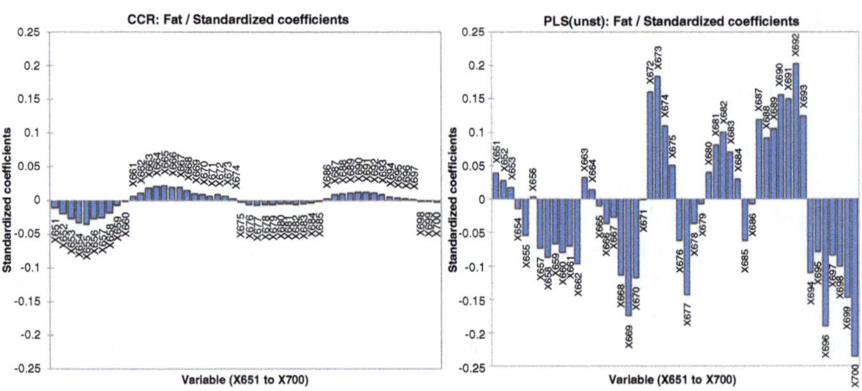

Fig. 2: Comparison of column plots of standardized coefficients for 50 highest wavelengths based on the CCR (*left*) vs. PLS-R estimated with unstandardized predictors (*right*)

To test the hypothesis that these higher wavelengths tend to be unimportant, we re-estimated the models after omitting these variables. Table 3b shows that for all three models, the CV-R^2 increases when these variables are omitted, supporting the hypothesis that these wavelengths are not important.

In order to compare the predictive performance of CCR with other regularization approaches, 100 simulated samples of size $N = 50$ were generated with 14 predictors according to the assumptions of OLS regression. An additional 14 extraneous predictors, correlated with the 14 true predictors, plus 28 irrelevant predictors, were also generated and included among the candidate predictors. The results indicated that CCR outperformed PLS-R, Elastic Net, and sparse PLS with respect to mean squared error, and several other criteria. All methods were tuned using an independent validation sample of size 50 (for more details, see [6]).

5 Extension of CCR to Logistic Regression, Linear Discriminant Analysis and Survival Analysis

When the dependent variable is dichotomous, the CCR algorithm generalizes directly to CCR-LOGISTIC and CCR-LDA respectively depending upon whether no assumptions are made about the predictor distributions, or whether the normality assumptions from linear discriminant analysis are made. In either case, the generalization involves replacing Y by $Logit(Y)$ on the left side of the linear equations. Thus, for example, under CCR-LOGISTIC and CCR-LDA Eq. 2 becomes:

$$Logit(Y) = \alpha + \gamma_{1.g}^{(2)} S_1 + \lambda_g^{(2)} X_g \tag{3}$$

where parameter estimation in each regression equation is performed by use of the appropriate ML algorithm (for logistic regression or LDA).

M-fold cross-validation continues to be used for tuning, but CV-R^2 is replaced by the more appropriate statistics CV-Accuracy and CV-AUC, AUC denoting the Area Under the ROC Curve. Accuracy is most useful when the distribution of the dichotomous Y is approximately uniform, about 50% of the sample being in each group. When Y is skewed, accuracy frequently results in many ties and thus is not as useful. In such cases AUC can be used as a tie breaker with Accuracy as the primary criterion or in the case of large skew, AUC can replace accuracy as primary.

For survival data, Cox regression and other important log-linear hazard models can be expressed as Poisson regression models since the likelihood functions are equivalent [7]. As such, CCR can be employed using the logit equation above where Y is a dichotomous variable indicating the occurrence of a rare event. In this case since Y has an extreme skew, the AUC is used as the primary criterion.

Similar to the result for CCR-linear regression, predictions obtained for the saturated CCR model for dichotomous Y are equivalent to those from the corresponding traditional model (logistic regression, LDA and Poisson regression).[8] In addition, for

[8] In general, the saturated model occurs when $K \geq minimum(P, N-1)$.

dichotomous Y the 1-component CCR model is equivalent to Naïve Bayes, which is also called diagonal discriminant analysis [8] in the case of CCR-LDA.

In a surprising result reported in [9], for high dimensional data (small samples and many predictors) generated according to the LDA assumptions, traditional LDA does not work well, and is outperformed by Naïve Bayes. Because of the equivalences described above, this means that the *1-component* CCR model should outperform the *saturated* CCR model under such conditions. However, we know that the Naïve Bayes model will not work well if predictors include 1 or more important suppressor variables, since suppressor variables tend to have 0 loadings on the first component and require at least two components for their effects to be captured in the model [10]. Thus, a CCR model with two components should outperform Naïve Bayes whenever important suppressor variables are included among the predictors.

Despite extensive literature documenting the enhancement effects of suppressor variables (e.g. [11, 12]), most pre-screening methods omit suppressor variables prior to model development, resulting in suboptimal models.[9] Since suppressor variables are commonplace and often are among the most important predictors in a model [10], such screening is akin to "throwing out the baby with the bath water."

In order to compare the predictive performance of CCR with other sparse modeling methods in a realistic high dimensional setting, data were simulated according to LDA assumptions to reflect the relationships among real world data for prostate cancer patients and normals where at least one important suppressor variable was among the predictors. The simulated data involved 100 samples each with N = 25 cases in each group, the predictors including 28 valid predictors plus 56 that were irrelevant. The sparse methods included CCR, sparse PLS-R [13, 14] and the penalized regression methods Lasso and Elastic Net [15–17]. For tuning purposes, cross-validation with five folds was used with accuracy as the criterion for all methods.

Results showed that CCR with typically 4–10 components outperformed the other methods with respect to accuracy (82.6% vs. 80.9% for sparse PLS-R, and under 80% for Lasso and Elastic Net), and fewest irrelevant predictors (3.4 vs. 6.2 for Lasso, 11.5 for Elastic Net and 13.1 for sparse PLS-R). The most important variable, which was a suppressor variable, was captured in the CCR model in 91 of the 100 samples compared to 78 for sparse PLS-R, 61 for elastic net and only 51 for Lasso. For further details of this and other simulations see [6].

6 Extension to Latent Class Models

In practice, sample data often reflects two or more distinct subpopulations (latent segments), with different intercepts and/or different regression coefficients, possibly due to different key drivers or at least different effects for the key drivers. In this section we describe a 2-step hybrid approach for identifying the latent segments without use of the predictors (step 1) and then using CCR to develop a predictive

[9] For a rare exception, ISIS (see [19]) corrects for the exclusion of suppressor variables by the popular SIS screening. CCR has been shown to outperform ISIS in a simulation study [10].

model based on a possibly large number of predictors (step 2). If the predictors are characteristics of the respondents, then the dependent variable (Y) would be the latent classes, while if the predictors were attributes of objects being rated, Y would be taken as the ratings.

As an example of the first case where the latent segments have different intercepts, in step 1 a latent class (LC) survival analysis was conducted on a sample of patients with late stage prostate cancer. The LC model identified both long-term and short term survival groups [18]. The goal in that study was to use gene expression measurements to predict whether patients belong to the longer or shorter survival class. Since the relevant genes were not known beforehand, the large number of available candidate predictors (genes) ruled out use of traditional methods.

In this case, CCR can be used to simultaneously select the appropriate genes and develop reliable predictions of LC membership based on the selected genes. One way to perform this task is to predict the dichotomy formed by the two groups of patients classified according to the LC model. However, this approach is suboptimal because the classifications contain error due to modal assignment. That is, assigning patients with a posterior probability of say 0.6 of being a long term survivor to this class (with probability 1) ignores the 40% expected misclassification error ($1 - 0.6 = 0.4$). The better way is to perform a weighted logistic (or LDA) CCR regression, where posterior probabilities from the LC model serve as case weights.

Table 4: Results from CCR showing that $P = 3$ of the 16 attributes were selected for inclusion in the model together with the random intercept CFactor1

Results for segment 1		Results for segment 2	
Variable	Standardized coefficient	Variable	Standardized coefficient
CFactor1	0.425	CFactor1	0.555
Fructose	−0.128	Sweeteningpower	−0.169
Sweeteningpower	0.238	Smellintensity	−0.129
Acidity	−0.325	Acidity	0.214

As an example of the second case, consider ratings on 6 different orange juice (OJ) drinks provided by 96 judges [20]. Based on these ratings, in step 1 a LC regression determines that there are two latent segments[10] exhibiting different OJ preferences. In step 2, separate weighted least squares CCR regressions are performed for each class to predict ratings based on the 16 OJ attributes. For a given class, posterior membership probabilities for that class are used as case weights.

For this application CCR is needed because traditional regression can include no more than six attributes in the model due to the fact that the attributes describe the six juices rather than the respondents. In addition, since these data consist of multiple records (6) per case, residuals from records associated with the same case are correlated, a violation of the independent observations assumption. This violation

[10] The number of classes was determined based on the Bayesian Information Criterion. For further details of this methodology, see [21].

is handled in step 1 by the LC model satisfying the "local independence" assumption. In step 2, the cross-validation is refined by assigning records associated with the same case to the same fold. Separate CCR models are developed for each LC segment, and then combined to obtain predicted ratings, providing substantial improvement over the traditional regression (CV-R^2 increases from 0.28 to 0.48). Results of step 2 are summarized in Table 4, showing that the most important attribute for both segments is acidity since it has the highest standardized coefficient magnitude. Segment 1 tends to prefer juices with low acidity (negative coefficient) and high sweetening power (positive coefficient) while the reverse is true for segment 2. Details of this analysis are provided in tutorials from www.statisticalinnovations.com.

Acknowledgements The author gratefully acknowledges the contributions of Yiping Yuan in conducting the simulations and providing the proof in the Appendix, Michel Tennenhaus for contributing the car price and orange juice example datasets, programming assistance and ongoing support from Jeremy Magland, Alexander Ahlstom, William Barker, and helpful comments from Harald Martens, Vincenzo Vinzi, Anthony Babinec, Gary Bennett, Karl Wassmann, David Rindskoff, Nicole Krämer, and an anonymous reviewer.

Appendix

Claim: OLS predictions based on X are equivalent to predictions based on $S = XA$, where A is a nonsingular matrix.

Proof:

- Predictions based on X:

$$\hat{Y} = X\hat{\beta} = X(X'X)^{-1}X'Y.$$

- Predictions based on S:

$$\begin{aligned}\hat{Y} &= S\hat{\gamma} \\ &= S(S'S)^{-1}S'Y = XA((XA)'XA)^{-1}(XA)'Y \\ &= XA(A'X'XA)^{-1}A'X'Y = XAA^{-1}(X'X)^{-1}A'^{-1}A'X'Y \\ &= X(X'X)^{-1}X'Y.\end{aligned}$$

Equations 4 and 5 above follow from standard operations with square matrices:

$$(BC)' = C'B' \quad \text{and} \quad (BC)^{-1} = C^{-1}B^{-1}.$$

It also follows that the OLS regression coefficients for X are identical to those obtained from CCR with a saturated model (i.e., $K = P$).

References

[1] J. Magidson, "COREXpress User's Guide: Manual for COREXpress", Belmont, MA: Statistical Innovations Inc., 2011.

[2] B. Osbourne, T. Fearn, A. Miller, and S. Douglas, "Application of near infrared reflectance spectroscopy to compositional analysis of biscuits and biscuit dough," *Journal of Science and Food Agriculture*, 35, 99–105, 1984.

[3] L. Eriksson, E. Johansson, N. Kettaneh-Wold, and S. Wold, "Introduction to multi- and megavariate data analysis using projection methods (PCA & PLS)," *Umetrics*, pp. 213–225, 1999.

[4] N. Kraemer, and A. Boulesteix, "Penalized Partial Least Squares (PPLS)," R Package, V. 1.05, Aug. 2011.

[5] P.J. Brown, T. Fearn, and M. Vannucci, "Bayesian wavelet regression on curves with application to a spectroscopic calibration problem," *Journal of the American Statistical Association*, 96, 398–408, 2001.

[6] J. Magidson, "Correlated Component Regression: A Prediction/Classification Methodology for Possibly Many Features," *JSM Proceedings of the American Statistical Association*, pp. 4372–4386, 2010.

[7] N. Laird and D. Oliver, "Covariance analysis of censored survival data using log-linear analysis techniques," *Journal of the American Statistical Association*, 76, pp. 231–240, 1981.

[8] T.R. Golub, D.K.Slonim, P. Tamayo, C. Huard, M. Gaasenbeek, J.P. Mesirov, H. Coller, M.L. Loh, J.R. Downing, M.A. Caligiuri, C.D. Bloomfield, and E.S. Lander, "Molecular classification of cancer: class discovery and class prediction by gene expression monitoring," *Science*, 286, pp. 531–537, Oct. 1999.

[9] P. Bickel and E. Levina, "Some theory for Fisher's linear discriminant function, 'naïve Bayes' and some alternatives when there are many more variables than observations," *Bernoulli*, 10, 989–1010, 2004.

[10] J. Magidson and K. Wassmann, "The Role of Proxy Genes in Predictive Models: An Application to Early Detection of Prostate Cancer," *2010 JSM Proceedings of American Statistical Association, Biometrics Section*, pp. 2739–2753, 2010.

[11] P. Horst, "The role of predictor variables which are independent of the criterion," *Social Science Research Bulletin*, 48, pp. 431–436, 1941.

[12] H. Lynn, "Suppression and Confounding in Action," *The American Statistician*, 57, pp. 58–61, 2003.

[13] H. Chun and S. Keleş, "Sparse partial least squares regression for simultaneous dimension reduction and variable selection," University of Wisconsin, Madison, 2009.

[14] H. Chun and S. Keleş, "Sparse Partial Least Squares Classification for High Dimensional Data," *Statistical Applications in Genetics and Molecular Biology*, 9, 17, 2010.

[15] R. Tibshirani, "Regression Shrinkage and Selection via the Lasso," *Journal of the Royal Statistical Society Series B (Methodological)*, 58, pp. 267–288, 1996.

[16] J. Friedman, T.Hastie, and R.Tibshirani, "Regularization Paths for Generalized Linear Models via Coordinate Descent," *Journal of Statistical Software*, 33, pp. 1–22, 2010.

[17] J. Friedman, T. Hastie, and R. Tibshirani, "Lasso and elastic-net regularized generalized linear models," Version 1.3, Jstatsoft.org, April 25, 2010.

[18] R. Ross, M. Galsky, H. Scher, J. Magidson, K. Wassmann, G. Lee, L. Katz, S. Subudhi, A. Anand, M. Fleisher, P. Kantoff, and W. Oh, "A whole-blood RNA transcript-based prognostic model in men with castration-resistant prostate cancer: a prospective study," *Lancet Oncology*, 2012; http://dx.doi.org/10.1016/S1470-2045(12)70263-2.

[19] J. Fan, Samworth, and W. Yichao, "Ultrahigh Dimensional Feature Selection: Beyond the Linear Model," *Journal of Machine Learning Research*, 10, pp. 2013–2038, 2009.
[20] M Tenenhaus, M., Pagès, J., Ambroisine L. and C. Guinot, "PLS methodology for studying relationships between hedonic judgments and product characteristics," *Food Quality and Preference*, 16, pp. 315–325, 2005.
[21] J. Magidson, and J. Vermunt, "Latent Class Models," in D.Kaplan (Ed.), *The Sage Handbook of Quantitative Methodology for the Social Sciences*, pp. 175–198. Thousand Oaks: Sage Publications, 2004.

Part II
Large Datasets and Genomics

Integrating Partial Least Squares Correlation and Correspondence Analysis for Nominal Data

Derek Beaton, Francesca Filbey, and Hervé Abdi

Abstract We present an extension of PLS—called partial least squares correspondence analysis (PLSCA)—tailored for the analysis of nominal data. As the name indicates, PLSCA combines features of PLS (analyzing the information common to two tables) and correspondence analysis (CA, analyzing nominal data). We also present inferential techniques for PLSCA such as bootstrap, permutation, and χ^2 omnibus tests. We illustrate PLSCA with two nominal data tables that store (respectively) behavioral and genetics information.

Key words: Partial least squares, Correspondence analysis, Multiple correspondence analysis, Chi-square distance, Genomics

1 Introduction

With the advent of relatively inexpensive genome-wide sequencing it is now possible to obtain large amounts of detailed genetic information on large samples of participants, and, so, several large sample studies are currently under way whose main goal is to relate genetics to behavior or clinical status. In these studies, the genetic information of each participant is a long list of pairs (one per chromosome) of DNA nucleotides (A, T, C, and G)—which could occur in $2^4 = 16$ different configurations—grouped in 23 chromosomes. However, only genomic locations that show enough variability in a population are used. These locations of variability are called single nucleotide polymorphisms (SNPs). Each SNP has a major allele (e.g., A), which is the most frequent nucleotide (in a population), and a minor allele (e.g., T; rare in a population but required to be found in at least 5% of the population to

D. Beaton (✉) • F. Filbey • H. Abdi
The University of Texas at Dallas, School of Behavioral and Brain Sciences.
MS: Gr.4.1 800 West Campbell Road, Richardson, TX 75080-3021, USA
e-mail: beaton@utdallas.edu; filbey@utdallas.edu; herve@utdallas.edu

H. Abdi et al. (eds.), *New Perspectives in Partial Least Squares and Related Methods*, Springer Proceedings in Mathematics & Statistics 56, DOI 10.1007/978-1-4614-8283-3_4,

be considered "relevant"). Thus, in practice only three variants for each location are used: the major homozygote (e.g., *AA*), the minor homozygote (e.g., *TT*), and the heterozygote (e.g., *AT*).

Multivariate data sets of SNPs are most often *re*-coded through a process of counting alleles: 0, 1, or 2. While 1 is always the heterozygote, 0 and 2 could be ambiguous. For example, minor homozygotes can be coded according to two different schemes: (1) having 2 minor alleles [1] or (2) having 0 major alleles [2]. In most analyses, the SNPs are treated as quantitative data because most statistical methods used rely upon quantitative measures [3–5]. Some multivariate approaches for SNPs include independent components analysis (ICA) [6], sparse reduced-rank regression (sRRR) [7], multivariate distance matrix regression (MDMR) [8, 9], and PLS regression (PLSR) [10, 11]. It should be noted that both sRRR and MDMR are PLSR-like techniques. However, these methods depend on the allele counting approach that assumes a uniform linear increase *for all* SNP*s* from 0 to 1 and from 1 to 2, but SNPs do not identify *how much* of an allele is present, *only which* allele (i.e., nucleotide variation) is present. Because the assumptions of a quantitative coding scheme seem unrealistic, we have decided to use a *qualitative* coding scheme and to consider that the values 0, 1, and 2 represent three different levels of a nominal variable (e.g., $0 = AA$, $1 = AT$, and $2 = TT$). In studies relating genetics and behavior, behavior is evaluated by surveys or questionnaires that also provide qualitative answers. So the problem of relating genetics and behavior reduces to finding the information common to two tables of qualitative data. Partial least square correlation (PLSC, see [12, 14]) would be an obvious solution to this "two-table problem" but it works only for quantitative data. An obvious candidate to analyze one table of qualitative data is correspondence analysis (CA), which generalizes principal component analysis (PCA) to qualitative data. In this paper, we present partial least squares-correspondence analysis (PLSCA): A generalization of PLSC—tailored for qualitative data—that integrates features of PLSC and CA. We illustrate PLSCA with an example on genetics and substance abuse.

2 PLSC and PLSCA

2.1 Notations

Matrices are denoted by bold face upper-case letters (e.g., $\mathbf{X}$), vectors by bold face lower case letters (e.g., $\mathbf{m}$). The identity matrix is denoted $\mathbf{I}$. The transpose operation is denoted ^{T} and the inverse of a square matrix is denoted $^{-1}$. The diag{} operator transforms a vector into a diagonal matrix when applied to a vector and extracts the diagonal element of a matrix when applied to a matrix.

2.2 PLSC: A Refresher

Partial least square correlation [12, 13] is a technique whose goal is to find and analyze the information common to two data tables collecting information on the same observations. This technique seems to have been independently (re)discovered by multiple authors and therefore, it exists under different names such as "inter-battery analysis" (in 1958 and probably the earliest instance of the technique, [15]), "PLS-SVD" [12, 17, 18], "intercorrelation analysis," "canonical covariance analysis," [19], "robust canonical analysis" [20], or "co-inertia analysis" [21]. In PLSC, $\mathbf{X}$ and $\mathbf{Y}$ denote two I by J and I by K matrices that describe the I observations (respectively) by J and K quantitative variables. The data matrices are, in general, pre-processed such that each variable has zero mean and unitary norm; the pre-processed data matrices are denoted $\mathbf{Z_X}$ and $\mathbf{Z_Y}$. The first step of PLSC is to compute the correlation matrix $\mathbf{R} = \mathbf{Z_X}^{\mathsf{T}}\mathbf{Z_Y}$, whose singular value decomposition (SVD, [22–24]) is $\mathbf{R} = \mathbf{U_X}\boldsymbol{\Delta}\mathbf{U_Y}^{\mathsf{T}}$. The matrices $\mathbf{U_X}$ and $\mathbf{U_Y}$ contain (respectively) the left and right singular vectors of $\mathbf{R}$. In PLSC parlance, the singular vectors are called *saliences* [25]. The diagonal matrix $\boldsymbol{\Delta}$ stores the singular values of $\mathbf{R}$: each singular value expresses how much a pair of singular vectors "explains $\mathbf{R}$." To express the saliences relative to the observations described in $\mathbf{Z_X}$ and $\mathbf{Z_Y}$, these matrices are projected onto their respective saliences. This creates two sets of *latent variables*—which are linear combinations of the original variables— which are denoted $\mathbf{L_X}$ and $\mathbf{L_Y}$, and are computed as:

$$\mathbf{L_X} = \mathbf{Z_X}\mathbf{U_X} \text{ and } \mathbf{L_Y} = \mathbf{Z_Y}\mathbf{U_Y}. \tag{1}$$

A pair of latent variables (i.e., one column from $\mathbf{L_X}$ and one column $\mathbf{L_Y}$) is denoted $\boldsymbol{\ell}_{\mathbf{X},\ell}$ and $\boldsymbol{\ell}_{\mathbf{Y},\ell}$ and together these two latent variables reflect the relationship between $\mathbf{X}$ and $\mathbf{Y}$ where the singular value associated to a pair of latent variables is equal to their covariance (see, e.g., [12]).

2.2.1 What Does PLSC Optimize?

The goal of PLSC is to find pairs of latent vectors $\boldsymbol{\ell}_{\mathbf{X},\ell}$ and $\boldsymbol{\ell}_{\mathbf{Y},\ell}$ with maximal covariance under the constraints that pairs of latent vectors of different indices are uncorrelated and coefficients of latent variables are normalized [15, 16]. Formally, we want to find:

$$\boldsymbol{\ell}_{\mathbf{X},\ell} = \mathbf{Z_X}\mathbf{u}_{\mathbf{X},\ell} \quad \text{and} \quad \boldsymbol{\ell}_{\mathbf{Y},\ell} = \mathbf{Z_Y}\mathbf{u}_{\mathbf{Y},\ell} \quad \text{such that} \quad \boldsymbol{\ell}_{\mathbf{X},\ell}^{\mathsf{T}}\boldsymbol{\ell}_{\mathbf{Y},\ell} = \max \tag{2}$$

under the constraints that

$$\boldsymbol{\ell}_{\mathbf{X},\ell}^{\mathsf{T}}\boldsymbol{\ell}_{\mathbf{Y},\ell'} = 0 \text{ when } \ell \neq \ell' \tag{3}$$

(note that $\boldsymbol{\ell}_{\mathbf{X},\ell}^{\mathsf{T}}\boldsymbol{\ell}_{\mathbf{X},\ell'}$ and $\boldsymbol{\ell}_{\mathbf{Y},\ell}^{\mathsf{T}}\boldsymbol{\ell}_{\mathbf{Y},\ell'}$ are *not* required to be null) and

$$\mathbf{u}_{\mathbf{X},\ell}^{\mathsf{T}}\mathbf{u}_{\mathbf{X},\ell} = \mathbf{u}_{\mathbf{Y},\ell}^{\mathsf{T}}\mathbf{u}_{\mathbf{Y},\ell} = 1\ . \tag{4}$$

2.3 PLSCA

In PLSC, **X** and **Y** are I by J and I by K matrices that describe the same I observations with (respectively) N_X and N_Y nominal variables. These variables are expressed with a 0/1 group coding (i.e., a nominal variable is coded with as many columns as it has levels and a value of 1 indicates that the observation has this level, 0 if it does not). The centroid of **X** (resp., **Y**) is denoted $\bar{\mathbf{x}}$ (resp., $\bar{\mathbf{y}}$), the relative frequency for each column of **X**, (resp., **Y**) is denoted $\mathbf{m_X}$ (resp. $\mathbf{m_Y}$). These centroids are computed as:

$$\mathbf{m_X} = \left(\mathbf{X}^\mathsf{T}\mathbf{1}\right) \times N_X^{-1} \text{ and } \mathbf{m_Y} = \left(\mathbf{Y}^\mathsf{T}\mathbf{1}\right) \times N_Y^{-1}. \tag{5}$$

In PLSCA, each variable is weighted according to the information it provides. Because a rare variable provides more information than a frequent variable, the weight of a variable is defined as the inverse of its relative frequency. Specifically, the weights of **X** (resp **Y**) are stored as the diagonal elements of the diagonal matrix $\mathbf{W_X}$ (resp. $\mathbf{W_Y}$) computed as: $\mathbf{W_X} = \operatorname{diag}\{\mathbf{m_X}\}^{-1}$ and $\mathbf{W_Y} = \operatorname{diag}\{\mathbf{m_Y}\}^{-1}$. The first step in PLSCA is to normalize the data matrices such that their sum of squares is equal to respectively $\frac{1}{N_X}$ and $\frac{1}{N_Y}$. Then the normalized matrices are centered in order to eliminate their means. The centered and normalized matrices are denoted $\mathbf{Z_X}$ and $\mathbf{Z_Y}$ and are computed as: $\mathbf{Z_X} = \left(\mathbf{X} - \mathbf{1}\bar{\mathbf{x}}^\mathsf{T}\right) \times I^{-\frac{1}{2}} N_X^{-1}$ and $\mathbf{Z_Y} = \left(\mathbf{Y} - \mathbf{1}\bar{\mathbf{y}}^\mathsf{T}\right) \times I^{-\frac{1}{2}} N_Y^{-1}$. Just like in PLSC, the next step is to compute the matrix J by K matrix **R** as $\mathbf{R} = \mathbf{Z_X}^\mathsf{T}\mathbf{Z_Y}$. The matrix **R** is then decomposed with the *generalized* SVD as:

$$\mathbf{R} = \mathbf{U_X}\boldsymbol{\Delta}\mathbf{U_Y^\mathsf{T}} \text{ with } \mathbf{U_X^\mathsf{T}}\mathbf{W_X}\mathbf{U_X} = \mathbf{U_Y^\mathsf{T}}\mathbf{W_Y}\mathbf{U_Y} = \mathbf{I}\,. \tag{6}$$

In PLSCA the saliences, denoted $\mathbf{S_X}$ and $\mathbf{S_Y}$, are slightly different from the singular vectors and are computed as $\mathbf{S_X} = \mathbf{W_X}\mathbf{U_X}$ and $\mathbf{S_Y} = \mathbf{W_Y}\mathbf{U_Y}$. Note that

$$\mathbf{S_X}^\mathsf{T}\mathbf{W_X}^{-1}\mathbf{S_X} = \mathbf{I} \text{ and } \mathbf{S_Y}^\mathsf{T}\mathbf{W_Y}^{-1}\mathbf{S_Y} = \mathbf{I}. \tag{7}$$

To express the saliences relative to the observations described in $\mathbf{Z_X}$ and $\mathbf{Z_Y}$, these matrices are projected onto their respective saliences. This creates two sets of *latent variables*—which are linear combinations of the original variables—that are denoted $\mathbf{L_X}$ and $\mathbf{L_Y}$ and are computed as:

$$\mathbf{L_X} = \mathbf{Z_X}\mathbf{S_X} = \mathbf{Z_X}\mathbf{W_X}\mathbf{U_X} \text{ and } \mathbf{L_Y} = \mathbf{Z_Y}\mathbf{S_Y} = \mathbf{Z_Y}\mathbf{W_Y}\mathbf{U_Y}\,. \tag{8}$$

2.4 What Does PLSCA Optimize?

In PLSCA, the goal is to find linear combinations of $\mathbf{Z_X}$ and $\mathbf{Z_Y}$ called *latent variables* $\boldsymbol{\ell}_{\mathbf{X},\ell}$ and $\boldsymbol{\ell}_{\mathbf{Y},\ell}$ which have maximal covariance under the constraints that pairs of latent vectors with different indices are uncorrelated and that the coefficients of each latent variables are normalized to unit length. Formally, we want to find

$$\boldsymbol{\ell}_{\mathbf{X},\ell} = \mathbf{Z_X}\mathbf{W_X}\mathbf{u}_{\mathbf{X},\ell} \quad \text{and} \quad \boldsymbol{\ell}_{\mathbf{Y},\ell} = \mathbf{Z_Y}\mathbf{W_Y}\mathbf{u}_{\mathbf{Y},\ell} \text{ such that } \quad \boldsymbol{\ell}_{\mathbf{X},\ell}^\mathsf{T}\boldsymbol{\ell}_{\mathbf{Y},\ell} = \max, \tag{9}$$

under the constraints that

$$\boldsymbol{\ell}_{\mathbf{X},\ell}^{\mathsf{T}}\boldsymbol{\ell}_{\mathbf{Y},\ell'} = 0 \text{ when } \ell \neq \ell' \tag{10}$$

and

$$\mathbf{u}_{\mathbf{X},\ell}^{\mathsf{T}}\mathbf{W_X}^{-1}\mathbf{u}_{\mathbf{X},\ell} = \mathbf{u}_{\mathbf{Y},\ell}^{\mathsf{T}}\mathbf{W_Y}^{-1}\mathbf{u}_{\mathbf{Y},\ell} = 1. \tag{11}$$

It follows from the properties of the generalized SVD [22] that $\mathbf{u}_{\mathbf{X},\ell}$ and $\mathbf{u}_{\mathbf{Y},\ell}$ are singular vectors of $\mathbf{R}$. Specifically, the product of the matrix of latent variables can be rewritten as (from Eq. 8):

$$\mathbf{L_X^{\mathsf{T}}L_Y} = \mathbf{U_X^{\mathsf{T}}W_XZ_X^{\mathsf{T}}Z_YW_YU_Y} = \mathbf{U_XW_X^{\mathsf{T}}RW_YU_Y} = \mathbf{U_XW_X^{\mathsf{T}}U_X}\boldsymbol{\Delta}\mathbf{U_YW_YU_Y} = \boldsymbol{\Delta}. \tag{12}$$

As a consequence, the covariance of a pair of latent variables $\boldsymbol{\ell}_{\mathbf{X},\ell}$ and $\boldsymbol{\ell}_{\mathbf{Y},\ell}$ is equal to their singular value:

$$\boldsymbol{\ell}_{\mathbf{X},\ell}^{\mathsf{T}}\boldsymbol{\ell}_{\mathbf{Y},\ell} = \delta_\ell\ . \tag{13}$$

So, when $\ell = 1$, we have the largest possible covariance between the pair of latent variables. Also, the orthogonality constraint for the optimization is automatically satisfied because the singular vectors constitute an orthonormal basis for their respective matrices. So, when $\ell = 2$ we have the largest possible *covariance* for the latent variables under the constraints that the latent variables are uncorrelated with the first pair of latent variables and so on for larger values of ℓ. So PLSCA and CA differ mostly by how they scale salience vs. factors scores and latent variables vs. supplementary factor scores. Correspondence analysis lends itself to biplots because the scaling scheme of factors/saliences and factor scores/latent variables allows all of them to be plotted on the same graph as they both have the same scale.

2.4.1 Links to Correspondence Analysis

In this section we show that PLSCA can be implemented as a specific case of correspondence analysis (CA) which, itself, can be seen as a generalization of PCA to nominal variables ([26, 27], for closely related approaches see [21, 28, 29]). Specifically, CA was designed to analyze contingency tables. For these tables, a standard descriptive statistic is Pearson's φ^2 coefficient of correlation whose significance is traditionally tested by the χ^2 test (recall that the coefficient φ^2 is equal to the table's independence χ^2 divided by the number of elements of the contingency table). In CA, φ^2—which, in this context, is often called the *total inertia* of the table—is decomposed into a series of orthogonal components called factors. In the present context, CA will first create, from $\mathbf{X}$ and $\mathbf{Y}$, a J by K contingency table denoted $\mathbf{S}^*$ and computed as: $\mathbf{S}^* = \mathbf{X}^{\mathsf{T}}\mathbf{Y}$. This contingency table is then transformed into a *correspondence* matrix (i.e., a matrix with nonnegative elements whose sum is equal to 1) denoted $\mathbf{S}$ and computed as $\mathbf{S} = \mathbf{S}^* s_{++}^{-1}$ (with s_{++} being the sum of all the elements of $\mathbf{S}^*$). The factors of CA are obtained by performing a generalized SVD on the double centered $\mathbf{S}$ matrix obtained as: $\left(\mathbf{S} - \mathbf{m_X}\mathbf{m_Y}^{\mathsf{T}}\right)$. Simple algebraic manipulation shows that this matrix is, in fact, equal to matrix **R** of PLSCA. Correspondence analysis then performs the SVD described in Eq. 6. The factor scores for the **X** and **Y** set are computed as

$$\mathbf{F_X} = \mathbf{W_X U_X} \boldsymbol{\Delta} \text{ and } \mathbf{F_Y} = \mathbf{W_Y U_Y} \boldsymbol{\Delta} \,. \tag{14}$$

For each set, the factor scores are pairwise orthogonal (under the constraints imposed by $\mathbf{W_X}^{-1}$ and $\mathbf{W_Y}^{-1}$) and the variance of the columns (i.e., a specific factor) of each set is equal to the square of its singular value. Specifically:

$$\mathbf{F_X}^{\mathsf{T}} \mathbf{W_X}^{-1} \mathbf{F_X} = \mathbf{F_Y}^{\mathsf{T}} \mathbf{W_Y}^{-1} \mathbf{F_Y} = \boldsymbol{\Delta}^2 \,. \tag{15}$$

The original **X** and **Y** matrices can be projected as *supplementary elements* on their respective factor scores. These supplementary factors scores denoted respectively $\mathbf{G_X}$ and $\mathbf{G_Y}$ are computed as

$$\mathbf{G_X} = N_X^{-1} \mathbf{X F_X} \boldsymbol{\Delta}^{-1} = N_X^{-1} \mathbf{X W_X U_X} \text{ and } \mathbf{G_Y} = N_Y^{-1} \mathbf{Y F_Y} \boldsymbol{\Delta}^{-1} = N_Y^{-1} \mathbf{Y W_Y U_Y} \,. \tag{16}$$

Note that the pre-multiplication by N_X and N_Y transforms the data matrices such that each row represents frequencies (this is called a *row profile* in correspondence analysis) and so each row now sums to one. This last equation shows that an observation is positioned as the barycenter of the coordinates of its variables. These projections are very closely related to the latent variables (see Eqs. 8 and 16) and are computed as

$$\mathbf{G_X} = I^{\frac{1}{2}} \mathbf{L_X} \text{ and } \mathbf{G_Y} = I^{\frac{1}{2}} \mathbf{L_Y}. \tag{17}$$

Both PLS and CA contribute to the interpretation of PLSCA. PLS shows that the latent variables have maximum covariance, CA shows that factors scores have maximal variance and that this variance "explains" a proportion of the φ^2 associated to the contingency table. Traditionally CA is interpreted with graphs plotting one dimension against the other. For these graphs, using the factor scores is preferable to the saliences because these plots preserve the similarity between elements. In CA, it is also possible to plot the factor scores of **X** and **Y** in the same graph (because they have the same variance) which is called a *symmetric* plot. If one set is privileged, it is possible to use an *asymmetric* plot in which the factor scores of the privileged set have a variance of one and the factor scores of the other set have a variance of δ^2.

2.5 Inference

Later in this paper, we present with an example three inferential methods of PLSCA: (1) a permutation test of the data for an omnibus χ^2 test to determine if, overall, the structure of the data is not due to chance, (2) a permutation test of the data to determine what, if any factors are not due to chance, and (3) a bootstrap test to determine which measures contribute a significant amount of variance.

3 Illustration

To illustrate how PLSCA works and how to interpret the results, we have created a small example from a subset of data to be analyzed. The data come from a study on the individual and additive role of specific genes and substance abuse in marijuana users [30]. Here, our (toy) hypothesis is that marijuana abusing participants ($I = 50$) with specific genotypes are more likely to frequent additional substances (i.e., certain genotypes *predispose* people to be polysubstance users).

3.1 Data

Each participant is given a survey that asks if they do or do not use certain (other) drugs—specifically, ecstasy (e), crack/cocaine (cc) or crystal meth (cm). Additionally, each participant is genotyped for COMT (which inactivates certain neurotransmitters) and FAAH (modulates fatty acid signals). The data are arranged in matrices **X** (behavior) and **Y** (SNPs; see Table 1).

Table 1: Example of nominal coding of drug use (*left*) and genotype (*right*). (**a**) Drug use (**b**) Genotypes

(a)

	CC		CM		E	
	yes	no	yes	no	yes	no
*Subj.*1	1	0	1	0	1	0
*Subj.*2	1	0	0	1	0	1
…	…	…	…	…	…	…
*Subj.*49	0	1	1	0	0	1
*Subj.*50	1	0	0	1	1	0

(b)

	COMT			FAAH		
	AG	AA	GG	CA	AA	CC
*Subj.*1	1	0	0	1	0	0
*Subj.*2	0.56	0.20	0.22	1	0	0
…	…	…	…	…	…	…
*Subj.*49	1	0	0	1	0	0
*Subj.*50	1	0	0	0	1	0

Sometimes genotype data cannot be obtained (e.g., COMT for Subject 2). This could happen if, for example, the saliva sample were too degraded to detect which nucleotides are present. Instances of missing data receive the average values from the whole sample. From **X** and **Y** we compute **R** (Table 2), which is a contingency table with the measures (columns) of **X** on the rows and the measures (columns) of **Y** on the columns. The **R** matrix is then decomposed with CA.

Table 2: The contingency table produced from **X** and **Y**

	COMT			FAAH		
	AG	AA	GG	CA	AA	CC
cc.yes	18.705	5.614	6.682	15.927	3.366	11.707
cc.no	9.705	4.614	4.682	13.341	0.293	5.366
cm.no	19.841	7.023	9.136	20.098	1.512	14.39
cm.yes	8.568	3.205	2.227	9.171	2.146	2.683
e.yes	10.000	1.000	9.000	10.171	2.146	7.683
e.no	18.409	9.227	2.364	19.098	1.512	9.39

3.2 PLSCA Results

With factor scores and factor maps, we can now interpret the results. The factor map is made up of two factors (1 and 2), which are displayed as axes. As in all SVD-based techniques, each factor explains a certain amount of variance within the dataset. Factor 1 (horizontal) explains 69% of the variance; Factor 2 explains 21%. Plotted on the factor map we see the rows (survey items, purple) and the columns (SNPs, green) from the **R** matrix (after decomposition). In CA, the distances between row items are directly interpretable. Likewise, the distances between column items are directly interpretable. However, the distances between row items and column items are not directly interpretable; the distances are *relative*. That is, "e.yes" is *more likely* to occur with COMT.GG *than other responses*.

In Fig. 1 on Factor 1, we see an interesting dichotomy. Marijuana users who have used crystal meth (cm.yes) are unlikely to use other drugs (e.no, cc.no); whereas marijuana users who have not used crystal meth (cm.no) may have used other drugs (e.yes, cc.yes). One explanation for this dichotomy is that ecstasy and cocaine could be considered more "social" drugs, whereas crystal meth is, socially, considerably frowned upon. But on Factor 2 we see that all "yes" responses occur above 0, where all "no" responses occur below 0. In this case, we can call Factor 1 "social drug use", and Factor 2 "any drug use". It is important to note that items (both rows and columns) near the origin occur in high frequency and therefore are considered "average." Items that are *not* average help with interpretation. Additionally, we see SNPs with our responses on the factor map. From this map, we know that FAAH.AA, COMT.GG and COMT.AA are rare (small frequency). Furthermore, we can see that FAAH.AA is more likely to occur with other drug use (besides marijuana) *than no drug use*, compared to other SNPs.

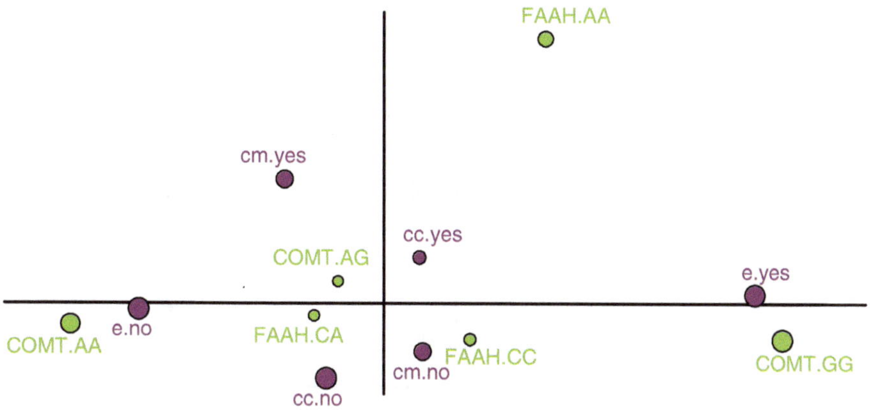

Fig. 1: Factors 1 (*horizontal*: 69% of variance) and 2 (*vertical*: 21% of variance). From the relative distances between SNPs and other drug use, we can infer that FAAH.AA is more likely to occur with other drug use (besides marijuana) *than no drug use*, compared to other SNPs; or, the *AA* allele of FAAH may predispose individuals to *polysubstance* abuse

3.3 Latent Variables

In the PLS framework, we compute latent variables from the singular vectors. The latent variables of **X** ($\mathbf{L_X}$) and **Y** ($\mathbf{L_X}$) are computed in order to show the relationships of participants with respect to SNPs (**X**; Fig. 2a) and behaviors (**Y**; Fig. 2b). In the latent variable plots, the circle size grows as more individuals are associated to it. That is, for example, in Fig. 2a, the large circle on the bottom left, with the number 13 in it, represents 13 individuals. This dot indicates that 13 individuals have the same patterns of responses to drug use.

3.4 Inferential Results

3.4.1 Permutation Tests

A permutation test of the data can test the omnibus null hypothesis. This test is performed by computing the χ^2 value (or alternatively, the total inertia) of the entire table for each permutation. The original table has a χ^2 value of 19.02, which falls outside the 95 %-ile for 1,000 permutations (which is 18.81) and this indicates that the overall structure of the data is significant (see Fig. 3).The same permutation tests are used to determine which components contribute more variance than due to chance. We test the components with the distribution of the eigenvalues. From the

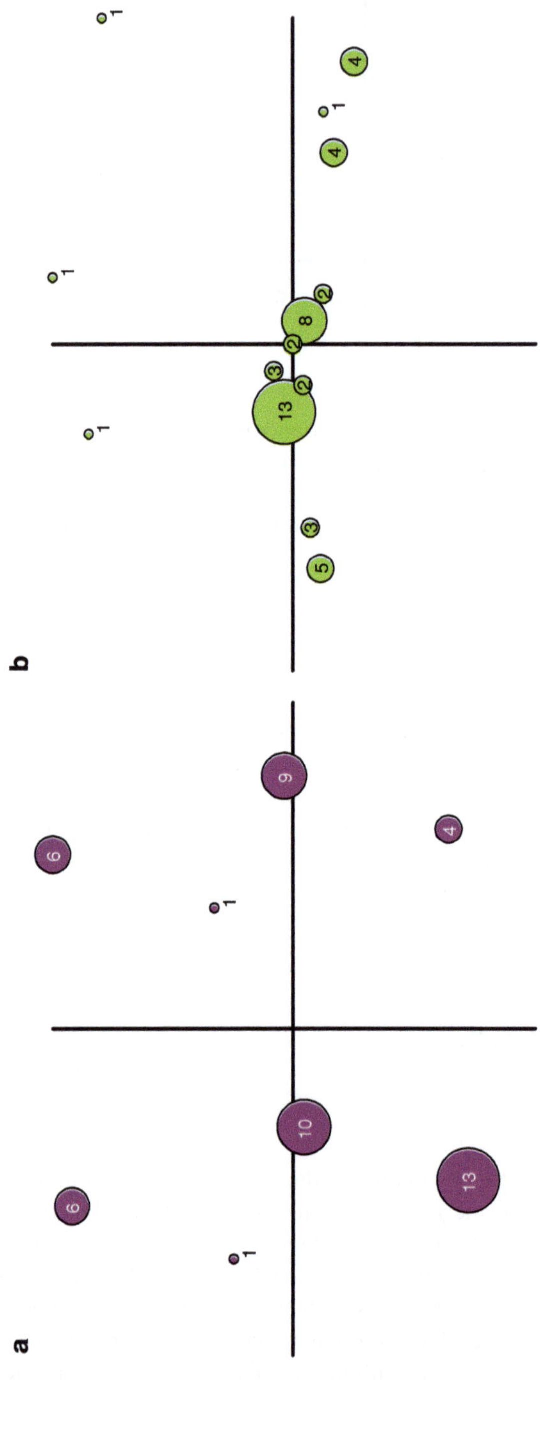

Fig. 2: Participants' latent variables for Factors 1 and 2. **(a)** (*left*) drug use **(b)** (*right*) genotype. The numbers in or near the circles give the number of participants and the size of the circles is proportional to the number of participants

toy example, only the third component (not shown above, see Fig. 4) contributes a significant amount of variance (note that this implementation of the permutation test is likely to give correct values only for the first factor, because the inertia extracted by the subsequent factors depend in part upon the inertia extracted by earlier factors; a better approach would be to recompute the permutation test for a given factor after having partialled out the inertia of all previous factors from the data matrices).

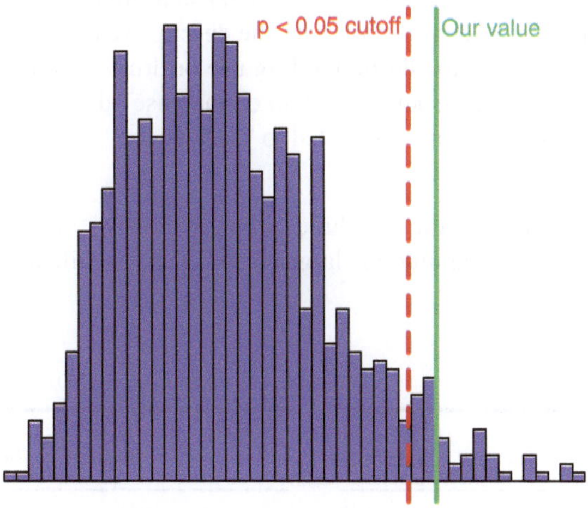

Fig. 3: The distribution for the omnibus χ^2 test. The *red line* shows the 95 ‰ (i.e., $p < 0.05$) for 1,000 permutations and the *green line* is the computed inertia value from our data. The overall structure of our data is significant ($p = 0.027$)

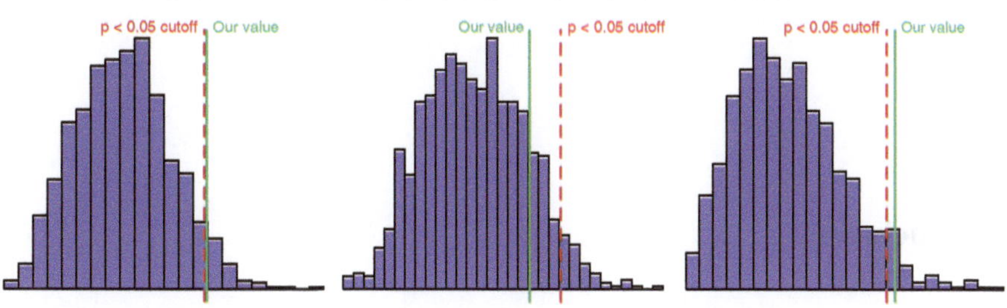

Fig. 4: Distributions for the permutation tests for each factor (1, 2, and 3, respectively). The *red lines* show the 95 ‰ (i.e., $p < 0.05$) for 1,000 permutations and the *green lines* are the eigenvalues of the factors. Factors 1 and 3 reach significance ($p = 0.048$ and $p = 0.033$, respectively) but Factor 2 does not ($p = 0.152$)

3.4.2 Bootstrap Ratios

Bootstrap resampling [31] of the observations provides distributions of how each of the measures (behavior and SNPs) changes with resampling. These distributions are used to build bootstrap ratios (also called bootstrap intervals t). When a value falls in the tail of a distribution (e.g., a bootstrap ratio of magnitude > 2), it is considered significant at the appropriate α level (e.g., $p < 0.05$). Table 3 shows that COMT (AA and GG) and ecstasy use (and non-use) contribute significantly to Factor 1.

The bootstrap tests, in conjunction with the descriptive results, indicate that certain genotypes are related to additional drug use or drug avoidance. More specifically, COMT.AA is more associated to "no ecstasy use" than any other allele and, oppositely, COMT.GG is more associated to "ecstasy use" than any other allele.

Table 3: Bootstrap ratios for the first three factors of the PLSCA. *Bold values* indicate bootstrap ratios whose magnitude is larger than 2 (i.e. "significant"). (**a**) Drug use (**b**) Genotypes

(a)

	Factor 1	Factor 2	Factor 3
cc.yes	0.291	0.714	−0.767
cc.no	−0.480	**−2.978**	0.879
cm.no	0.308	−1.434	−0.475
cm.yes	−0.786	1.036	0.697
e.yes	2.458	0.133	0.232
e.no	**−3.175**	−0.157	−0.266

(b)

	Factor 1	Factor 2	Factor 3
COMT.AG	−0.531	0.336	−0.430
COMT.AA	**−2.797**	−0.218	0.039
COMT.GG	**3.982**	−0.499	0.403
FAAH.CA	−0.858	−0.216	0.834
FAAH.AA	0.535	1.724	−0.033
FAAH.CC	0.693	−0.549	−1.367

4 Conclusion

In this paper, we presented PLSCA, a new method tailored to the analysis of genetics, behavioral and brain imaging data. PLSCA stands apart from current methods, because it directly analyzes SNPs as qualitative variables. Furthermore, PLSCA is particularly suited for the concomitant analysis of genetics and high-level behaviors as explored, for example, with surveys. Surveys are essential for the analysis of genetics and behavior as they are often designed and refined to capture the specific behaviors of given populations or psychological constructs. This way, these survey data work as an "anchor" to provide variance for genetics data. PLSCA, being the ideal tool to analyze the relationship between survey and genetic data, will help to better understand the genetic underpinnings of brains, behavior, and cognition.

References

[1] J. de Leon, J. C. Correa, G. Ruaño, A. Windemuth, M. J. Arranz, and F. J. Diaz, "Exploring genetic variations that may be associated with the direct effects of some antipsychotics on lipid levels," *Schizophrenia Research* **98**, pp. 1–3, 2008.

[2] C. Cruchaga, J. Kauwe, K. Mayo, N. Spiegel, S. Bertelsen, P. Nowotny, A. Shah, R. Abraham, P. Hollingworth, D. Harold, *et al.*, "SNPs associated with cerebrospinal fluid phospho-tau levels influence rate of decline in Alzheimer's disease," *PLoS Genetics* **6**, 2010.

[3] D. Y. Lin, Y. Hu, and B. E. Huang, ' 'Simple and efficient analysis of disease association with missing genotype data," *American Journal of Human Genetics* **82**, pp. 444–452, 2008.

[4] C. Lippert, J. Listgarten, Y. Liu, C. M. Kadie, R. I. Davidson, and D. Heckerman, "FaST linear mixed models for genome-wide association studies," *Nature Methods* **8**, pp. 833–835, 2011.

[5] C. J. Hoggart, J. C. Whittaker, M. De Iorio, and D. J. Balding, "Simultaneous analysis of all SNPs in Genome-Wide and Re-Sequencing association studies," *PLoS Genetics* **4**, p. e1000130, 2008.

[6] J. Liu, G. Pearlson, A. Windemuth, G. Ruano, N. I. Perrone-Bizzozero, and V. Calhoun, "Combining fMRI and SNP data to investigate connections between brain function and genetics using parallel ICA," *Human Brain Mapping* **30**, pp. 241–255, 2009.

[7] M. Vounou, T. E. Nichols, and G. Montana, "Discovering genetic associations with high-dimensional neuroimaging phenotypes: A sparse reduced-rank regression approach," *NeuroImage* **53**, pp. 1147–1159, 2010.

[8] M. A. Zapala and N. J. Schork, "Multivariate regression analysis of distance matrices for testing associations between gene expression patterns and related variables," *Proceedings of the National Academy of Sciences* **103**, pp. 19430–19435, 2006.

[9] C. S. Bloss, K. M. Schiabor, and N. J. Schork, "Human behavioral informatics in genetic studies of neuropsychiatric disease: Multivariate profile-based analysis," *Brain Research Bulletin* **83**, pp. 177–188, 2010.

[10] G. Moser, B. Tier, R. E. Crump, M. S. Khatkar, and H. W. Raadsma, "A comparison of five methods to predict genomic breeding values of dairy bulls from genome-wide SNP markers," *Genetics Selection Evolution* **41**, p. 56, 2009.

[11] J. Poline, C. Lalanne, A. Tenenhaus, E. Duchesnay, B. Thirion, and V. Frouin, "Imaging genetics: bio-informatics and bio-statistics challenges," in *19th International Conference on Computational Statistics*, Y. Lechevallier and G. Saporta, (eds.), (Paris, France), 2010.

[12] A. Krishnan, L. J. Williams, A. R. McIntosh, and H. Abdi, "Partial least squares (PLS) methods for neuroimaging: A tutorial and review," *NeuroImage* **56**, pp. 455–475, 2011.

[13] A. McIntosh, F. Bookstein, J. Haxby, and C. Grady, "Spatial pattern analysis of functional brain images using partial least squares," *NeuroImage* **3**, pp. 143–157, 1996.

[14] A. Krishnan, N. Kriegeskorte, and H. Abdi, "Distance-based partial least squares analysis," in *New perspectives in Partial Least Squares and Related Methods*, H. Abdi, W. Chin, V. Esposito Vinzi, G. Russolilo, and L. Trinchera, (eds.), New York, Springeer Verlag, pp. 131–145.

[15] L.R., Tucker, "An inter-battery method of factor analysis." *Psychometrika* **23**, pp. 111–136, 1958.

[16] H. Abdi and L.J. Williams, "Partial least squares methods: Partial least squares correlation and partial least square regression," in: *Methods in Molecular Biology: Computational Toxicology*, B. Reisfeld and A. Mayeno (eds.), pp. 549–579. New York: Springer Verlag. 2013.

[17] F.L. Bookstein, P.L. Sampson, A.P. Streissguth, and H.M. Barr, "Exploiting redundant measurements of dose and developmental outcome: New methods from the behavioral teratology of alcohol," *Developmental Psychology* **32**, pp. 404–415, 1996.

[18] P.D. Sampson, A.P. Streissguth, H.M. Barr, and F.S. Bookstein, "Neurobehavioral effect of prenatal alcohol: Part II, partial least square analysis," *Neurotoxicology and Teratology* **11**, pp. 477–491, 1989

[19] A. Tishler, D. Dvir, A. Shenhar, and S. Lipovetsky, "Identifying critical success factors in defense development projects: A multivariate analysis," *Technological Forecasting and Social Change* **51**, pp. 151–171, 1996.

[20] A. Tishler, and S. Lipovetsky, "Modeling and forecasting with robust canonical analysis: method and application ," *Computers and Operations Research* **27**, pp. 217–232, 2000.

[21] S. Dolédec, and D. Chessel, "Co-inertia analysis: an alernative method for studying species-environment relationships." *Freshwater Biology* **31**, pp. 277–294, 1994.

[22] H. Abdi, "Singular value decomposition (SVD) and generalized singular value decomposition (GSVD)," in *Encyclopedia of Measurement and Statistics*, N. Salkind, ed., pp. 907–912, Thousand Oaks (CA): Sage, 2007.

[23] M. Greenacre, *Theory and Applications of Correspondence Analysis,* London, Academic Press, 1984.

[24] H. Yanai, K. Takeuchi, and Y. Takane, *Projection Matrices, Generalized Inverse Matrices, and Singular Value Decomposition*, New York, Springer, 2011.

[25] F. Bookstein, "Partial least squares: a dose–response model for measurement in the behavioral and brain sciences," *Psycoloquy* **5**, 1994.

[26] H. Abdi and L. J. Williams, "Correspondence analysis," in *Encyclopedia of Research Design*, pp. 267–278, Thousand Oaks, (CA), Sage, 2010 .

[27] H. Abdi and D. Valentin, "Multiple correspondence analysis," in *Encyclopedia of Measurement and Statistics*, pp. 651–657, Thousand Oaks, (CA),Sage, 2007.

[28] A. Leclerc, "L'analyse des correspondances sur juxtaposition de tableaux de contingence," *Revue de Statistique Appliquée* **23**, pp. 5–16

[29] L. Lebart, M. Piron, and A. Morineau, *Statistiques Exploratoire Multidimensionnelle: Visualisations et Inférences en Fouille de Données*, Paris, Dunod, 2006.

[30] F. M. Filbey, J. P. Schacht, U. S. Myers, R. S. Chavez, and K. E. Hutchison, "Individual and additive effects of the CNR1 and FAAH genes on brain response to marijuana cues," *Neuropsychopharmacology* **35**, pp. 967–975, 2009.

[31] T. Hesterberg, "Bootstrap," *Wiley Interdisciplinary Reviews: Computational Statistics* **3**, pp. 497–526, 2011.

Clustered Variable Selection by Regularized Elimination in PLS

Tahir Mehmood and Lars Snipen

Abstract Variable selection is a crucial issue in many sciences, including modern biology, where an example is the selection of genomic markers for classification (diagnosing diseases, recognizing pathogenic bacteria, etc.). This becomes complicated as biological variables are in general correlated. For example, genes may be easily correlated, if they provide common biological functions. Variable selection may dissolve the group effects and mislead the focus onto a specific variable instead of a variable cluster. We study the selection and estimation properties of variable clusters in high dimensional settings when the number of variables exceeds the sample size. To address the issue a regularized elimination procedure in multiblock-PLS (mbPLS) is used, where highly correlated variables are clustered together, and whole groups are selected if they establish a relation with the response.

Key words: Regularization, High-dimension, Collinearity, Clustering, Power, Parameter estimation

1 Introduction

Multivariate approaches have the potential to provide superior statistical power, increased interpretability of the results and a deeper functional understanding of the multivariate relationship, without resulting in an excessive number of hypotheses to test. Hence, it could provide decisive statistical and biological advantages over classical univariate analysis [3, 4, 8, 14]. However, in applying multivariate analysis to the problem of recovering complex relationships much is still left to improve. In particular, multivariate approaches are sensitive to parameter estimation and parameter estimation remains a serious challenge, partially because variables tend to show

T. Mehmood (✉) • L. Snipen
Biostatistics, Department of Chemistry, Biotechnology and Food Sciences,
Norwegian University of Life Sciences, Norway
e-mail: tahir.mehmood@umb.no; lars.snipen@umb.no

H. Abdi et al. (eds.), *New Perspectives in Partial Least Squares and Related Methods*, Springer Proceedings in Mathematics & Statistics 56, DOI 10.1007/978-1-4614-8283-3_5,

extensive co-linearity, which can destroy the asymptotic consistency of the estimators for univariate responses [2]. A possible solution to the challenge is to include variable selection with these approaches. For biological interpretation of the results, focus in relational studies is shifting towards the selection of cluster of variables instead of variables itself [7]. Since genes having common pathways or similar biological functions tend to have high correlations [15], it is natural to cluster correlated variables [20]. This is important in a modeling perspective, where incorporating the biological information on genes with similar function increases the accuracy of detecting the reality behind a phenomenon [9, 12].

For the selection of a cluster of variables, one possibility is to represent a cluster of correlated variables by averaging the variables, then using this representative member' only in the model with variable selection [13]. Averaging variables can include bias in the variable selection [21]. The best subset selection will result in an unbiased model, and the lasso [16] method is a possible approach, but this tends to select a single variable instead of a cluster. Yuan and Lin [22] suggest first to select the relevant variables within each group through lasso and then select the relevant clusters by using 'group lasso'. Another possibility is an elastic net to select the cluster of variables, and if one variable in a cluster is selected the whole cluster will be selected automatically [23]. However, in this way cluster of highly connected variables can only be selected if their regression coefficients tend to be equal [21]. This motivates for a powerful structure extraction tool for cluster selection.

In high dimensional data, Partial Least Squares (PLS) is a popular solution to handle the situation where the number of variables exceeds the sample size. There are many advances in PLS to deal with different structures of the data. Multiblock PLS (mbPLS) is a way to deal with the current situation [1, 10, 19], where cluster of variables can be considered as blocks of variables. For cluster (block) selection, two main possibilities exist. One is a statistical significance test of block coefficients [17] and the other are based on block importance on prediction (BIP) [18]. A recently conducted study proposes to select clusters on the basis of stability to gain better interpretation [6]. Considering this we have used a modification of regularized stepwise procedure [11], where a significant number of clusters can be eliminated at the cost of a nonsignificant increase in model RMSE (root mean square error), which results in better understandability of the model and higher stability of the selection. The suggested procedure ranks the clusters based on the BIP measure in a regularized stepwise procedure. Here, we exemplify the applicability of this procedure in a search for codon/di-codon usage related to optimal growth temperatures in prokaryote.

2 Approach

2.1 Data

Genome sequences for 44 *Actinobacteria* genomes and the respective growth temperature information were obtained from NCBI Genome Projects (http://www.ncbi.

nlm.nih.gov/genomes/lproks.cgi). Optimal growth temperature, of each bacterium, is the response variable **y** in data set. For each genome, genes were found by the gene-finding software Prodigal [5]. For each genome, we collected the frequencies of each codon and each di-codon over all genes. The predictor variables thus consists of relative frequencies for all codons and di-codons, giving a predictor matrix **X** with a total of $p = 64 + 64^2 = 4,160$ variables (columns).

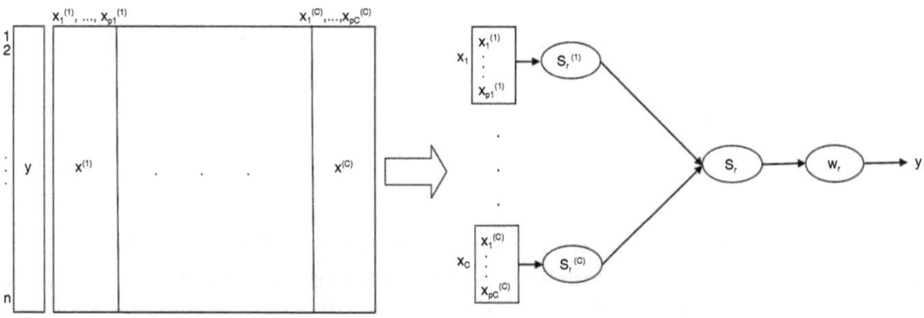

Fig. 1: An overview of the relation between clusters and latent variables in mbPLS

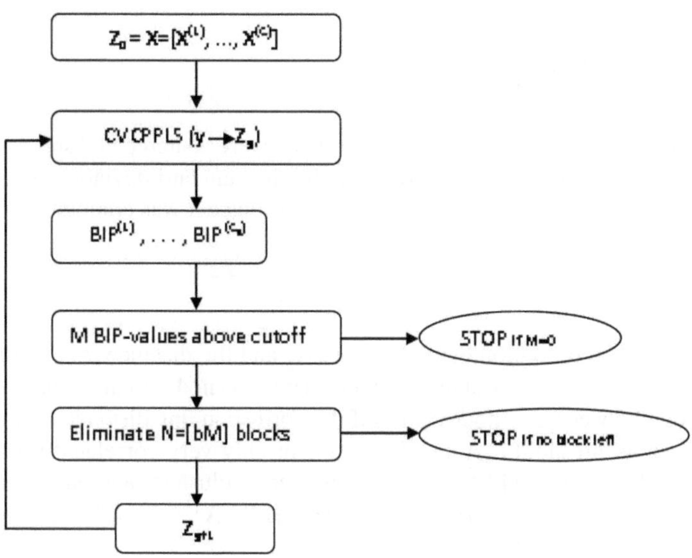

Fig. 2: An overview of the block elimination used in our stepwise elimination procedure

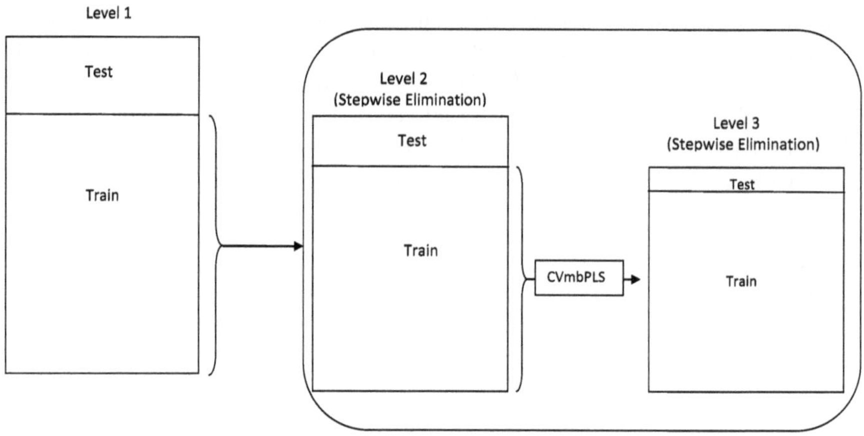

Fig. 3: An overview of the testing-training procedure used in this study. The *rectangles* illustrate the predictor matrix. At level 1 we split at random the data into a test set and training set (25/75). This was repeated 30 times. Inside our suggested method, the stepwise elimination, there are two levels of cross-validation. First a ten-fold cross-validation was used to optimize selection parameters 'a' and 'b', and at level 3 leave-one-out cross-validation was used to optimize the regularized mb-PLS method

2.2 *Clusters of Variables*

Prior to all model fitting, all variables in y and X were centered and standardized by subtracting the column mean and dividing by the standard deviation. For any two variables x_i and x_j, correlation based pairwise distance d was computed as,

$$d(x_i, x_j) = \frac{1 - cor(x_i, x_j)}{2}$$

where cor is the correlation between x_i and x_j and the distance d is a number between 0 and 1. Next, all variables were represented as nodes in an undirected graph, and an edge between two nodes exists if the corresponding distance between them is below some threshold t. A small t results in only very correlated variables being linked. This graph will form say C number of clusters, and each cluster $X^{(c)}$ (c= 1, ..., C) contains p_c variables. This defines $X=[X^{(1)},\ldots,X^{(C)}]$ having $p=\sum p_c$ columns.

2.3 *Multiblock-PLS (mbPLS)*

The association between the response y and the blocks $X=[X^{(1)},\ldots,X^{(C)}]$ is assumed to be linear. Since we have to deal with a 'small n large p' situation together with a block structure, this can be handled with mbPLS. The main focus in this algorithm is to seek scores for each block, s, which are used to generate combined scores t Fig. 1 illustrate this connection. Here, we have adopted the mbPLS procedure [10] with some modification, where each score s are also normalized by the number of variables in each cluster. Algorithm starts with $E_0 = X = [X^{(1)}, \dots, X^{(C)}]$ and $f_0 = y$.

Algorithm 1

---

for r=*1:R* **do**
  for c=*1:C* **do**
    $u_r^{(c)} = \frac{(E_{r-1}^{(c)})' f_{r-1}}{\|(E_{r-1}^{(c)})' f_{r-1}\|}$ $s_r^{(c)} = E_{r-1}^{(c)} u_r^{(c)}$ $l^{(c)}$ = number of columns of $E^{(c)}$
  end
  $S_r = [s_r^{(1)}/l^{(1)}|\dots|s_r^{(C)}/l^{(C)}]$ $w_r = \frac{(S_r)' f_{r-1}}{\|(S_r)' f_{r-1}\|}$ $t_r = S_r w_r$ $p_r = (t_r' t_r)^{-1} E_{r-1}' t_r$ $q_r = (t_r' t_r)^{-1} f_{r-1}' t_r$ $E_r = E_{r-1} - t_r p_r'$ $f_r = f_{r-1} - t_r q_r$ Extract each block $E_r^{(c)}$ from E_r
end

For prediction, model coefficients are stored; $U^{(c)} = [u_1^{(c)}, \dots, u_R^{(C)}]$, $P = [p_1, \dots, p_R]$, $Q = [q_1, \dots, q_R]$ and $W = [w_1, \dots, w_R]$, and for test data $N = [N^{(1)}, \dots, N^{(C)}]$ which is scaled as X with $\hat{y}$=0 and $E_0 = N$.

Algorithm 2

---

for r=*1:R* **do**
  for c=*1:C* **do**
    $s_r^{(c)} = E_{r-1}^{(c)} u_r^{(c)}$
  end
  $S_r = [s_r^{(1)}|\dots|s_r^{(C)}]$ $t_r = S_r w_r$ $E_r = E_{r-1} - t_r p_r'$ $\hat{y} = \hat{y} + t_r q_r$ Extract each block $E_r^{(c)}$ from E_r
end

2.4 *Algorithm for Cluster Selection*

Recently, we have suggested a stepwise estimation algorithm for parsimonious variable selection [11], where stability based variable selection procedure is adopted, and data have been split randomly in a predefined number of subsets (test and training). For each split, a stepwise procedure is adopted to select the variables. Stable variables that are being selected by stepwise elimination from all split of the data are selected finally. This algorithm was also implemented here, but feature selection was performed on clusters of variables instead of individual variables, where the 'worst' clusters were iteratively eliminated using a greedy algorithm. The algorithm requires a ranking of the blocks in X. For this, block importance on prediction (BIP) [18] is utilized and defined as

$$BIP^{(c)} = \sqrt{C\sum_{r=1}^{R} cov^2(y,t_r)w_r^{(c)^2} / \sum_{r=1}^{R} cov^2(y,t_r)}$$

where $w_r^{(c)^2}$ are the loading weights for cluster c, cov means covariance, C is the number of clusters and is included in above relation so that $\sum_{c=1}^{C}(BIP^{(c)})^2 = 1$. The BIP weights the contribution of each cluster according to the variance explained by each PLS component. Cluster c can be eliminated, if $BIP^c < a$ for some user-defined threshold $a \in [0,\infty)$. Defining a is a critical issue, here we have modified the stepwise algorithm [11] for cluster selection.

The stepwise elimination algorithm can be sketched as follows: Let $Z_0 = X = [X_1,\ldots,X_C]$.

1. For iteration g run y and Z_g through cross validated mbPLS. The matrix Z_g has p_g clusters, and we get the same number of criterion values, sorted in ascending order as $BIP^{(1)},\ldots,BIP^{(C_g)}$.
2. There are M criterion values below the cutoff a. If $M = 0$, terminate the elimination here.
3. Else, let $N = \lceil bM \rceil$ for some fraction $b \in \langle 0,1]$. Eliminate the clusters corresponding to the N most extreme criterion values.
4. If there are still more than one cluster left, let Z_{g+1} contain these clusters, and return to 1.

The fraction b determines the 'steplength' of the elimination algorithm, where an b close to '0' will only eliminate a few cluster in every iteration. An overview of block elimination is given in Fig. 2. The fraction b and threshold a can be obtained through cross validation.

From each iteration g of the elimination, we get a root mean square error (RMSE) and is denoted by L_g. RMSE value closer to '0' indicates satisfactory prediction ability. The number of influencing clusters decreases at each iteration, and L_g will often decrease until some optimum is achieved, and then increase again as we keep on eliminating. A potentially much simpler model can be achieved by a relatively small sacrifice of optimum RMSE [11]. This means we need a rejection level m, where for each iteration beyond optimum RMSE L^* we can compute the t-test p-value, to give a perspective on the trade-off between understandability of the model and RMSE.

2.4.1 The Split of Data into Test and Training and Parameter Tuning

We fixed a at the extreme value 10 and considered three levels of step length $b = (0.1, 0.5, 1)$. In the first regularization step, we tried different rejection levels $(m = (0.90, 0.99))$. For accurate model estimation, the data was split at three levels. Figure 3 gives a graphical overview. At level 1, we split the data into a test set containing 25% of the genomes and a training set containing the remaining 75%. This split was repeated 30 times, each time sampling the test set at random, i.e. the 30 test (and training) sets were partially overlapping. In each of the 30 instances, selected clusters were used for classifying the level 1 test set, and the RMSE on prediction (RMSEP) was computed. Inside the stepwise elimination, there are two levels of cross-validation as indicated by the right part of Fig. 3. First, a ten-fold cross-validation was used to optimize the fraction *b* and the rejection level *b* in the elimination part of the algorithm. At the final level, leave-one-out cross-validation was used to estimate all parameters in the mbPLS method. These two procedures together corresponds to a 'cross-model validation' [11].

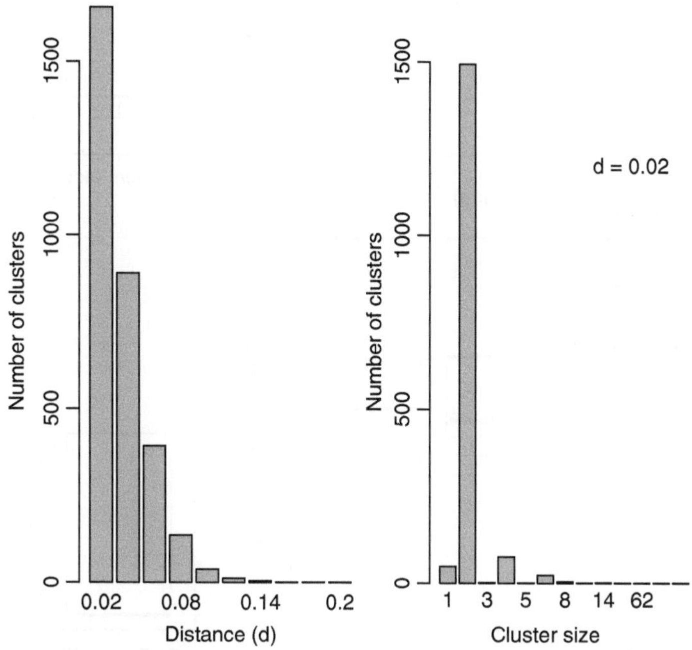

Fig. 4: The number of clusters obtained with the different thresholds on d are presented in *left panel*. The number of clusters decreases with the increase of threshold on 'd'The *right panel* shows the size distribution of the clusters using the smallest threshold d = 0.02

3 Results and Discussion

For codon and di-codons clusters identification which influence the variation in optimal growth temperature of *Actinobacteria*, we have first clustered these variables based on their correlation. Figure 4, left panel presents the number of clusters obtained with the different correlation thresholds *d*. The number of clusters decreases with the increase of threshold on *d*. For the clusters to be meaningful variables inside each cluster need to be highly correlated [20]; hence we have selected the

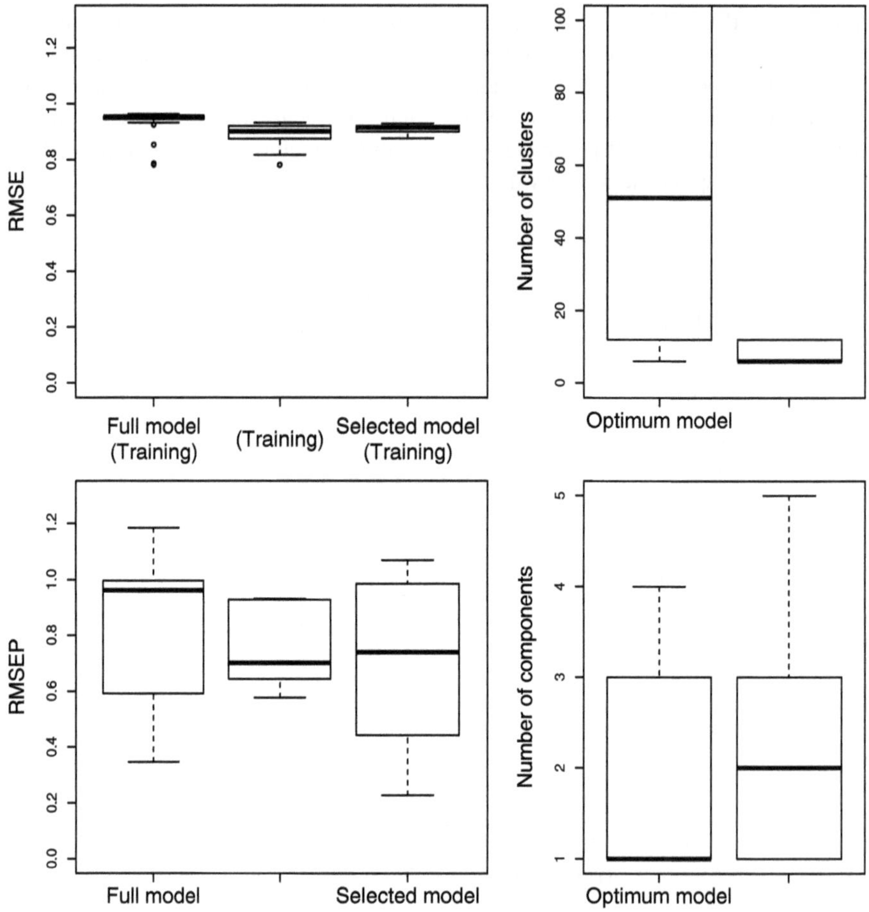

Fig. 5: The distribution of RMSE for a full model, optimum model and selected model on training data are displayed in the *upper left panel*, and RMSEP on test data in the *lower left panel*. In the *upper right panel* the number of selected clusters and in *lower right panel* the number of components from our selected model are compared with the optimum model

strict threshold $d = 0.02$. These clusters are consisting of highly correlated variables only and gives the maximum number of clusters. The distribution of cluster sizes is presented in the right panel of Fig. 4, indicating that most clusters are small (two members).

In suggested algorithm for cluster selection, it is possible to have results from the full model, the optimum model (smallest possible RMSE) and the selected model. The selected model allows for a small increase in RMSE to achieve a huge reduction in number of selected clusters. In Fig. 5 in the upper left panel, the distribution of RMSE for the full, optimum and selected model on training data is displayed.

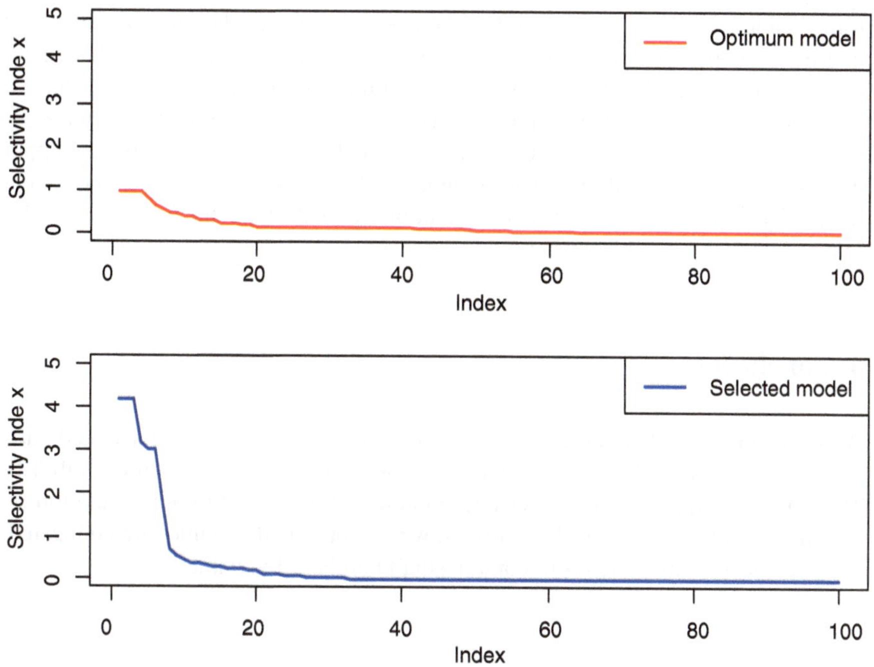

Fig. 6: The selectivity score is sorted in descending order for the optimum model in *upper panel* and for the selected model in *lower panel*. There are 1,656 cluster each having selectivity score, here only the first 100 values (out of 1,656) are shown

Training data indicates that, by elimination of some 'noice' clusters, we get decreased RMSE by going from the full model to the optimum model. Selected model has slightly poorer RMSE compared to the optimum model, which is a sacrifice we are willing to make in order to achieve the improved understandability of a reduced model. The upper right panel of Fig. 5 indicates the number of selected clusters for the optimum and selected model. On test data, we get more variation in RMSEP, as presented in the lower left panel of the Fig. 5, but the trend is same. Furthermore, we find the selected and optimum models are very simple in the sense that most of the models used only one/two components, as presented in lower right panel of Fig. 5.

Stability and selectivity of selected clusters is an important factor for any multivariate analysis, so the cluster selection on the bases of stability can bring better interpretation [6]. To evaluate model stability and selectivity, we recently introduced a simple *selectivity score* [11]: if a cluster is selected as one out of C cluster, it will get a score of $1/C$. Repeating the selection for each split of the data, we simply add up the scores for each cluster. Thus, a cluster having a large selectivity score tend to be repeatedly selected as one among a few clusters. In Fig. 6, the selectivity score is sorted in descending order and is presented for optimum model in upper left panel and for selected model in the lower panel. The selected model indeed found some stable cluster of codon and di-codon selection for explaining the variability in growth temperature of *Actinobacteria* prokaryote genomes and is a fundamental requirement for any further analysis. To fulfill this requirement, we need to have a rough idea of the 'null-distribution' of this selectivity score, for this we ran the selection on data where the response y was permuted at random. From this, the upper 0.1% of the null-distribution is determined, which approximately corresponds to the selectivity score above 0.5 for clusters. Out of 1,656 clusters, this procedure finally selects less than 20 stable clusters.

4 Conclusion

We have generalized variable selection procedure for the cluster of correlated variables selection through a stepwise backward block elimination procedure in mbPLS. The derived results give the better interpretation of modeled biological relation by selecting small number of stable clusters, where correlated variables are connected. Further selected cluster tends to contain small number of variables.

References

[1] Bougeard, S., Qannari, E.M., Rose, N.: Multiblock redundancy analysis- interpretation tools and application in epidemiology. Journal of Chemometrics **1**, (2011)

[2] Chun, H. and Keleş, S.: Sparse partial least squares regression for simultaneous dimension reduction and variable selection. Journal of the Royal Statistical Society: Series B (Statistical Methodology) **72**, 3–25 (2010)

[3] David B.A., Bonnie T., Pamela S.J., Robert C.E., Ming C.I., Nicholas J.S.: Multiple phenotype modeling in gene-mapping studies of quantitative traits- power advantages. Am. J. Hum. Genet. **63**, 1190–1201 (1998)

[4] Guri G., Patrick F., Jochen K., Michael P., Corey N., Daniel F.J., Angela M.C., Michael I.J. Adam P.A., Ronald W.D.: Chemogenomic profiling: identifying the functional interactions of small molecules in yeast. The National Academy of Sciences. **101**, 793–798 (2004)

[5] Hyatt, D., Chen, G. L., Locascio, P. F., Land, M. L., Larimer, F. W., Hauser, L. J.:Prodigal-prokaryotic gene recognition and translation initiation site identification. BMC Bioinformatics. **11**, 119 (2010)

[6] Jiang, J.H., Berry, R.J., Siesler, H.W., Ozaki, Y.: Wavelength interval selection in multicomponent spectral analysis by moving window partial least-squares regression with applications to mid-infrared and near-infrared spectroscopic data. Analytical chemistry **74**, 3555–3565 (2002)

[7] Jorgensen, K.M., Hjelle, S.M., Oye, O.K., Puntervoll, P., Reikvam, H., Skavland, J., Anderssen, E., Bruserud, O., Gjertsen, B.T.: Untangling the intracellular signalling network in cancer-a strategy for data integration in acute myeloid leukaemia. Journal of proteomics. **17**, (2010)

[8] Keleş, S. and Chun, H.: Comments on: Augmenting the bootstrap to analyze high dimensional genomic data. TEST. **17**, 1133–0686 (2008)

[9] Lan, H., Chen, M., Flowers, J.B., Yandell, B.S., others: Combined expression trait correlations and expression quantitative trait locus mapping. PLoS Genetics. **2**, e6 (2006)

[10] Lopes J.A., Menezes J.C., Westerhuis J.A., Smilde A.K.: Multiblock PLS analysis of an industrial pharmaceutical process. Biotechnology and bioengineering **80**, 419–427 (2002)
[11] Mehmood, T., Martens, H., Warringer, J., Snipen, L.: A Partial Least Squares based algorithm for parsimonious variable selection. Algorithms for Molecular Biology. **6**, 27 (2011)
[12] Pan, W.: Network-based multiple locus linkage analysis of expression traits. Bioinformatics. **25**, 1390–1401 (2009)
[13] Park, M.Y., Hastie, T., Tibshirani, R.: Averaged gene expressions for regression. Biostatistics **8**, 212 (2007)
[14] Peter K., Mariza A.: Group 6: Pleiotropy and multivariate analysis. Genetic Epidemiology. **25–1**, S50–56 (2003)
[15] Segal, M.R., Dahlquist, K.D., Conklin, B.R.: Regression approaches for microarray data analysis. Journal of Computational Biology **10**, 961–980 (2003)
[16] Tibshirani, R.: Regression shrinkage and selection via the lasso. Journal of the Royal Statistical Society. Series B (Methodological). **1**, 267–288 (1996)
[17] Vinzi, V.E., Chin, W.W. and Henseler, J.: Handbook of partial least squares: Concepts, methods and applications. Springer Verlag (2009)
[18] Vivien, M., Sabatier, R.: Generalized orthogonal multiple co-inertia analysis (–PLS): new multiblock component and regression methods. Journal of chemometrics **17**, 287–301 (2003)
[19] Wangen, L.E., Kowalski, B.R.: A multiblock partial least squares algorithm for investigating complex chemical systems. Journal of chemometrics **3**, 3–20 (1989)
[20] Wei, F. and Huang, J.: Consistent group selection in high-dimensional linear regression. Bernoulli. **16**, 1369–1384 (2010)
[21] Xie, J. and Zeng, L.: Group Variable Selection Methods and Their Applications in Analysis of Genomic Data. Frontiers in Computational and Systems Biology **1**, 231–248 (2010)
[22] Yuan, M. and Lin, Y.: Model selection and estimation in regression with grouped variables. Journal of the Royal Statistical Society. Series B (Methodological). **68**, 49–67 (2006)
[23] Zou, H. and Hastie, T.: Regularization and variable selection via the elastic net. Journal of the Royal Statistical Society. Series B (Methodological) **67**, 301–320 (2005)

[10] Lopes J.A., Menezes J.C., Westerhuis J.A., Smilde A.K., Multiblock PLS analysis of an industrial pharmaceutical process, Biotechnology and Bioengineering [illegible]

[11] Mehmood T., Martens H., Warringer J., Snipen L., A Partial Least Squares based [illegible] variable selection, Algorithms for Molecular Biology [illegible]

[illegible]

PLS Regression and Hybrid Methods in Genomics Association Studies

Antonio Ciampi, Lin Yang, Aurélie Labbe, and Chantal Mérette

Abstract Using data from a case-control study on schizophrenia, we demonstrate the use of PLS regression in constructing predictors of a phenotype from Single Nucleotide Polymorphisms (SNPs). We consider straightforward application of PLS regression as well as two hybrid methods, in which PLS regression scores are used as input for a tree-growing algorithm and a clustering algorithm respectively. We compare these approaches with other classic predictors used in statistical learning, showing that our PLS-based hybrid methods outperform both classic predictors and straightforward PLS regression.

Key words: PLS Regression, Bagging, SNP, GWAS

1 Introduction

Genome-wide association studies (GWAS) have raised great hopes. It is now possible to determine the value of virtually every Single Value Polymorphism (SNP) of a subject in a reasonable time frame and at rapidly decreasing costs. It is also possible to scan the whole genome of a fairly large number of subjects and assess the impact of each SNP on a particular phenotype, such as 'presence of a disease.' The remarkable advances that have made this possible should have empowered

A. Ciampi (✉) • A. Labbe
Department of Epidemiology, Biostatistics, and Occupational Health, Montréal, Canada
e-mail: antonio.ciampi@mcgill.ca; aurelie.labbe@mcgill.ca

L. Yang
Division of Clinical Epidemiology, McGill University Health Centre, Montréal, Canada
e-mail: lin.yang@mcgill.ca

C. Mérette
Faculty of Medicine, Department of Psychiatry and Neurosciences, Université Laval, Canada
e-mail: chantal.merette@laval.ca

H. Abdi et al. (eds.), *New Perspectives in Partial Least Squares and Related Methods*, Springer Proceedings in Mathematics & Statistics 56, DOI 10.1007/978-1-4614-8283-3_6,

scientists to obtain a deep understanding of the genetic causes of human diseases. Unfortunately this understanding remains elusive. Much has been written on the problem of "missing heritability," (i.e., the failure to find strong evidence at the genome level of well known patterns of disease transmission within families [1, 2]). Various complex hypotheses have been advanced to explain the paradox (see, e.g., [3, 4]), however none of these hypotheses has been universally accepted within the scientific community [5].

Less attention is paid to the fact that GWAS studies often use only univariate tests to identify associations between SNPs and phenotypes such as common diseases. In contrast, biology suggests that genes often cause common diseases through complex pathways, which should be studied with multivariate statistical models including both linear effects and interactions. Yet, there exist powerful but rarely-used methods originating from the statistical learning literature [6, 7] that may assist in identifying multi-SNP relationships and gaining understanding of SNP-trait associations.

This work originates from an attempt to identify genetic variants associated to schizophrenia from data collected in a case-control study. The data consisted of 237 cases and 132 controls. Preliminary univariate analyses had identified a number of candidate regions of the genome and 41 Single Nucleotide Polymorphisms (SNPs) belonging to these regions, covering 7 chromosomes. SNPs on the same chromosome were found to be correlated, but some correlations between SNPs on different chromosomes were also detected. Our goal was to construct a parsimonious predictive model for schizophrenia using the 41 SNPs which were found 'significant' in univariate analyses, or an appropriately selected subset of these.

In view of the correlations among SNPs, PLS regression appeared to be a method of choice for building such a model. At the same time, it seemed interesting to compare PLS regression with other 'classical' methods of data analysis and with methods from the statistical learning literature. The choice is vast, but we chose to limit ourselves to the most popular ones. Since PLS is a linear method, i.e. not naturally designed to detect interactions among predictors, we decided to include approaches that are designed to detect interactions. Moreover, in view of the unique features of each technique, we also decided to 'mix and match' various approaches, developing what might be seen as two novel hybrid methods.

The next section lists and briefly describes the methods we have used. Section 2 describes our approach for evaluating and comparing methods. Section 3 describes the results of our comparisons. Finally in Sect. 5 we draw some tentative conclusions and discuss ideas for future research.

2 Constructing Predictors from Data

We distinguish here between (generalized) linear and non-linear prediction methods for constructing predictors. Linear methods are known to be more stable and robust than non-linear methods; on the other hand non-linear methods may provide unique insight into a data set, revealing non-linear relationships and complex interactions, which a linear method would not uncover.

2.1 (Generalized) Linear Prediction

The linear prediction is at the intersection of "classical" and "statistical learning" methodology. Classical theory offers a solid theoretical justification for linear methods, given a vector of predictors. However, the choice of the predictors out of a potentially large pool can be seen as a task of statistical learning. Prediction of a binary variable, the focus of this work, can be based on logistic regression and Fisher's canonical discriminant analysis (CANDISC [6]). Given a binary outcome variable Y and a number of predictors $X_1, X_2, \ldots, X_k$. the logistic regression model has the (generalized) linear form:

$$\log(\frac{p}{1-p}) = \beta_0 + \beta_1 X_1 + \beta_2 X_2 + \ldots + \beta_k X_k, \tag{1}$$

where p is the probability that $Y = 1$ and $X_1, X_2, \ldots, X_k$ are the independent variables (predictors). The β's are known as the regression coefficients, which have to be estimated from the data, usually by likelihood maximization. Variable selection methods are numerous: we have chosen the computationally light approaches based on the AIC and BIC criteria [6]).

Fisher's CANDISC is equivalent to canonical correlation analysis between the predictor variables and a set of K class indicator variables, where K is the number of classes ($K = 2$ here). CANDISC derives a linear combination of the predictor variables that has the highest possible correlation with the class indicator variables. This maximum correlation is called the first canonical correlation. The coefficients of the linear combinations are the canonical coefficients. The variable defined by the linear combination is the first canonical variable. The second canonical correlation is obtained by finding the linear combination uncorrelated with the first canonical variable that has the highest possible multiple correlation with the groups. The process of extracting canonical variables can be repeated until the number of canonical variables equals the number of original variables or the number of classes minus one, whichever is smaller. Canonical variables are also called canonical components. For $K = 1$ CANDISC provides a 2-dimensional representation of the data, which gives a very useful visualization: if subjects belonging to different classes are represented by points of different color, one can easily see to what extent the canonical variables are able to separate the classes.

Partial Least Square (PLS) regression was developed to handle data sets with a large number P of correlated predictor variables, including the case $P \gg N$ (with N denoting the sample size). It can be seen as an alternative to variable selection. No variable is excluded from a PLS model, but variables are weighted differently, in such a way that unimportant variables receive very small weights. PLS regression is not likelihood based, but is close to CANDISC in that it is based on purely geometric ideas. The goal is to successively find PLS components, i.e. mutually orthogonal linear combinations of the predictor variables that maximize the product of the correlation of the predictor variables with the outcome and of the variance of the linear combination itself. PLS regression has also features that permit to discard some of the PLS components, so that an effective dimension reduction is achieved. Usually this step is carried out through cross-validation. It should be noted that PLS regression has been shown to be similar to other forms of shrinking and dimension reduction, such as ridge regression (RR) and principal component regression (PCR, []). Guided by this, by now, classical result, we have worked with PLS regression but have not included RR and PCR regression in the methods used in this paper.

2.2 Non-linear Prediction

We have limited ourselves to prediction trees and some methods that are built on prediction trees, known as ensemble methods as they combine predictors obtained from several subsets of the data. The approach to tree-growing used in this work is essentially the classical CART developed by Breiman et al. [8]. The technique is aimed at finding a rule which could predict the value of a dependent variable Y from known values of P explanatory variables X_p , $p = 1,\ldots,P$ (predictors). CART builds trees and formulates simple if/then rules for recursive partitioning (splitting) of all the observations into smaller subgroups. Each such step may give rise to new "branches." The goal of this process is to maximize homogeneity of the values of the dependent variable Y in the various subgroups. It implies that CART does not stop splitting until terminal nodes contain observations only of one class. Maximum trees may turn out to be of very high complexity and consist of many levels. Therefore, they have to be optimized to solve this problem. Tree optimization implies choosing the right size of tree-cutting off insignificant nodes and even subtrees. Two strategies can be used in practice: optimization by setting a minimum sample size of each terminal node and cross-validation based on optimal proportion between the complexity of the tree and misclassification error. In our study, the latter strategy was applied to find the optimal size of the pruned classification tree.

Bootstrap aggregating, shortened as bagging [8], is a method of averaging the predictions obtained by many decision trees over a collection of bootstrap samples, thereby reducing its variance to avoid over-fitting. Bagging consists of building a simple classifier using successively different bootstrap samples. In bagging, the bootstrap samples are based on the unweighted bootstrap and majority voting makes the predictions.

Boosting is a popular machine learning method for transforming a collection of weak classifiers or trees into one strong classifier in order to improve the accuracy of any given learning algorithm. It reduces over-fitting by weighting each classifier's contribution to the final fit. In boosting, the bootstrap samples are built iteratively using weights that depend on the predictions made in the last iteration. There are many Boosting algorithms differing in details. The AdaBoost algorithm, short for Adaptive Boosting (see [9] for a description) was used in this work.

Random Forest (introduced by Breiman [10]), use recursive partitioning to generate many trees and then aggregate the results. Each tree is independently constructed using a bootstrap sample of the data. The algorithm produces, in addition to predictionfor new data, measures of importance (in determining the prediction) for each variable.

2.3 Two Hybrid Methods

With the aim to combine the complementary features of linear and non linear methods, we briefly describe here two hybrid approaches based on PLS regression and trees. PLS regression identifies a subspace L of the predictor space of reduced dimension m, which contains (most of) the useful predictive information. This subspace is spanned by the first M PLS components: $PLS_1, PLS_2, \ldots, PLS_M$. Prediction is then based on a linear function of these PLS components. To capture non-linear predictive information, we could attempt to search for a model of the form:

$$\log\left(\frac{p}{1-p}\right) = g_1(PLS_1) + g_2(PLS_2) + \ldots + g_m(PLS_M), \tag{2}$$

where the g's are non-linear functions to be determined from the data. There are many possible choices. Here we limit ourselves to indicator functions of subsets of L.

Specifically, we present here two approaches:

(A) PLS + Prediction trees: Construct a prediction tree using the first m PLS components as predictors. This indicator functions are those associated to the leaves of the tree. A variant of this approach includes also the original SNPs variables in the tree construction.

(B) PLS + Clustering: Construct a partition of L using a clustering algorithm (e.g. K-means). The indicator functions are those associated to the sets of the partition.

It should be noted that both approaches are heavily data dependent. Data determine both the shape and the size of the tree; similarly, data determine both the definition and the size of the partition obtained through a clustering algorithm. If we wish to compare these two approaches with each other and with other ones, some careful cross-validation is needed.

3 Comparing Predictors

To compare the various predictors constructed from the data, we have used here three evaluation measures: the misclassification error, the Brier score and the C-statistics. The Brier's score [11] measures the distances between true and predicted binary outcome for a particular test set. So a lower Brier score represents higher accuracy of a predictive model. For binary outcomes, the C-statistic (C for concordance [12]), is defined, as the fraction of all pairs of subjects with predictions concordant with outcomes (Prediction and outcome are concordant for a pair of subjects if they have either the same observed and predicted outcome or different observed and predicted outcome). It can be shown that the C-statistic is identical to the area under the receiver operating characteristic (ROC) curve [13].

For a given predictor, these measures can be calculated on the same sample on which the predictor has been constructed: this is known as re-substitution. As is well known, however, re-substitution measures yield over-optimistic evaluations of predictors. Moreover, the comparison among different predictors cannot be trusted, since the methods to build them vary broadly in the amount of data dependence: For instance, a PLS predictor uses the data only to estimate the parameters of the underlying model, while a prediction tree is built by a very intensive search at each node. Therefore, to obtain fairer comparisons, we have adopted the same cross-validation approach as described in [8]. We proceeded as follows:

(i) From the original sample, we left out a random 10% of the subjects;
(ii) We constructed the predictors on the remaining 90% (learning sample) and then used the left-out 10% (test set) to calculate the evaluation measures;
(iii) The procedure was repeated 100 times and the cross-validated evaluation measures were calculated as the average over all misclassification errors thus obtained.

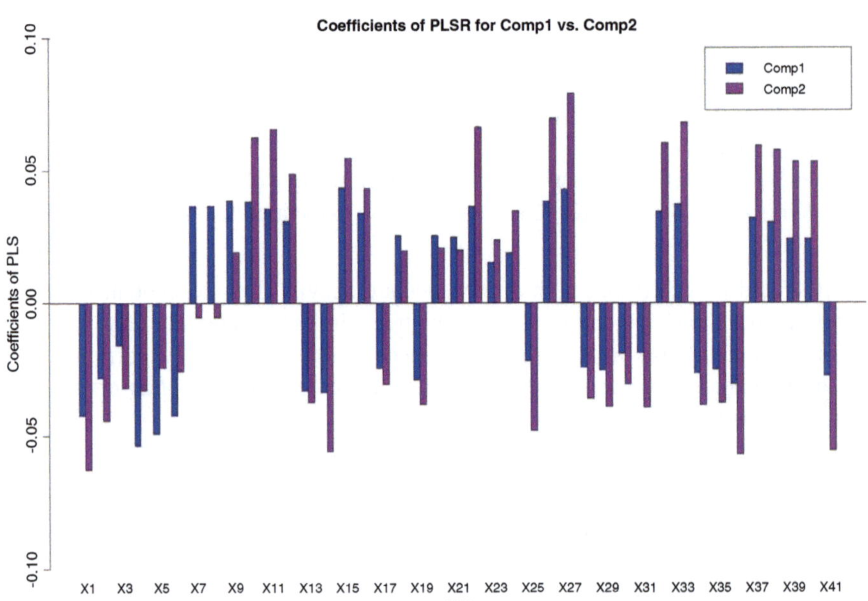

Fig. 1: PLSR coefficients

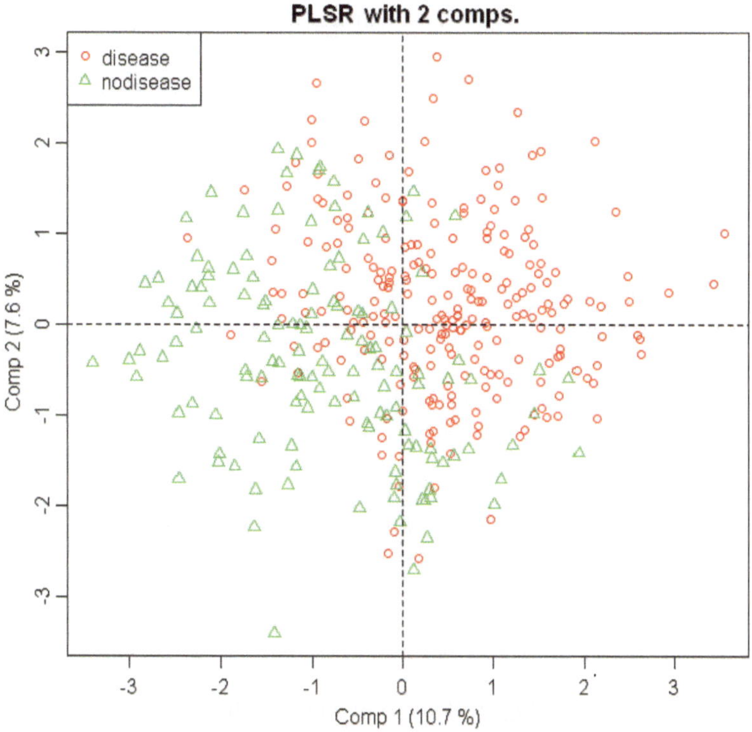

Fig. 2: The first two dimensions of the PLSR separating disease and non-disease groups

4 Results

We summarize the results obtained by applying 12 data analytic approaches on the schizophrenia data set in Table 1. These include CANDISC, four variants of logistic regression, PLS regression, three tree-based ensemble methods and two hybrid approaches (PLS + tree-growing, and PLS + K-means clustering).

For each predictor we calculated the re-substitution error rate, the cross-validated error, as well as the cross-validated Brier score and C-statistic.

The first column of the table contains the re-substitution error, and is interesting for illustrative purposes. As it could be expected, the re-substitution error is over-optimistic and favors models with a greater number of parameters and/or of higher complexity, e.g. we obtain 0 errors with Bagging and Boosting, and the smallest non-zero error with CANDISC and logistic regression including all 41 SNPs. The second column, containing the cross-validated error, tells a very different story: it shows that our two hybrid methods outperform all others; also, PLS regression outperforms the classical methods. The other two evaluation measures shown in the third and fourth column are essentially in agreement with the cross-validated error in ranking predictors.

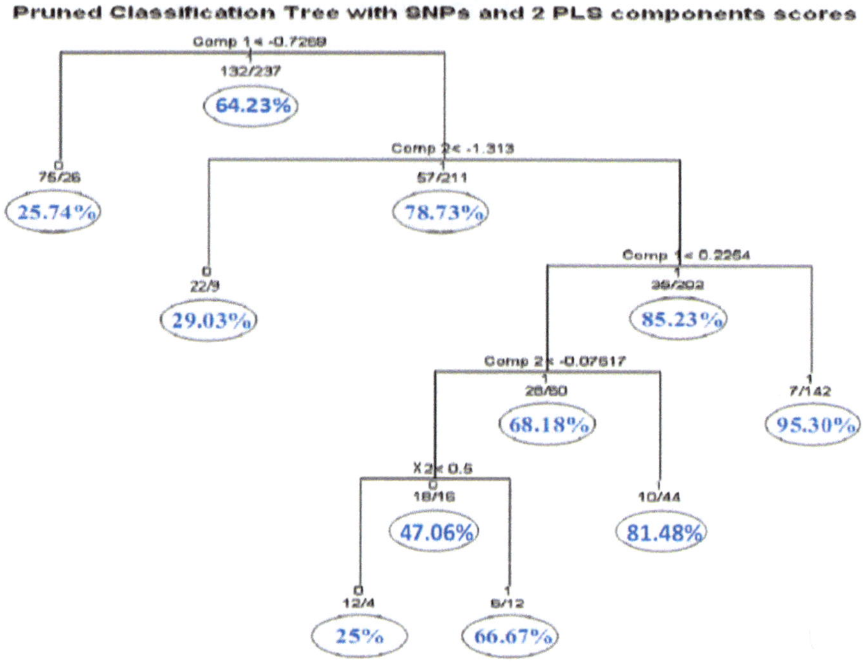

Fig. 3: Pruned classification tree

The two hybrid approaches give the best results, followed by PLS regression (with cross-validation to select the number of components). The performance of the Random forest is comparable with that of PLS, while the other tree-based ensemble methods perform worse than logistic regression.

The data are confidential; therefore we cannot disclose details leading to the identification of specific SNPs. However, Fig. 1 shows the impact of certain SNPs on prediction, as well as the fact that some SNPs are associated to risk and other to protection. Since the SNPs are numbered in an order that corresponds to their coordinates on the genome, it is also apparent that there are spatial patterns in the associations.

Figure 2 shows the discrimination between cases and controls obtained with the first two PLS components. The K-means algorithm simply identifies clusters in this plane, making the prediction sharper (data not shown because of space limitation). Finally in Fig. 3 we show the tree based on both SNPs and the PLS components.

Table 1: Comparison of predictors

Model	Resubstitution error rate	CV-error	Brier score	C-statistic
CANDISC	0.184	0.389	0.291	0.671
Logistic linear model with 41 SNPs	0.184	0.257	0.181	0.807
Logistic linear model with 21 SNPs (AIC)	0.201	0.238	0.169	0.825
Logistic linear model with 15 SNPs (BIC)	0.219	0.256	0.175	0.810
Logistic linear model with first three PCs	0.260	0.271	0.186	0.766
Pruned Tree with only 41SNPs	0.230	0.363	0.327	0.615
PLS regression model with two PLS Comps	0.205	0.232	0.165	0.826
Pruned Tree with 41 SNPs and 2 PLS Comps	0.189	0.210	0.163	0.836
K-means (nine clusters) with two PLS Comps	0.208	0.209	0.149	0.847
Random forest	0.244	0.239	0.177	0.801
Bagging	0	0.262	0.181	0.784
AdaBoosting	0	0.252	0.179	0.783

5 Conclusion

The superiority of PLS regression over the classical methods is not surprising, since it was designed for this type of situations: a large (relative to sample size) number of correlated predictors. More interesting is its superiority over the statistical learning methods (when cross-validation was properly used). This may be interpreted as evidence that highly correlated predictors, appropriately handled by PLS regression, may cause problems even for sophisticated learning approaches. On the other hand, the superiority of our hybrid methods may be due to the integration of the advantages of PLS regression and statistical learning methods. At least in this situation, the capability of handling correlation which distinguishes PLS regression, combines well with be ability to handle non-linear relationships (e.g., interactions in the case of trees) which characterized statistical learning methods. Therefore it seems useful to further explore hybrid methods. We plan to do so by analyzing a variety of real data sets and planning appropriate simulation studies.

References

[1] L.B. Maher, "Personal genomes, the case of the missing heritability," *Nature*, **456**, 18–21, 2008.

[2] T.A. Manolio, *et al.*, "Finding the missing heritability of complex diseases" *Nature*, **461**, 747–753, 2009.

[3] R.A. Fisher, *The Genetical Theory of Natural Selection*, Oxford University Press, Oxford, 1930.

[4] P.M. Visscher, W.G. Hill, and N. Wray, "Heritability in the Genomics era: Errors and misconceptions," *Nature Review Genetics*, **9**, 255–266, 2008.

[5] G. Gibson, "Rare and common variants: twenty arguments," *Nature Review Genetics*, **13**, 135–145, 2012.
[6] T. Hastie,T., R. Tibshirani, J.H., Friedman, *The elements of Statistical Learning* New York, Springer, 2008.
[7] I. Frank, J. Friedman, "A statistical view of some Chemometrics regression tools," *Technometrics* **35**, 109–135, 1993.
[8] L. Breiman, "Bagging Predictors" *Machine Learning*, **26**, 123–140, 1996.
[9] Y. Freund, R.E. Schapire, "A short introduction to boosting," *Journal of Japanese Society for Artificial Intelligence*, **14**, 771–780, 1999.
[10] L. Breiman, "Random forests," *Machine Learning*, **45**, 5–32, 2001.
[11] G.W. Brier, "Verification of forecasts expressed in terms of probability," *Monthly Weather Review*, 78, 1–3, 1950.
[12] E.M. Ohman, C.B. Granger, R.A. Harrington, K.L. Lee, "Risk stratification and therapeutic decision making in acute coronary syndromes," *The Journal of the American Medical Association,* **284**, 876–878, 2000.
[13] J.A. Hanley, B.J. McNeil, "The meaning and use of the area under a receiver operating characteristic (ROC) curve," *Radiology*, **143**, 29–36, 1982.

Globally Sparse PLS Regression

Tzu-Yu Liu, Laura Trinchera, Arthur Tenenhaus, Dennis Wei, and Alfred O. Hero

Abstract Partial least squares (PLS) regression combines dimensionality reduction and prediction using a latent variable model. It provides better predictive ability than principal component analysis by taking into account both the independent and response variables in the dimension reduction procedure. However, PLS suffers from over-fitting problems for few samples but many variables. We formulate a new criterion for sparse PLS by adding a structured sparsity constraint to the global SIMPLS optimization. The constraint is a sparsity-inducing norm, which is useful for selecting the important variables shared among all the components. The optimization is solved by an augmented Lagrangian method to obtain the PLS components and to perform variable selection simultaneously. We propose a novel greedy algorithm to overcome the computation difficulties. Experiments demonstrate that our approach to PLS regression attains better performance with fewer selected predictors.

Key words: Sparse PLS, Sparsity, Regularization, Over-fitting, Principal component analysis

T.-Y. Liu (✉) • D. Wei • A.O. Hero
Electrical Engineering and Computer Science Department,
University of Michigan, Ann Arbor, MI, USA
e-mail: joyliu@umich.edu; dlwei@eecs.umich.edu; hero@eecs.umich.edu

L. Trinchera
Rouen Business School, Rouen, France
e-mail: ltr@rouenbs.fr

A. Tenenhaus
Supélec, Gif-sur-Yvette, France
e-mail: arthur.tenenhaus@supelec.fr

H. Abdi et al. (eds.), *New Perspectives in Partial Least Squares and Related Methods*, Springer Proceedings in Mathematics & Statistics 56, DOI 10.1007/978-1-4614-8283-3_7,

1 Introduction

Partial least squares (PLS) regression combines dimensionality reduction and prediction using a latent variable model. It was first developed for regression analysis in chemometrics [5, 6, 8, 9, 11, 15, 17–19], and has been successfully applied to many different areas, including sensory science and more recently genetics. Since PLS-R does not require matrix inversion or diagonalization, it can be applied to problems with large numbers of variables. As predictor dimension increases, variable selection becomes essential to avoid over-fitting, to provide more accurate predictors and to yield more interpretable parameters. For this reason sparse PLS was developed by H. Chun and S. Keles [10]. The sparse PLS algorithm performs variable selection and dimension reduction simultaneously using an L_1 type variable selection penalty. However, the L_1 penalty used in [10] penalizes each variable independently and this can result in different sets of variables being selected for each PLS component leading to an excessively large number of variables. In this paper we propose a global variable selection approach that penalizes the total number of variables across all PLS components. Put another way, the proposed global penalty guarantees that the selected variables are shared among the PLS components. This results in improved PLS performance with fewer variables. We formulate PLS with global sparsity as a variational optimization problem with objective function equal to the univariate PLS criterion with added mixed norm sparsity constraint on the weight matrix. The mixed norm sparsity penalty is the L_1 norm of the L_2 norm on the subsets of variables used by each PLS component. A novel augmented Lagrangian method is proposed to solve the optimization problem and soft thresholding for sparsity occurs naturally as part of the iterative solution. Experiment results show that the modified PLS attains better performance (lower mean squared error, MSE) with many fewer selected predictor variables.

2 Partial Least Squares Regression

Partial Least Squares (PLS) methods embrace a suite of data analysis techniques based on algorithms belonging to the PLS family. These algorithms consist of various extensions of the Nonlinear estimation by Iterative PArtial Least Squares (NIPALS) algorithm that was proposed by Herman Wold [2] as an alternative algorithm for implementing a Principal Component Analysis (PCA) [3]. The NIPALS approach was slightly modified by Herman Wold son, Svante, and Harald Martens, in order to obtain a regularized component based regression tool, known as PLS Regression (PLS-R) [4, 9].

Suppose that the data consists of n samples of independent variables $X \in R^{n\times p}$ and dependent variables (responses) $Y \in R^{n\times q}$. In standard PLS Regression the aim is to define orthogonal latent components in R^p, and then use such latent components as predictors for Y in an ordinary least squares framework.The X weights used to compute the latent components can be specified by using iterative algorithms belonging to the NIPALS family or by a sequence of eigen-decompositions.

Moreover, in the univariate response case, it does not make sense to calculate components in the unidimensional response space. For the k-th component, the X weights can be directly computed as a function of Y. In particular, for the first component the X weights are defined such that the covariance between the predictors and the univariate response is maximized. In both the univariate and multivariate cases, the general underlying model behind the PLS Regression is $X = TP^T + E$ and $Y = TQ^T + F$, where T is the latent component matrix, P and Q are the loading matrices, E and F are the residual terms.

2.1 Univariate Response

We assume, without loss of generality, that all the variables have been centered in a pre-processing step. For univariate Y, i.e $q = 1$, PLS Regression, also often denoted as PLS1, successively finds X weights $R = [\mathbf{r}_1\ \mathbf{r}_2\ \dots\ \mathbf{r}_K]$ as the solution to the constrained optimization

$$\mathbf{r}_k = \arg\max_{\mathbf{r}}\{\mathbf{r}^T X_{(k-1)}^T Y_{k-1} Y_{k-1}^T X_{(k-1)} \mathbf{r}\} s.t.\ \mathbf{r}^T\mathbf{r} = 1 \tag{1}$$

where $X_{(k-1)}$ is the matrix of the residuals (i.e. the deflated matrix) from the regression of the X-variables on the first $k-1$ latent components, and $X_0 = X$. Due to the deflation on data after each iteration for finding the weight vector $\mathbf{r}_k$, the orthogonality constraint is satisfied by construction. These weights are then used to find the orthogonal latent components $T = X_{(k-1)}R$. Such components can be also expressed in terms of original variables (instead of deflated variables), i.e. as $T = XW$, where W is the matrix containing the weights to be applied to the original variables in order to exactly obtain the latent components [13].

For a fixed number of components, the response variable Y is predicted in an ordinary least squares regression model where the latent components play the role of the exogenous variables

$$\arg\min_{Q}\{||Y - TQ^T||_2\} = (T^T T)^{-1} T^T Y \tag{2}$$

This provides the regression coefficients $\hat{\beta}^{PLS} = W\hat{Q}^T$ for the model $Y = X\beta^{PLS} + F$.

Depending on the number of selected latent components the length $||\hat{\beta}^{PLS}||_2$ of the vector of the PLS coefficient estimators changes. In particular, de Jong [1] has shown that the sequence of these coefficient vectors have lengths that are strictly increasing as the number of component increases. This sequence converges to the ordinary least squares coefficient vector and the maximum number of latent components obtainable equals the rank of the X matrix. Thus, by using a number of latent components $K < rank(X)$, PLS-R performs a dimension reduction by shrinking the β vector. Hence, PLS-R is a suitable tool for problems with data containing many more variables p than observations n.

The objective function in (1) can be interpreted as maximizing the squared covariance between Y and the latent component: $corr^2(Y, X_{k-1}\mathbf{r}_k)\mathrm{var}(X_{k-1}\mathbf{r}_k)$. Because the response Y has been taken into account to formulate the latent matrix, PLS has better performance in prediction problems than principle component analysis(PCA) does [20]. This is one of the main difference between PLS and principle component analysis (PCA) [14].

2.2 Multivariate Response

Similarly to univariate response PLS-R, multivariate response PLS-R selects latent components in R^p and R^q, i.e. $\mathbf{t}_k$ and $\mathbf{v}_k$, such that the covariance between $\mathbf{t}_k$ and $\mathbf{v}_k$ is maximized. For a specific component, the sets of weights $\mathbf{r}_k \in R^p$ and $\mathbf{c}_k \in R^q$ are obtained by solving

$$\max\{\mathbf{t}^T\mathbf{v}\} = \max\{\mathbf{r}^T X_{k-1}^T Y_{k-1}\mathbf{c}\} s.t.\ \mathbf{r}^T\mathbf{r} = \mathbf{c}^T\mathbf{c} = 1 \tag{3}$$

where $\mathbf{t}_k = X_{(k-1)}\mathbf{r}_k$, $\mathbf{v}_k = Y_{(k-1)}\mathbf{c}_k$, and $X_{(k-1)}$ and $Y_{(k-1)}$ are the deflated matrices associated to X and Y. Notice that the optimal solution $\mathbf{c}_k$ should be proportional to $Y_{k-1}^T X_{k-1}\mathbf{r}_k$. Therefore, the optimization in (3) is equivalent to

$$\max_{\mathbf{r}}\{\mathbf{r}^T X_{k-1}^T Y_{k-1} Y_{k-1}^T X_{k-1}\mathbf{r}\} s.t.\ \mathbf{r}^T\mathbf{r} = 1 \tag{4}$$

For each component, the solution to this criterion can be obtained by using a so called PLS2 algorithm. A detailed description of the iterative algorithm as presented by Höskuldsson is in Algorithm 3 [7].

Algorithm 3 PLS2 algorithm

for *k=1:K* **do**
  initialize $\mathbf{r}$ $X = X_{new}$ $Y = Y_{new}$ **while** *solution has not converged* **do**
    $\mathbf{t} = X\mathbf{r}$ $\mathbf{c} = Y^T\mathbf{t}$ Scale $\mathbf{c}$ to length 1 $\mathbf{v} = Y\mathbf{c}$ $\mathbf{r} = X^T\mathbf{v}$ Scale $\mathbf{r}$ to length 1
  end
  loading vector $\mathbf{p} = X^T\mathbf{t}/(\mathbf{t}^T\mathbf{t})$ deflate $X_{new} = X - \mathbf{t}\mathbf{p}^T$ regression $\mathbf{b} = Y^T\mathbf{t}/(\mathbf{t}^T\mathbf{t})$ deflate $Y_{new} = Y - \mathbf{t}\mathbf{b}^T$ $\mathbf{r_k} = \mathbf{r}$
end

In 1993 de Jong proposed a variant of the PLS2 algorithm, called Straightforward Implementation of a statistically inspired Modification of PLS (SIMPLS), which calculates the PLS latent components directly as linear combinations of the original variables [12]. The SIMPLS was first developed as an optimality problem and solve the optimization

$$\begin{aligned} \mathbf{w}_k &= \arg\max_{\mathbf{w}}(\mathbf{w}^T X^T Y Y^T X\mathbf{w}) \\ s.t.\ \mathbf{w}^T\mathbf{w} &= 1,\ \ \mathbf{w}^T X^T X\mathbf{w}_j = 0\ for\ j = 1,\ldots,k-1. \end{aligned} \tag{5}$$

Ter Braak and de Jong [21] provided a detailed comparison between the objective functions for PLS2 in (4) and SIMPLS in (5) and shown that the successive weight vectors $\mathbf{w}_k$ can be derived either from the deflated data matrices or original variables in PLS2 and SIMPLS respectively. Let W^+ be the Moore-Penrose inverse of $W = [\mathbf{w}_1\ \mathbf{w}_2\ \ldots\ \mathbf{w}_{k-1}]$. The PLS2 algorithm (Algorithm 3) is equivalent to solving the optimization

$$\mathbf{w}_k = \arg\max_{\mathbf{w}}(\mathbf{w}^T X^T Y Y^T X \mathbf{w})$$
$$s.t. \mathbf{w}^T (I - WW^+)\mathbf{w} = 1, \mathbf{w}^T X^T X \mathbf{w}_i = 0 \; for \; i = 1,\ldots,k-1. \tag{6}$$

Both NIPALS and SIMPLS have the same objective function but each are maximized under different constraints. NIPALS and SIMPLS are equivalent when Y is univariate, but provide slightly different weight vectors in multivariate scenarios. The performance depends on the nature of the data, but SIMPLS appears easier to interpret since it does not involve deflation of the data sets [12]. However NIPALS can manage missing data when SIMPLS needs complete data. We develop our globally sparse PLS based on the SIMPLS optimization formulation.

3 Globally Sparse PLS Regression

One approach to sparse PLS is to add the L_1 norm of the weight vector, a sparsity inducing penalty, to (5). The solution for the first component would be obtained by solving

$$\mathbf{w}_1 = \arg\max_{\mathbf{w}}(\mathbf{w}^T X^T Y Y^T X \mathbf{w}) \; s.t. \; \mathbf{w}^T \mathbf{w} = 1, \; ||\mathbf{w}||_1 \leq \lambda. \tag{7}$$

The addition of the L_1 norm is similar to SCOTLASS (simplified component lasso technique), the sparse PCA proposed by Jolliffe [16]. However, the solution of SCOTLASS is not sufficiently sparse, and the same issue remains in (7). Chun and Keles [10] reformulated the problem, promoting the exact zero property by imposing the L_1 penalty on a surrogate of the weight vector instead of the original weight vector [10], as shown in (8). For the first component, they solve the following optimization by alternating between updating $\mathbf{w}$ and $\mathbf{z}$ (block coordinate descent). The L_2 norm addresses the potential singularity problem when solving for $\mathbf{z}$.

$$\mathbf{w}_1, \mathbf{z}_1 = \arg\min_{\mathbf{w},\mathbf{z}}\{-\kappa \mathbf{w}^T X^T Y Y^T X \mathbf{w} + (1-\kappa)(\mathbf{z}-\mathbf{w})^T X^T Y Y^T X(\mathbf{z}-\mathbf{w}) + \lambda_1||\mathbf{z}||_1 + \lambda_2||\mathbf{z}||_2^2\}$$
$$s.t. \; \mathbf{w}^T \mathbf{w} = 1 \tag{8}$$

As mentioned in the Introduction, this formulation penalizes the variables in each PLS component independently. This paper proposes an alternative in which variables are penalized simultaneously over all directions. First, we define the global weight matrix, consisting of the K weight vectors, as

$$W = \begin{bmatrix} | & | & & | \\ \mathbf{w}_1 & \mathbf{w}_2 & \cdots & \mathbf{w}_K \\ | & | & & | \end{bmatrix} = \begin{bmatrix} - & \mathbf{w}_{(1)}^T & - \\ - & \mathbf{w}_{(2)}^T & - \\ & \vdots & \\ - & \mathbf{w}_{(p)}^T & - \end{bmatrix}$$

Notice that the elements in a particular row of W, i.e. $\mathbf{w}_{(j)}^T$, are all associated with the same predictor variable $\mathbf{x}_j$. Therefore, rows of zeros correspond to variables that are not selected. To illustrate the drawbacks of penalizing each variable independently, as in [10], suppose that each entry in W is selected independently with probability p_1. The probability that the $(j)_{th}$ variable is not selected becomes $(1-p_1)^K$, and the probability that all the variables are selected for at least one weight vector is $[1-(1-p_1)^K]^p$, which increases as the number of weight vectors K increases. This suggests that for large K the local variable selection approach of [10] may not lead to an overall sparse and parsimonious PLS model. In such cases a group sparsity constraint is necessary to limit the number of selected variables. The globally sparse PLS variable selection problem is to find the top K weight vectors that best relate X to Y, while using limited number of variables.

$$W = \arg\min_{W} -\frac{1}{n^2}\sum_{k=1}^{K}\mathbf{w}_k^T X^T Y Y^T X \mathbf{w}_k + \lambda \sum_{j=1}^{p} ||\mathbf{w}_{(j)}||_2 \tag{9}$$
$$s.t.\ \mathbf{w}_k^T\mathbf{w}_k = 1\ \forall\, k\ and\ \mathbf{w}_k^T X^T X \mathbf{w}_i = 0\ \forall\, i \neq k$$

The objective function (9) is the summation of the first K terms in the SIMPLS objective. Instead of the sequential greedy solution in PLS2 algorithm, the proposed globally sparse PLS must solve for the K weight vectors simultaneously. The L_2 norm of each row of W promotes grouping entries in W that relate to the same predictor variable, whereas the L_1 norm promotes a small number of groups, as in (7).

We propose to solve the optimization (9) by augmented Lagrangian methods, which allows one to solve (9) by variable splitting iterations. Augmented Lagrangian methods introduce a new variable M, constrained such that $M = W$, such that the row vectors $\mathbf{m}_{(j)}$ of M obey the same structural pattern as the rows of W:

$$\min_{W,M} -\frac{1}{n^2}\sum_{k=1}^{K}\mathbf{w}_k^T X^T Y Y^T X \mathbf{w}_k + \lambda \sum_{j=1}^{p} ||\mathbf{m}_{(j)}||_2 \tag{10}$$
$$s.t.\ \mathbf{w}_k^T\mathbf{w}_k = 1\ \forall\, k\ ,\ \mathbf{w}_k^T X^T X \mathbf{w}_i = 0\ \forall\, i \neq k,\ and\ M = W$$

The optimization (10) can be solved by replacing the constrained problem by an unconstrained one with an additional penalty on the Frobenius norm of the difference $M - W$. This penalized optimization can be iteratively solved by a block coordinate descent method that alternates between optimizing over W and over M (See Algorithm 4). We initialize the Algorithm 4 with $M(0)$ equals to the solution of standard PLS, and $D(0)$ equals to the zero matrix. Once the algorithm converges, the final PLS regression coefficients are obtained by applying the standard PLS regression on the selected variables keeping the same number of components K. The optimization over W can be further simplified to a secular equation problem, whereas the optimization over M can be shown to reduce to solving a soft thresholding operation. As described later in the experimental comparisons section, the parameters λ and K are decided by cross validation.

Algorithm 4 Exact solution of the global PLS variable selection problem using the augmented Lagrangian method

set $\tau = 0$, choose $\mu > 0, M(0), W(0), D(0)$ **while** *stopping criterion is not satisfied* **do**

$$W(\tau+1) = \arg\min_{W} -\frac{1}{n^2}\sum_{k=1}^{K} \mathbf{w}_k^T X^T Y Y^T X \mathbf{w}_k + \frac{\mu}{2}||W - M(\tau) - D(\tau)||_F^2$$

$$s.t.\quad \mathbf{w}_k^T \mathbf{w}_k = 1 \;\; \forall k, \quad \mathbf{w}_k^T X^T X \mathbf{w}_i = 0 \;\; \forall\; i \neq k \quad M(\tau+1) = \arg\min_{M} \lambda \sum_{j=1}^{p} ||\mathbf{m}_{(j)}||_2 + \frac{\mu}{2}||W(\tau+1) - M - D(\tau)||_F^2 \quad D(\tau+1) = D(\tau) - W(\tau+1) + M(\tau+1)$$

end

4 Experimental Comparisons

In this section we show experimental results obtained by comparing standard PLS-R, L_1 penalized PLS-R [10], our proposed globally sparse PLS-R, and Correlated Component Regression [22]. All the methods have been applied on the Octane data set (see [13]). The Octane data is a real data set consisting of 39 gasoline samples for which the digitized Octane spectra have been recorded at 225 wavelengths (in nm). The aim is to predict the Octane number, a key measurement of the physical properties of gasoline, using the spectra as predictors. This is of major interest in real applications, because the conventional procedure to calculate the Octane number is time consuming and involves expensive and maintenance-intensive equipment as well as skilled labor.

The experiments are composed of 150 trials. In each trial we randomly split the 39 samples into 26 training samples and 13 test samples. The regularization parameter λ and number of components K are selected by 2-fold cross validation on the training set, while μ is fixed to 2,000. The averaged results over the 150 trials are shown in Table 1. All the methods but CCR perform reasonably in terms of MSE on the test set. We further show the variable selection frequencies for the first three PLS methods over the 150 trials superimposed on the octane data in Fig. 1. In chemometrics, the rule of thumb is to look for variables that have large amplitudes in first derivatives with respect to wavelength. Notice that both L_1 penalized PLS-R and globally sparse PLS have selected variables around 1,200 and 1,350 nm, and the selected region in the latter case is more confined. Box and Whisker plots for comparing the MSE, number of selected variables, and number of components of these three PLS formulations are shown in Fig. 2. Comparing our proposed globally sparse PLS with standard PLS and L_1 penalized PLS [10], we see that PLS with global variable selection attains better performance in terms of MSE, the number of predictors, and the number of components.

Table 1: Performance of the PLS with global variable selection compared with standard PLS and L_1 penalized PLS

methods	MSE	number of var.	number of comp.
PLS-R	0.0564	225	5.5
L_1 penalized PLS-R	0.0509	87.3	4.5
globally sparse PLS-R	0.0481	38.5	3.8
CCR	0.8284	19.1	6

5 Conclusion

The formulation of the SIMPLS objective function with an added group sparsity penalty greatly reduces the number of variables used to predict the response. This suggests that when multiple components are desired, the variable selection technique should take into account the sparsity structure for the same variables among all the components. Our proposed globally sparse PLS algorithm is able to achieve as good or better performance with fewer predictor variables and fewer components as compared to competing methods. It is useful for performing dimension reduction and variable selection simultaneously in applications with large dimensional data but comparatively few samples ($n < p$). In future work, we will apply globally sparse PLS algorithms to multivariate response datasets.

Acknowledgements We would like to give special thanks to Douglas Rutledge, professor in AgroParisTech, for his expert knowledge in chemometrics to interpret the selected variables in octane data.

References

[1] de Jong, S. (1995). PLS shrinks. *Journal of Chemometrics*, **9**(4), 323–326.
[2] Wold, H. (1966). Nonlinear estimation by iterative least squares procedures. *Research papers in statistics*, 411–444.

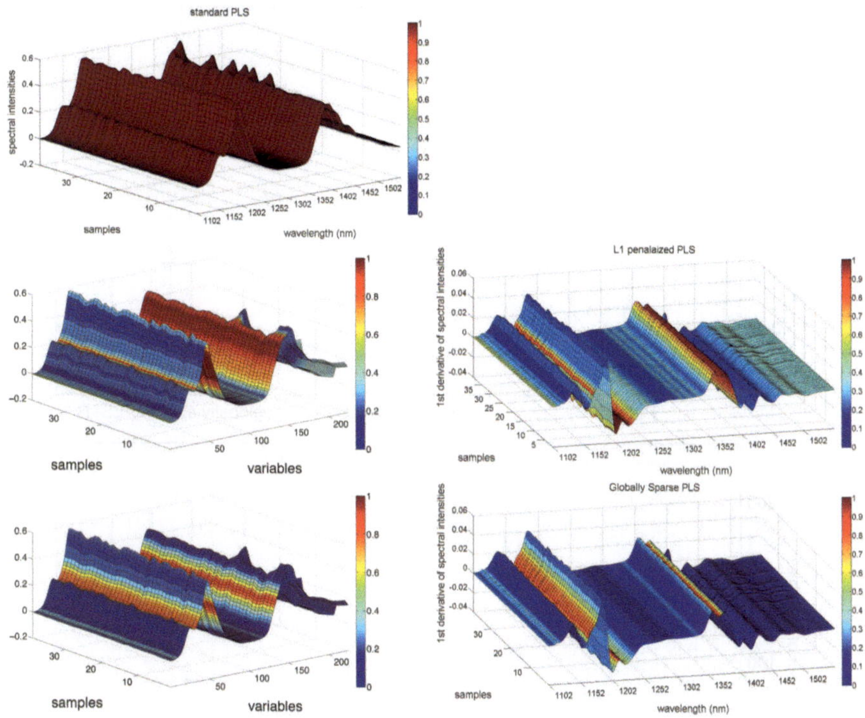

Fig. 1: Variable selection frequency superimposed on the octane data: The hight of the surfaces represents the exact value of the data over 225 variables for the 39 samples. The *color* of the surface shows the selection frequency of the variables as depicted on the colorbar

[3] Hotelling, H. (1933). Analysis of a complex of statistical variables into principal components. *Journal of educational psychology*, **24**(6), 417.

[4] Wold, S., Ruhe, A., Wold, H., and Dunn III, W. J. (1984). The collinearity problem in linear regression. The partial least squares (PLS) approach to generalized inverses. *SIAM Journal on Scientific and Statistical Computing*, **5**(3), 735–743.

[5] Martens, H., and Naes, T. (1989). *Multivariate calibration*. Wiley.

[6] Rossouw, D., Robert-Granié, C., and Besse, P. (2008). A sparse PLS for variable selection when integrating omics data. *Genetics and Molecular Biology*, **7**(1), 35.

[7] Höskuldsson, A. (1988). PLS regression methods. *Journal of Chemometrics*, **2**(3), 211–228.

[8] Wold, H. (1975). Soft modelling by latent variables: the non-linear iterative partial least squares (NIPALS) approach. *Perspectives in Probability and Statistics, In Honor of MS Bartlett*, 117–144.

[9] Wold, S., Martens, H., and Wold, H. (1983). The multivariate calibration problem in chemistry solved by the PLS method. *Proceedings of the Conference on Matrix Pencils. Lectures Notes in Mathematics*, 286–293.

[10] Chun, H., and KeleÅÿ, S. (2010). Sparse partial least squares regression for simultaneous dimension reduction and variable selection. *Journal of the Royal Statistical Society: Series B (Statistical Methodology)*, **72**, 3–25.

[11] Bach, F. R. (2008). Consistency of the group Lasso and multiple kernel learning. *The Journal of Machine Learning Research*, **9**, 1179–1225.

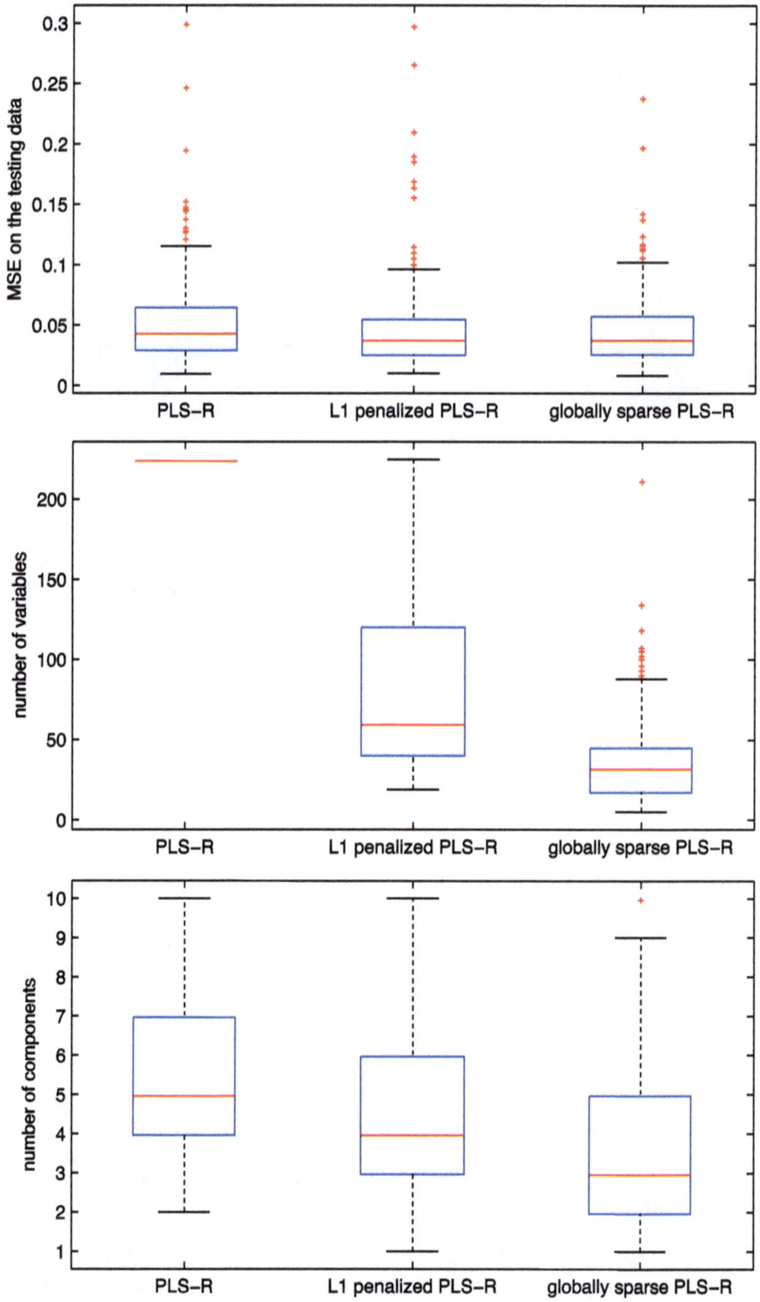

Fig. 2: The Box and Whisker plot for comparing MSE, and number of selected variables, and number of components on the test samples

[12] de Jong, S. (1993). SIMPLS: an alternative approach to partial least squares regression. *Chemometrics and Intelligent Laboratory Systems*, **18**(3), 251–263.
[13] Tenenhaus, M. (1998). *La Régression PLS: théorie et pratique*. Editions Technip.
[14] Boulesteix, A. L., and Strimmer, K. (2007). Partial least squares: a versatile tool for the analysis of high-dimensional genomic data. *Briefings in bioinformatics*, **8**(1), 32–44.
[15] ter Braak, C. J., and de Jong, S. (1998). The objective function of partial least squares regression. *Journal of chemometrics*, **12**(1), 41–54.
[16] Jolliffe, I. T., Trendafilov, N. T., and Uddin, M. (2003). A modified principal component technique based on the LASSO. *Journal of Computational and Graphical Statistics*, **12**(3), 531–547.
[17] Gander, W., Golub, G. H., and von Matt, U. (1989). A constrained eigenvalue problem. *Linear Algebra and its applications*, **114**, 815–839.
[18] Beck, A., Ben-Tal, A., and Teboulle, M. (2006). Finding a global optimal solution for a quadratically constrained fractional quadratic problem with applications to the regularized total least squares. *SIAM Journal on Matrix Analysis and Applications*, **28**(2), 425–445.
[19] Jolliffe, I. (2002). *Principal component analysis*. John Wiley & Sons, Ltd.
[20] de Jong, S. (2005). PLS fits closer than PCR. *Journal of chemometrics*, **7**(6), 551–557.
[21] ter Braak, C. J., and de Jong, S. (1998). The objective function of partial least squares regression. *Journal of chemometrics*, **12**(1), 41–54.
[22] Magidson, J. (2010). Correlated Component Regression: A Prediction/Classification Methodology for Possibly Many Features. *Proceedings of the American Statistical Association*.

Part III
Brain Imaging

Distance-Based Partial Least Squares Analysis

Anjali Krishnan, Nikolaus Kriegeskorte, and Hervé Abdi

Abstract Distances matrices are traditionally analyzed with statistical methods that represent distances as maps such as Metric Multidimensional Scaling (MDS), Generalized Procrustes Analysis (GPA), Individual Differences Scaling (INDSCAL), and DISTATIS. MDS analyzes only one distance matrix at a time while GPA, INDSCAL and DISTATIS extract similarities between several distance matrices. However, none of these methods is predictive. Partial Least Squares Regression (PLSR) predicts one matrix from another, but does not analyze distance matrices. We introduce a new statistical method called DIstance-based Partial Least Squares Regression (DISPLSR), which predicts one distance matrix from another. We illustrate DISPLSR with data obtained from a neuroimaging experiment, which explored semantic categorization.

Key words: Partial least squares, Regression, Correlation, Distance, MDS, DISTATIS

1 Introduction

Distance matrices are ubiquitous in the social sciences and several multivariate descriptive methods have been developed to analyze them. Specifically, methods such as metric multidimensional scaling (MDS), generalized Procrustes analysis (GPA),

A. Krishnan (✉)
Institute of Cognitive Science, University of Colorado Boulder, Boulder, CO, USA
e-mail: anjali.krishnan@colorado.edu

N. Kriegeskorte
MRC Cognition and Brain Sciences Unit, University of Cambridge, Cambridge, UK
e-mail: nikolaus.kriegeskorte@mrc-cbu.cam.ac.uk

H. Abdi
School of Behavioral and Brain Sciences, The University of Texas at Dallas,
MS: Gr.4.1 800 West Campbell Road, Richardson, TX 75080-3021, USA
e-mail: herve@utdallas.edu

H. Abdi et al. (eds.), *New Perspectives in Partial Least Squares and Related Methods*,
Springer Proceedings in Mathematics & Statistics 56, DOI 10.1007/978-1-4614-8283-3_8,

individual differences scaling (INDSCAL), and DISTATIS, all display distances as points on a map. MDS transforms a distance matrix into a cross-product matrix (akin to a variance-covariance matrix) in order to compute a suitable coordinate system such that the original distances between elements are represented as accurately as possible as Euclidean distances [1, 2]. GPA analyzes the similarities between more than two distance matrices obtained on the same elements [3]. INDSCAL analyzes multiple distance matrices that each store distances measured in the same observations [4]. DISTATIS analyzes multiple distance matrices simultaneously by computing an optimal compromise between all the distance matrices ([5–7]). While MDS only analyzes one distance matrix at a time, and GPA, INDSCAL and DISTATIS analyze several distance matrices, none of these methods predicts one distance matrix from another.

Partial Least Squares Regression (PLSR) predicts a set of dependent variables (predictee) from a set of independent variables (predictor). PLSR belongs to the family of PLS methods which also includes Partial Least Squares Correlation (PLSC) (also called inter-battery-analysis [9], PLS-SVD [10, 11], inter-correlation analysis, canonical covariance analysis, [12], robust canonical analysis [13], or co-inertia analysis [14]), a correlation technique that analyzes associations between two matrices. While PLSC has mostly been used in neuroimaging research [8], PLSR has been applied in many fields such as Econometrics and Chemometrics [15–17]. The basis for PLSC and PLSR is the singular value decomposition (SVD) of a matrix [18–20]. PLSR requires an iterative application of the SVD, in order to find latent variables that model the independent variables and simultaneously predict the dependent variables. Each iteration of the SVD produces orthogonal (i.e., uncorrelated) latent variables and corresponding regression weights for prediction. PLSR displays the latent variables in the form of maps, which describe the relation between the predictor and predictee.

There are a few methods that are related to PLSR and similar to DISPLSR. Sample-based Partial Least Squares analysis (SAMPLS), developed for comparative molecular field analysis [21], extracts latent variables from inter-sample distances (transformed into covariances) to predict a single response variable. A variation of SAMPLS was later developed [22] with the modification that inter-sample distances were calculated intrinsically (i.e., distances were calculated for each pair of elements separately and then aggregated for the SAMPLS analysis). Distance-based Redundancy Analysis (DB-RDA), developed for ecological research [23], combines MDS and redundancy analysis (related to multiple linear regression with more than one dependent variable) to predict a set of dependent variables from a distance matrix. Multivariate Distance Matrix Regression (MDMR), developed for genomic research [24], assesses the relationship between a distance matrix and non-distance responses. This method was performed without using dimensional analysis and instead defined a modified F statistic similar to the F statistic in ANOVA (i.e., the ratio between explained and unexplained variance). Although SAMPLS, DB-RDA, and MDMR are relevant methods to analyze distance matrices, they do not predict one distance matrix from another.

2 Methodology

In order to reduce the long computation time taken to iteratively derive latent variables in PLSR, a kernel PLSR algorithm was developed to condense large data matrices before the PLSR step [25]. Kernel PLSR computes association matrices (akin to variance-covariance matrices) for the predictor and predictee separately. The product of these association matrices, called the *kernel* matrix, is used to compute the latent variables for the PLSR step. While PLSR does not analyze distance matrices, the structure of the association matrices in Kernel PLSR is comparable to the cross-product matrix generated by MDS for distances. Therefore, the properties of both Kernel PLSR and MDS were adapted to develop DIstance-based Partial Least Squares Regression (DISPLSR) and DIstance-based Partial Least Squares Correlation (DISPLSC). Both versions of DISPLS are discussed here with mathematical details and illustrations.

2.1 Distance-Based Partial Least Squares Regression

The main algorithm for DISPLSR is derived from Kernel PLSR. All vectors in the Kernel PLSR algorithm can be computed with the eigenvalue decomposition of *kernel* matrices [26], which in turn are computed as the product of two association matrices. For Kernel PLSR, the association matrix for an $I \times J$ matrix $\mathbf{X}$ is computed as $\mathbf{XX}^\mathsf{T}$ (i.e., $\mathbf{S_X}$), and the association matrix for $I \times K$ matrix $\mathbf{Y}$ is computed as $\mathbf{YY}^\mathsf{T}$ (i.e., $\mathbf{S_Y}$). For DISPLSR, the data are in the form of distances: an $I \times I$ predictor distance matrix $\mathbf{D_X}$ and an $I \times I$ predicted distance matrix $\mathbf{D_Y}$. These distance matrices are converted into cross-product matrices $\mathbf{S_X}$ and $\mathbf{S_Y}$. In order to convert a distance matrix d into a cross-product matrix we first define a mass vector, whose elements are all positive and whose sum is equal to 1. When the masses for the rows are equal, the value of each element is $\frac{1}{I}$. The masses are stored in a vector $\mathbf{m}$ so that:

$$\mathbf{m}^\mathsf{T}\mathbf{1} = 1. \tag{1}$$

Then, an $I \times I$ conformable *centering* matrix is defined as:

$$\Xi = \mathbf{I} - \mathbf{1m}^\mathsf{T}, \tag{2}$$

where $\mathbf{I}$ is a conformable identity matrix. The cross-product matrix is obtained by double-centering the rows and columns of the distance matrix as:

$$\mathbf{S} = -\frac{1}{2}\Xi\mathbf{D}\Xi^\mathsf{T}. \tag{3}$$

Both $\mathbf{D_X}$ and $\mathbf{D_Y}$ are transformed into $\mathbf{S_X}$ and $\mathbf{S_Y}$, respectively, which are renamed as $\mathbf{S_{X0}}$ and $\mathbf{S_{Y0}}$ as the input for the first iteration of the DISPLSR algorithm.

The first latent variable for the predictor is determined from the solution of this singular value decomposition problem:

$$(\mathbf{S_{X0}}\mathbf{S_{Y0}})\,\mathbf{t}_1 = \delta_1 \mathbf{t}_1, \tag{4}$$

where $\mathbf{S_{X0}}\mathbf{S_{Y0}}$ is the kernel matrix for the first iteration of DISPLSR, and δ_1 and $\mathbf{t}_1$ are respectively the first singular value and the first right singular vector of $\mathbf{S_{X0}}\mathbf{S_{Y0}}$. The first latent variable for the predictee is computed as:

$$\mathbf{u}_1 = \mathbf{S_{Y0}}\mathbf{t}_1. \tag{5}$$

Because the Kernel PLSR latent variables (i.e., $\mathbf{t}$ and $\mathbf{u}$) are calculated on the cross-product matrices, their lengths are not comparable with the latent variables that are computed in the original PLSR algorithm. Therefore, latent variables have to be rescaled to the original PLSR algorithm in order to get comparable prediction [25]. First, we scale $\mathbf{u}_1$ to $\mathbf{u}_{\text{temp}}$ as:

$$\mathbf{u}_{1_{\text{temp}}} = \frac{\mathbf{u}_1}{\mathbf{t}_1^{\mathsf{T}}\mathbf{u}_1}. \tag{6}$$

Then, we obtain the weights to rescale $\mathbf{t}_1$ as:

$$\mathbf{t}_{1_{\text{weight}}} = \mathbf{u}_{1_{\text{temp}}}^{\mathsf{T}}\mathbf{S_X}\mathbf{u}_{1_{\text{temp}}}. \tag{7}$$

Finally, we rescale both $\mathbf{t}_1$ and $\mathbf{u}_{\text{temp}}$ as:

$$\mathbf{t}_{1_{\text{scaled}}} = \mathbf{t}_1\sqrt{\mathbf{t}_{1_{\text{weight}}}}, \tag{8}$$

and

$$\mathbf{u}_{1_{\text{scaled}}} = \mathbf{u}_{1_{\text{temp}}}\sqrt{\mathbf{t}_{1_{\text{weight}}}}. \tag{9}$$

Once the first latent variables have been computed, the matrices $\mathbf{S_{X0}}$ and $\mathbf{S_{Y0}}$ are deflated. Because we have now condensed the original matrices into cross-product matrices, the deflation is done directly on the cross-product matrices by multiplying them with an *updating* matrix $\mathbf{G}$ (i.e., $\mathbf{G}_0$ for the first iteration) computed as:

$$\mathbf{G}_0 = \mathbf{I} - \mathbf{t}_{1_{\text{scaled}}}\mathbf{t}_{1_{\text{scaled}}}^{\mathsf{T}}. \tag{10}$$

The cross-product matrices $\mathbf{S_{X0}}$ and $\mathbf{S_{Y0}}$ are deflated as:

$$\mathbf{S_{X1}} = \mathbf{G}_0\mathbf{S_{X0}}\mathbf{G}_0^{\mathsf{T}}, \tag{11}$$

and

$$\mathbf{S_{Y1}} = \mathbf{G}_0\mathbf{S_{Y0}}\mathbf{G}_0^{\mathsf{T}}. \tag{12}$$

This process is continued until all the latent variables have been computed and rescaled to the original PLSR algorithm length. The latent variables of the predictor are stored in matrix $\mathbf{T}$ and the latent variables of the predictee are stored in matrix $\mathbf{U}$. To derive the weights (i.e., $\mathbf{W}$) and loadings (i.e., $\mathbf{C}$ and $\mathbf{P}$) for DISPLSR, the original matrices $\mathbf{S_X}$ and $\mathbf{S_Y}$ are projected onto the space of $\mathbf{T}$ and $\mathbf{U}$. We correct for scaling with the pseudo-inverse of the square root of $\mathbf{S_X}$ and $\mathbf{S_Y}$ (both being square matrices) as:

$$\mathbf{W} = \left(\mathbf{S_X}^{\frac{1}{2}}\right)^{+} \mathbf{S_X}\mathbf{U}. \tag{13}$$

$$\mathbf{C} = \left(\mathbf{S_Y}^{\frac{1}{2}}\right)^{+} \mathbf{S_Y}\mathbf{T}, \tag{14}$$

$$\mathbf{P} = \left(\mathbf{S_X}^{\frac{1}{2}}\right)^{+} \mathbf{S_X}\mathbf{T}, \tag{15}$$

The weights $\mathbf{W}$ and the loadings of $\mathbf{S_Y}$ on $\mathbf{T}$, (i.e., $\mathbf{C}$), are normalized so that the sum of squares equals one. The regression weights $\mathbf{B}_{\text{PLS}}$ are computed as in Kernel PLSR:

$$\mathbf{B}_{\text{PLS}} = \mathbf{W}\left(\mathbf{P}^{\top}\mathbf{W}\right)^{-1}\mathbf{C}^{\top}. \tag{16}$$

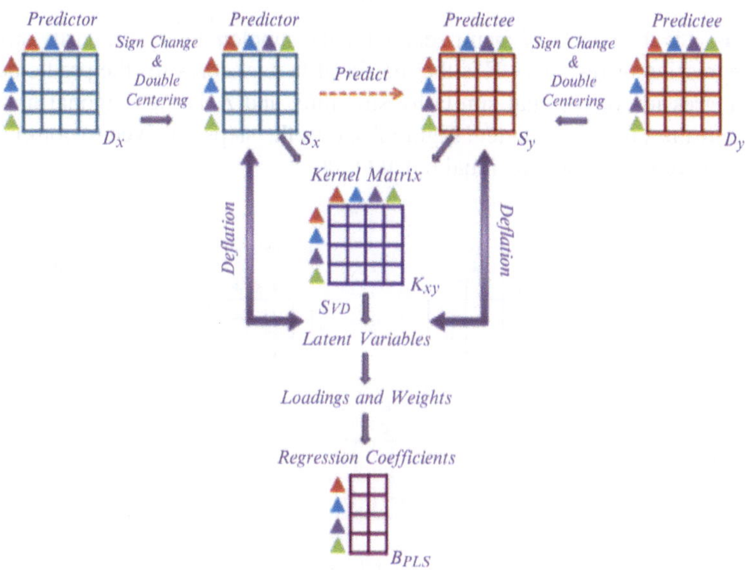

Fig. 1: Steps for DISPLS regression: compute cross-product matrices; compute kernel matrix; iteratively compute latent variables; compute weights, loadings and regression coefficients

Figure 1 shows the steps involved to generate latent variables and regression weights in DISPLSR. Once the latent variables and regression weights have been extracted, the predicted similarity structure is computed as:

$$\hat{\mathbf{S}}_{\mathbf{Y}} = \frac{1}{2\sqrt{I}}\left(\mathbf{S_X}\mathbf{B}_{\text{PLS}}\mathbf{B}_{\text{PLS}}^{\top}\mathbf{S_X}^{\top}\right), \tag{17}$$

where $\frac{1}{2\sqrt{I}}$ is a scaling factor with I being the number of rows (and columns) of $\mathbf{D_X}$ and $\mathbf{D_Y}$. The residual similarity structure is obtained by subtracting the predicted similarity structure from the original predictee:

$$\tilde{\mathbf{S}}_{\mathbf{Y}} = \mathbf{S}_{\mathbf{Y}} - \hat{\mathbf{S}}_{\mathbf{Y}} \,. \tag{18}$$

The quality of the DISPLSR model is evaluated by the R_V coefficient, which is similar to a squared coefficient of correlation [27, 28], computed as:

$$R_V = \frac{\mathsf{trace}\{\mathbf{S_Y}^{\mathsf{T}}\hat{\mathbf{S}}_{\mathbf{Y}}\}}{\sqrt{\left(\mathsf{trace}\{\mathbf{S_Y}^{\mathsf{T}}\mathbf{S_Y}\}\right)\left(\mathsf{trace}\{\hat{\mathbf{S}}_{\mathbf{Y}}^{\mathsf{T}}\hat{\mathbf{S}}_{\mathbf{Y}}\}\right)}} \,. \tag{19}$$

The original predictee is displayed on an MDS map. The regression and residual are projected as supplementary structures on this map to show the additivity of the regression model:

$$\mathbf{F}_{\mathsf{sup}} = \mathbf{S}_{\mathsf{sup}}^{\mathsf{T}}\mathbf{F}\boldsymbol{\Lambda}^{-1}, \tag{20}$$

where $\mathbf{F}_{\mathsf{sup}}$ is the matrix of factor scores for the supplementary similarity structure (i.e., regression or residual), $\mathbf{F}$ is the matrix of factor scores for the predictee, $\mathbf{S}_{\mathsf{sup}}$ is the regression or residual similarity structure, and $\boldsymbol{\Lambda}$ is the diagonal matrix of the eigenvalues of the predictee. Figure 2 shows the steps involved to compute and display the regression and residual in DISPLSR.

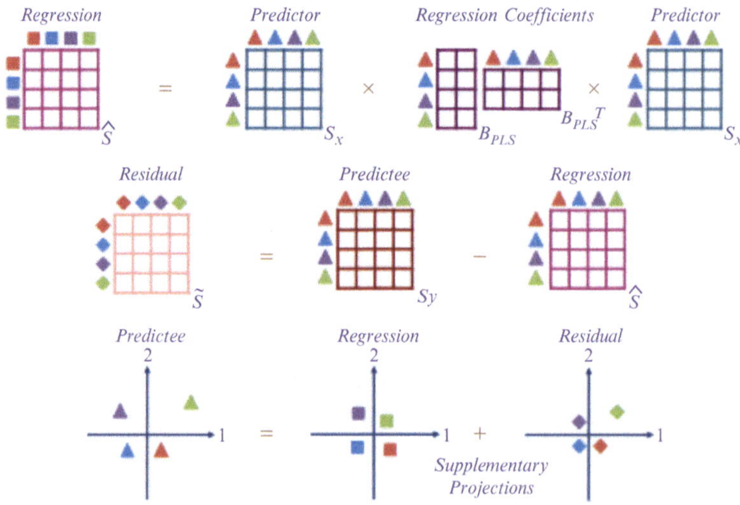

Fig. 2: Steps for DISPLS regression: compute regression; compute residual; project regression and residual as supplementary elements onto map of predictee

2.1.1 Cross-Validation for DISPLSR

A permutation test on the R_V coefficient can be used for hypothesis testing. In a permutation test, a new data set—called a *permuted* sample—is obtained by randomly reordering the labels of the rows and columns of one distance matrix and leaving the other distance matrix unchanged. The DISPLSR model is then recomputed for the permuted sample to obtain a new R_V coefficient. This procedure is repeated for a large number of permuted samples, say 1,000 or 10,000. The set of all the R_V coefficients provides a sampling distribution of the R_V coefficient under the null hypothesis. The resulting null distribution of the R_V is conditional on the original distance matrices (see [29] for more details). While the sampling distribution of the R_V coefficient for permuted positive semi-definite matrices has been documented [27, 30], the sampling distribution of the R_V coefficients for permuted *distance* matrices has not been fully explored.

2.2 *Distance-Based Partial Least Squares Correlation*

The original application of DISPLSR as a predictive method is ideal only when there exist a clear predictor and a clear predictee. When both the distance matrices are dependent variables, a correlation technique will better capture the similarity between the two matrices. The idea behind DIStance-based Partial Least Squares Correlation (DISPLSC; derived from PLSC [8]) is to extract the commonalities between two distance matrices. The difference between the algorithms of DISPLSR and DISPLSC is that for DISPLSC the latent variables are computed in one iteration of the SVD. Also, because there is no prediction step, the weights $\mathbf{W}$, loadings $\mathbf{P}$ and $\mathbf{C}$, and regression weights $\mathbf{B}_{\text{PLS}}$ are not computed for DISPLSC. Specifically, the kernel matrix K is given by:

$$K = \mathbf{S_X S_Y}\,. \tag{21}$$

The SVD of K is given by:

$$K = \mathbf{U}\boldsymbol{\Phi}\mathbf{V}^{\mathsf{T}}, \tag{22}$$

where $\mathbf{U}$ is the matrix of right singular vectors, $\mathbf{V}$ is the matrix of left singular vectors, and $\boldsymbol{\Phi}$ is the diagonal matrix of singular values. The latent variables for $\mathbf{S_X}$ (and $\mathbf{D_X}$) are given by:

$$\mathbf{L_{DX}} = \mathbf{S_X U}, \tag{23}$$

and describe the relationship between the distances in $\mathbf{D_X}$ with respect to the distances in $\mathbf{D_Y}$. The latent variables for $\mathbf{S_Y}$ (and $\mathbf{D_Y}$) are given by:

$$\mathbf{L_{DY}} = \mathbf{S_Y V}\,, \tag{24}$$

and describe the relationship between the distances in $\mathbf{D_Y}$ with respect to the distances in $\mathbf{D_X}$. Figure 3 shows the steps involved in DISPLSC.

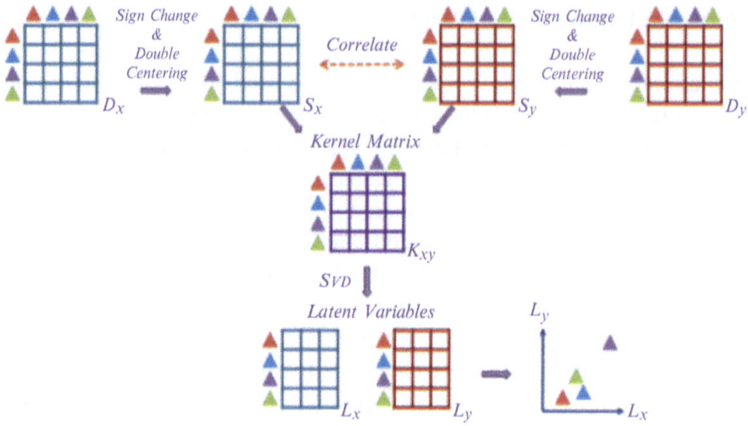

Fig. 3: Steps for DISPLS correlation: compute cross-product matrices; compute kernel matrix; compute latent variables

Just lile the quality of the DISPLSR model The quality of the DISPLSC model is evaluated by the R_V coefficient computed between $\mathbf{L_X}$ and $\mathbf{L_Y}$ (see Eq. 19).

2.2.1 Cross-Validation for DISPLSC

A permutation test on the R_V coefficient between the latent variables can be used for hypothesis testing. As mentioned earlier, a permuted sample is created by randomly reordering the labels of the rows and columns of one distance matrix, leaving the other distance matrix unchanged. The set of all the R_V coefficients computed from say 1,000 or 10,000 permutations, provides a sampling distribution of the R_V coefficient under the null hypothesis, which is conditional on the original distance matrices [29].

3 Illustration

We illustrate DISPLSR with data from a neuroimaging experiment, which explored how the brain semantically categorizes objects [31]. Functional Magnetic Resonance Imaging (fMRI) measures were obtained from the inferior temporal cortex of four human participants while they viewed pictures of 92 real-world objects. The similarity between patterns of brain activation elicited for each pair of pictures was measured as a correlation coefficient r (ranging from -1 to $+1$), and the correlation distance between these patterns of brain activation was quantified as $1-r$. The authors derived a stimulus-by-stimulus correlation distance matrix from the fMRI data for each participant and averaged the distance matrices across all four participants to get a mean distance matrix. The authors also obtained an averaged stimulus-by-stimulus correlation distance matrix from single-cell data measured from the inferior temporal cortex of two monkeys [32]. In addition, the authors computed a stimulus-by-stimulus correlation distance matrix from the actual pictures used in the experiment. The pictures were modeled using Gabor filters that generate a computational model of images based on texture and shape, which closely mimics the Gabor model of neuronal function in the primary visual cortex of mammalian brains [33].

An MDS was first performed on each of the distance matrices to display the objects on a map (only the first and second dimensions are displayed). Figure 4a shows the original map of the stimuli as represented by a Gabor model of the pictures. On the whole we see no categorical structure defined by the Gabor model. Figure 4b shows the original map of the stimuli as measured by *f*MRI from the human participants and Fig. 4c shows the original map of the stimuli as measured by single-cell recordings from the monkeys. We see categories such as scenes, human faces, monkeys and other animals in these maps. The single-cell data from the monkeys were more robust than the *f*MRI data from human participants and revealed some of the categories more clearly.

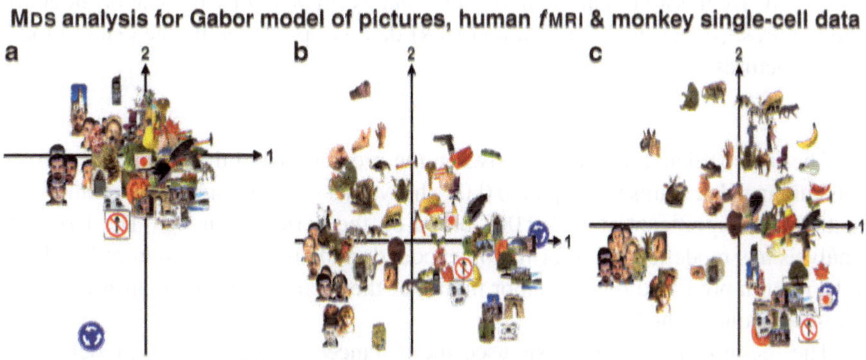

Fig. 4: (**a**) MDS map for the Gabor model of pictures; (**b**) MDS map for *f*MRI data from human participants; (**c**) MDS map for single-cell data from monkeys

We performed three separate DISPLSR analyses for the data. The first DISPLSR analysis used the distances from the Gabor model of the pictures (Fig. 4a) as the predictor and the distances from the *f*MRI data from the human participants (Fig. 4b) as the predictee. Figure 5a shows the map of the stimuli represented by the *f*MRI data as *predicted* from the Gabor model of the pictures (which appears to be linear because the predictor and predictee might have only one dimension in common). If we compare this map with the original map of stimuli derived from the *f*MRI data (see Fig. 4b), we see that DISPLSR predicts the face category, oblong objects and the roundabout on the first and second dimensions. If we subtract the map of the regression (Fig. 5a) from the original map (Fig. 4b), we obtain the residual map (Fig. 5b), which shows what is unique to brain activation in the inferior temporal cortex (i.e., cannot be predicted from the Gabor model of the pictures) such as semantic categories of faces (both human and monkey), animals, vegetables and scenes.

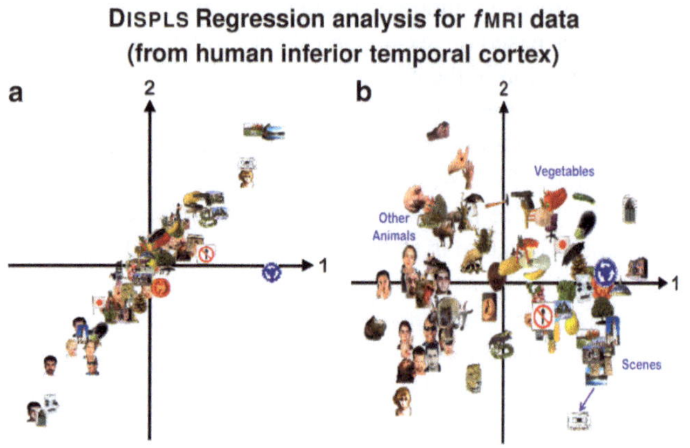

Fig. 5: (**a**) Regression map: human *f*MRI data as *predicted* by the Gabor model of the pictures; (**b**) residual map: human *f*MRI data *not* predicted by the Gabor model of the pictures

The R_V coefficient computed between the prediction and the original data from the human participants was equal to 0.08, this value—even though quite small—was statistically significant at $p < 0.001$ based on the permutation test. The DISPLSR analysis has modelled the information encoded in the brain as the sum of low-level information predicted by the Gabor model of the pictures and the high-level information unique to the brain.

The second DISPLSR analysis used the distances from the Gabor model of the pictures (Fig. 4a) as the predictor and the distances from single-cell data from the monkeys (Fig. 4c) as the predictee. Figure 6a shows the map of the stimuli represented by the single-cell data as *predicted* from the Gabor model of the pictures. If we compare this map with the original map of stimuli derived from the single-cell data (see Fig. 4c), we see the separation of oblong or rectangular objects from the circular objects on the first and second dimensions. If we subtract the map of the regression (Fig. 6a) from the original map (Fig. 4c), we obtain the residual map (Fig. 6b), which isolates semantic categories (Fig. 4c). Because the single-cell data are more robust, the residual did not isolate much of the categorical structure. The R_V coefficient between the prediction and the original data from the monkeys was not statistically significant based on a permutation test.

Because of the similarity between primate and human vision [34], we can reasonably speculate that some of the basic semantic categories in humans could be traced back to semantic categorization in monkeys. The third DISPLSR analysis used the

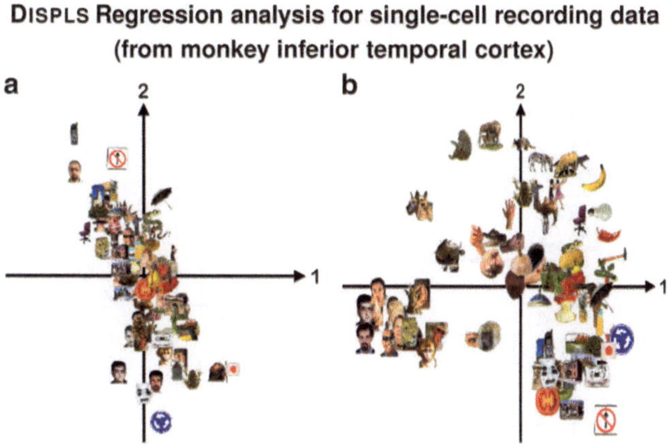

Fig. 6: (**a**) Regression map: monkey single-cell data as *predicted* by the Gabor model of the pictures; (**b**) residual map: monkey single-cell data *not* predicted by the Gabor model of the pictures

residual from the DISPLSR analysis for the monkeys (Fig. 6b) as the predictor and the residual from the DISPLSR analysis for the human participants (Fig. 5b) as the predictee in order to predict the semantic categorical structure in humans from the semantic categorical structure in monkeys (after removing perceptual information modeled by the Gabor filters).

Figure 7a shows the map of the stimuli represented by the *f*MRI data (not predicted by Gabor model) from the humans as *predicted* from the single-cell data (not predicted by Gabor model) from the monkeys. The first dimension separates the basic categories of animate and inanimate objects, and the second dimension separates the non-human and human related categories. This could imply that these monkeys (who were accustomed to a human environment) and human participants share information about these categories. If we subtract the map of the regression (Fig. 7b) from the original map (Fig. 5b), we obtain the residual map (Fig. 7b), which isolates semantic categories unique to human participants such as the natural and artificial objects. The R_V coefficient between the prediction and the residual data from the DISPLSR analysis for the human participants was not statistically significant based on a permutation test.

Lastly, we performed a DISPLS Correlation analysis to determine the commonalities between the data from the human participants and the monkeys. Figure 8a shows the first dimension of the latent similarity structure for both the human participants and the monkeys. We see that the first dimension separates the animate and inanimate objects. Figure 8b shows the second dimension of the latent similarity structure for both the human participants and the monkeys.

We see that the second dimension separates the natural and artificial objects. It is important to note that each of these maps only depict one dimension and therefore

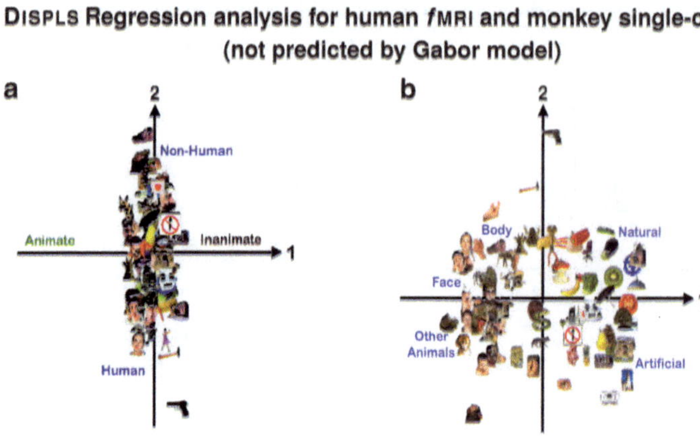

Fig. 7: (**a**) Regression map: non-Gabor *f*MRI data from the human participants as *predicted* by the non-Gabor single-cell data from the monkeys; (**b**) residual map: non-Gabor *f*MRI data from the human participants *not* predicted by by non-Gabor single-cell data from the monkeys

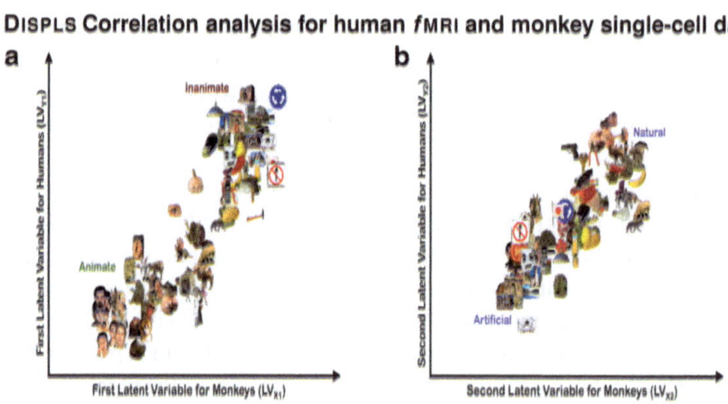

Fig. 8: (**a**) First dimension of latent similarity structures for monkeys and human participants; (**b**) second dimension of latent similarity structures for monkeys and human participants

objects that are be better represented on other orthogonal dimensions might appear to be mis-categorized. The R_V coefficient of 0.75 between the latent variables for the human participants and the latent variables for the monkeys was found to be statistically significant at $p < 0.001$ based on a permutation test.

4 Discussion

DISPLSR is a regression method, and so the residual will contain the unexplained part of experimental variance, which may confound the results of a DISPLSR analysis. DISPLSR (in combination with DISTATIS) can also be used to describe the directional dependencies of three or more sets of variables for path modeling or multi-block analyses. The reliability of the DISPLSR maps can be tested with the bootstrap method to generate confidence intervals for the predictee, predictor and residual [35]. A limitation of DISPLSR is that bootstrap maps are only generated when the original data, which were used to derive the distances between categories of observations, are available. Although other techniques exist that find commonalities between multiple distance matrices (e.g., GPA, INDSCAL, DISTATIS), it is worth exploring the potential of DISPLS Correlation (DISPLSC) to capture the information from one distance matrix in relation to the information from another distance matrix. More research is required to investigate the permutation of distance matrices in order to obtain a random sampling distribution, although a permutation test with the R_V coefficient for both DISPLSR and DISPLSC is currently possible. The prediction-based approach for distance matrices offered by DIStance-based Partial Least Squares Regression (DISPLSR), and the association-based approach for distance matrices offered by DIStance-based Partial Least Squares Correlation (DISPLSC) provide essential links between various domains of research.

References

[1] H. Abdi, "Metric multidimensional scaling," in *Encyclopedia of Measurement and Statistics*, N. J. Salkind, ed., pp. 598–605, Thousand Oaks (CA): Sage, 2007.

[2] W. S. Torgerson, "Multidimensional scaling: I. Theory and method," *Psychometrika* **17**, pp. 401–419, 1952.

[3] J. Gower and G. Dijksterhuis, *Procrustes Problems*, New York: Oxford University Press, 2004.

[4] J. D. Carroll and J.-J. Chang, "Analysis of individual differences in multidimensional scaling via an N-way generalization of Eckart-Young decomposition," *Psychometrika* **35**, pp. 283–319, 1970.

[5] H. Abdi, D. Valentin, A. J. O'Toole, and B. Edelman, "DISTATIS: The analysis of multiple distance matrices," in *Proceedings of the* IEEE *Computer Society: International Conference on Computer Vision and Pattern Recognition*, pp. 43–47, 2005.

[6] H. Abdi, J.P. Dunlop, and L.A. Williams, "How to compute reliability estimates and display confidence and tolerance intervals for pattern classiffiers using the Bootstrap and 3-way multidimensional scaling DISTATIS," in *NeuroImage* **17**, pp. 89–95, 2009.

[7] H. Abdi, D. Valentin, S. Chollet, and C. Chrea, " Analyzing assessors and products in sorting tasks: DISTATIS, theory and applications," in *Food Quality and Preference,* **18**, pp. 627–640, 2007.

[8] A. Krishnan, L. J. Williams, A. R. McIntosh, and H. Abdi, "Partial least squares (PLS) methods for neuroimaging: A tutorial and review," *NeuroImage* **56**, pp. 455–475, 2011.

[9] L.R., Tucker, "An inter-battery method of factor analysis." *Psychometrika* **23**, pp. 111–136, 1958.

[10] F.L. Bookstein, P.L. Sampson, A.P. Streissguth, and H.M. Barr, "Exploting redundant measurements of dose and developmental outcome: New methods from the behavioral teratology of alcohol," *Developmental Psychology* **32**, pp. 404–415, 1996.
[11] P.D. Sampson, A.P. Streissguth, H.M. Barr, and F.S.Bookstein, "Neurobehavioral effect of prenatal alcohol: Part II, partial least square analysis," *Neurotoxicology and Teratology* **11**, pp. 477–491.
[12] A. Tishler, D. Dvir, A. Shenhar, and S. Lipovetsky, "Identifying critical success factors in defense development projects: A multivariate analysis," *Technological Forecasting and Social Change* **51**, pp. 151–171, 1996.
[13] A. Tishler, and S. Lipovetsky, "Modelling and forecasting with robust canonical analysis: method and application ," *Computers and Operations Research* **27**, pp. 217–232, 2000.
[14] S. Dolédec, and D. Chessel, "Co-inertia analysis: an alernative methods for studying sepcies-environment relationships." *Fresehwater Biology* **31**, pp. 277–294.
[15] H. Abdi, "Partial least squares regression and projection on latent structure regression (PLS Regression)," *WIREs Computational Statistics* **2**, pp. 97–106, 2010.
[16] H. Wold, "Soft modelling: The basic design and some extensions," in *Systems under indirect observation: Causality-structure-prediction Part II*, K. Jöreskog and H. Wold, eds., pp. 1–54, Amsterdam: North-Holland Publishing Company, 1982.
[17] S. Wold, M. Sjöström, and L. Eriksson, "PLS-regression: A basic tool of chemometrics," *Chemometrics and Intelligent Laboratory Systems* **58**, pp. 109–130, 2001.
[18] H. Abdi, "Singular value decomposition (SVD) and generalized singular value decomposition (GSVD)," in *Encyclopedia of Measurement and Statistics*, N. Salkind, ed., pp. 907–912, Thousand Oaks (CA): Sage, 2007.
[19] M. Greenacre, *Theory and Applications of Correspondence Analysis*, Academic Press, London, 1984.
[20] H. Yanai, K. Takeuchi, and Y. Takane, *Projection Matrices, Generalized Inverse Matrices, and Singular Value Decomposition*, New York, Springer, 2011.
[21] B. L. Bush and R. B. Nachbar, Jr., "Sample-distance partial least squares: PLS optimized for many variables, with application to COMFA," *Journal of Computer-aided Molecular Design* **7**, pp. 587–619, 1993.
[22] Y. C. Martin, C. T. Lin, C. Hetti, and J. DeLazzer, "PLS analysis of distance matrices to detect nonlinear relationships between biological potency and molecular properties," *Journal of Medicinal Chemistry* **38**, pp. 3009–3015, 1995.
[23] P. Legendre and M. J. Anderson, "Distance-based redundancy analysis: Testing multispecies responses in multifactorial ecological experiments," *Ecological Monographs* **69**, pp. 1–24, 1999.
[24] M. A. Zapala and N. J. Schork, "Multivariate regression analysis of distance matrices for testing associations between gene expression patterns and related variables," *Proceedings of the National Academy of Sciences* **103**, pp. 19430–19435, 2006.
[25] S. Rännar, F. Lindgren, P. Geladi, and S. Wold, "A PLS kernel algorithm for data sets with many variables and fewer objects. Part I: Theory and algorithm," *Journal of Chemometrics* **8**, pp. 111–125, 1994.
[26] A. Höskuldsson, "PLS regression methods," *Journal of Chemometrics* **2**, pp. 211–228, 1988.
[27] H. Abdi, "Congruence: Congruence coefficient, R_V coefficient and Mantel coefficient," in *Encyclopedia of Research Design*, N. Salkind, D. D.M., and B. Frey, eds., pp. 222–229, Thousand Oaks (CA): Sage, 2010.
[28] H. Abdi, "R_V coefficient and congruence coefficient," in *Encyclopedia of Measurement and Statistics*, N. Salkind, ed., pp. 849–853, Thousand Oaks (CA): Sage, 2007.
[29] E. J. Dietz, "Permutation tests for association between two distance matrices," *Systematic Zoology* **32**, pp. 21–26, 1983.
[30] J. Josse, J. Pagès, and F. Husson, "Testing the significance of the R_V coefficient," *Computational Statistics & Data Analysis* **53**, pp. 82–91, 2008.
[31] N. Kriegeskorte, M. Mur, and P. Bandettini, "Representational similarity analysis connecting the branches of systems neuroscience," *Frontiers in Systems Neuroscience* **2**, p. doi:10.3389/neuro.06.004.2008, 2008.

[32] R. Kiani, H. Esteky, K. Mipour, and K. Tanaka, "Object category structure in response patterns of neuronal population in monkey inferior temporal cortex," *Journal of Neurophysiology* **97**, pp. 4296–4309, 2007.
[33] J. Daugman, "How iris recognition works," IEEE *Transactions on Circuits and Systems for Video Technology* **14**, pp. 21–30, 2004.
[34] G. Orban, D. van Essen, and W. Vanduffel, "Comparative mapping of higher visual areas in monkeys and humans," *Trends in Cognitive Sciences* **8**, pp. 315–324, 2004.
[35] B. Efron and R. Tibshirani, "Bootstrap methods for standard errors, confidence intervals, and other measures of statistical accuracy," *Statistical Science* **1**, pp. 54–77, 1986.

Dimension Reduction and Regularization Combined with Partial Least Squares in High Dimensional Imaging Genetics Studies

Edith Le Floch, Laura Trinchera, Vincent Guillemot, Arthur Tenenhaus, Jean-Baptiste Poline, Vincent Frouin, and Edouard Duchesnay

Abstract In the imaging genetics field, the classical univariate approach ignores the potential joint effects between genes or the potential covariations between brain regions. In this paper, we propose instead to investigate exploratory multivariate methods, namely partial least squares regression or canonical correlation analysis, in order to identify a set of genetic polymorphisms covarying with a set of neuroimaging phenotypes. However, in high-dimensional settings, such multivariate methods may encounter overfitting issues. Thus, we investigate the use of different strategies of regularization and dimension reduction, combined with PLS or CCA, to face the very high dimensionality of imaging genetics studies. We propose a comparison study of the different strategies on a simulated dataset. We estimate the generalisability of the multivariate association with a cross-validation scheme and assess the capacity of good detection. Univariate selection seems necessary to reduce the dimensionality. However, the best results are obtained by combining univariate filtering and L_1-regularized PLS, which suggests that discovering meaningful genetic associations calls for a multivariate approach.

Key words: Canonical correlation analysis, SNPs, GWAS, L_1 and L_2 regularization

E. Le Floch (✉) • V. Guillemot • J.-B. Poline • V. Frouin • E. Duchesnay
CEA, Saclay, Paris, France
e-mail: edith.lefloch@gmail.com;vincent.guillemot@cea.fr;jean-baptiste.poline@cea.fr; vincent.frouin@cea.fr;edouard.duchesnay@cea.fr

L. Trinchera
Rouen Business School, Rouen, France
e-mail: ltr@rouenbs.fr

A. Tenenhaus
Supélec, Gif-sur-Yvette, Paris, France
e-mail: arthur.tenenhaus@supelec.fr

H. Abdi et al. (eds.), *New Perspectives in Partial Least Squares and Related Methods*, Springer Proceedings in Mathematics & Statistics 56, DOI 10.1007/978-1-4614-8283-3_9,

1 Introduction

Brain imaging is increasingly recognized as an intermediate phenotype in understanding the complex path between genetics and behavioral or clinical phenotypes. In this context, a first goal is to propose methods with good sensitivity to identify the part of genetic variability that explains some neuroimaging variability.

Classical approaches rely on massive univariate linear modeling (MULM) [11], ignoring the potential interactions between genes or between brain regions. To overcome this limitation, we investigate exploratory multivariate methods in order to identify a set of Single Nucleotide Polymorphisms (SNPs) covarying with a set of neuroimaging phenotypes, derived from functional Magnetic Resonance Imaging (*f*MRI) data.

Partial least squares (PLS) regression [16] and canonical correlation analysis (CCA) [6] appear to be good candidates in order to look for associations between two blocks of data, as they extract pairs of covarying/correlated latent variables (one linear combination of the variables for each block). Another approach has also been proposed by [1] based on parallel independent component analysis in order to combine functional MRI data and SNPs from candidate regions. Nevertheless, all these multivariate methods encounter critical overfitting issues due to the very high dimensionality of the data. To face these issues, methods based on dimension reduction or regularization can be used.

Dimension reduction is essentially based on two paradigms: feature extraction and feature selection. Feature extraction looks for a low-dimensional representation of the data that explains most of its variability, such as principal components analysis (PCA). Feature selection methods may be divided into two categories: some univariate methods, which select relevant features independently from each other, and some multivariate methods, which consider feature inter-relations to select a subset of variables [5].

As for regularization, a sparse (L_1-regularized) version of PLS [2, 7, 9, 14, 15] and an L_2-regularised CCA [10] have recently been shown to provide good results in correlating two blocks of data such as transcriptomic and metabolomic data, gene expression levels and gene copy numbers, or gene expression levels and SNP data. Here we propose to transpose this idea to the SNP versus imaging context. One may note that such sparse multivariate methods based on L_1 penalization actually perform variable selection. Vounou et al. [13] also introduced a promising similar method, called sparse reduced-rank regression and based on L_1 penalization, that they applied to a simulated dataset made of thousands of SNPs and brain imaging data. This method is equivalent to sparse PLS in our high dimensional settings, since they make the classical approximation that in this case the covariance matrix of each block may be replaced by its diagonal elements.

We propose a comparison study, on a simulated dataset, of various regularization and preliminary dimension reduction strategies combined with PLS or CCA to deal with an increasing number of irrelevant SNPs in the training dataset. This work is complementary to ongoing studies conducted on experimental datasets since, knowing ground truth, it provides a new and essential insight into the different methods.

2 Simulated Dataset

A realistic dataset mimicking SNP and fMRI data was simulated in order to study the behavior of the different methods of interest, while knowing ground truth. A dataset **Y** of 500 samples with 34 imaging phenotypes was simulated from a multivariate normal distribution with parameters estimated from experimental data.

In order to simulate genotyping data with a genetic structure similar to that of our real data, we considered a simulation method that uses the HapMap CEU panel. We used the *gs* algorithm proposed by [8] with the phased (phase III) data for CEU unrelated individuals for Chromosome one; we only consider the genotype simulation capability of this software that may also generate linked phenotypes. We generated a dataset **X** consisting in 85,772 SNPs and 500 samples, using the extension method of the algorithm. We randomly selected 10 SNPs (out of 85,772) having a MAF equal to 0.20 and 8 imaging phenotypes (out of 34). We induced two independent causal patterns: for the first pattern we associated the first 5 SNPs with the first 4 imaging phenotypes; the second pattern was created associating the 5 remaining SNPs with the 4 last phenotypes. For each causal pattern $i \in \{1,2\}$, we induced a genetic effect using an additive genetic model involving the average of the causative SNPs (x_{ik}): $\bar{x}_i = \sum_{k=1}^{5} \frac{1}{5} x_{ik}$. Then each imaging phenotype y_{ij} ($j \in [1,\ldots,4]$) of the pattern i was affected using a linear model:

$$y^{\star}_{ij} = y_{ij} + \beta_{ij}\bar{x}_i \tag{1}$$

The parameter β_{ij} was set by controlling for the correlation (at a value of 0.50) between the jth affected imaging phenotype ($y^{\star}_{ij}$) and the causal SNPs ($\bar{x}_i$) i.e.: $\mathrm{corr}(y^{\star}_{ij}, \bar{x}_i) = 0.50$. Such control of the correlation (or the explained variance) is equivalent to the control of the effect size while controlling for the variances of SNPs ($\mathrm{var}(\bar{x}_i)$) and (unaffected) imaging phenotypes ($\mathrm{var}(y_{ij})$, as well as any spurious covariance between them ($\mathrm{cov}(y_{ij}, \bar{x}_i)$). We favor such control over a simple control for the effect size since the later may result in arbitrary large or weak associations depending on the genetic/imaging variances ratios.

SNPs whose r^2 coefficient with any of the causal SNPs is at least 0.80 are also considered as causal. Such linkage disequilibrium (LD) threshold, commonly used in the literature [3], led to 56 causal SNPs: 32 in "Pattern 1" and 24 in "Pattern 2." We use these SNPs as "ground truth" of truly causal SNPs to compute the power of good detection of the learning methods. Finally, we stripped off 10 blocks of SNPs around the 10 causal SNPs, from the whole genetic dataset, considering that neighboring SNPs were in LD with the marker if their r^2 were at least 0.20. The 5 first (resp. last) blocks, of Pattern 1 (resp. 2), are made of 127 (resp. 71) SNPs and contain all the 32 (resp. 24) SNPs that were declared as causal. The stripped blocks were concatenated and moved at the beginning of the dataset leading to 198 ($127+71$) informative features followed by 85,574 (i.e., $85{,}772-198$) non-informative (noise) features. Such a dataset organization provides a simple way to study the methods' performances while the dimensionality of the input dataset increases from 200 (mostly made of informative features) to 85,772 mostly made of noise.

3 Methods

3.1 Partial Least Squares and Canonical Correlation Analysis

PLS regression or CCA are used to model the associations between two blocks of variables (**X** and **Y**) hypothesizing that they are linked through unobserved latent variables. The latent variables (or components) are linear combinations of the observed variables, obtained by finding two weights vectors (**u** and **v**).

PLS builds successive and orthogonal latent variables for each block such that at each step the covariance between the pair of latent variables is maximal.

$$\max_{||\mathbf{u}_h||_2=||\mathbf{v}_h||_2=1} \mathbf{u}_h'\mathbf{X}_{h-1}'\mathbf{Y}_{h-1}\mathbf{v}_h \tag{2}$$

with $\mathbf{X}_0=\mathbf{X}$ and $\mathbf{Y}_0=\mathbf{Y}$ (whose columns have been standardized), and where $\mathbf{X}_{h-1}$ and $\mathbf{Y}_{h-1}$ are the deflated matrices obtained by subtracting, from blocks **X** and **Y**, the effects of their own previous latent variables respectively. This optimization problem is solved using the iterative algorithm of two-block PLS Regression (PLS2). Please note that PLS regression usually refers to an asymmetrical deflation of the two blocks of data (by the same latent variable: the one corresponding to block **X**), instead of the symmetrical deflation mode used here, called canonical mode.

CCA differs from PLS in that it maximizes the correlation (instead of the covariance) between latent variables at each step h:

$$\max_{||\mathbf{u}_h||=||\mathbf{v}_h||=1} \frac{\mathbf{u}_h'\mathbf{X}'\mathbf{Y}\mathbf{v}_h}{\sqrt{\mathbf{u}_h'\mathbf{X}'\mathbf{X}\mathbf{u}_h}\sqrt{\mathbf{v}_h'\mathbf{Y}'\mathbf{Y}\mathbf{v}_h}} \tag{3}$$

The solution may be obtained by computing the SVD of $(\mathbf{X}'\mathbf{X})^{-1/2}\,\mathbf{X}'\mathbf{Y}\,(\mathbf{Y}'\mathbf{Y})^{-1/2}$. For numerical issues, we use the dual formulation of CCA based on a linear kernel: Kernel CCA (KCCA).

3.2 Regularization Techniques

3.2.1 L2 Regularization of CCA

In order to solve the overfitting issue, regularization based on L_2 penalization may be introduced within CCA, by replacing the matrices $\mathbf{X}'\mathbf{X}$ and $\mathbf{Y}'\mathbf{Y}$ by $\mathbf{X}'\mathbf{X}+\lambda_2\mathbf{I}$ and $\mathbf{Y}'\mathbf{Y}+\lambda_2\mathbf{I}$. We call this method rKCCA.

However in high-dimensional settings, an extreme regularization is required such that regularized CCA becomes equivalent to PLS. Indeed, the approximation is often made that the covariance matrices $\frac{1}{n-1}\mathbf{X}'\mathbf{X}$ and $\frac{1}{n-1}\mathbf{Y}'\mathbf{Y}$ may be replaced by identity matrices.

3.2.2 L1 Regularization of PLS

Pushing further regularization, authors [7, 13] have proposed an approach that includes variable selection in PLS regression, based on L_1 penalization of the SNP weight vector $\mathbf{u}_h$ and leading to sparse PLS (sPLS):

$$\min_{||\mathbf{u}_h||_2=||\mathbf{v}_h||_2=1} -\mathbf{u}_h'\mathbf{X}_{h-1}'\mathbf{Y}_{h-1}\mathbf{v}_h + \lambda_{1X}||\mathbf{u}_h||_1 \tag{4}$$

where λ_{1X} is the L_1-penalization parameter for the weight vector of block $\mathbf{X}$. The sPLS criterion is bi-convex in $\mathbf{u}_h$ and $\mathbf{v}_h$ and may still be solved iteratively for $\mathbf{u}_h$ fixed or $\mathbf{v}_h$ fixed, using soft-thresholding of $\mathbf{u}_h$ within the inner loop of the PLS2 algorithm. Indeed, for $\mathbf{v}_h$ fixed,

$$\widehat{\mathbf{u}_h} = g_{\lambda_{1X}}(\mathbf{X}_{h-1}'\mathbf{Y}_{h-1}\mathbf{v}_h) \tag{5}$$

where $g_\lambda(y) = \text{sign}(y)(|y| - \lambda)_+$ is the soft-thresholding function.

For the sake of simplicity, in the rest of the paper, we replaced the penalization term by the sPLS selection rate: $s_{\lambda_{1X}}$, as the number of selected SNPs (with non null weights) out of the total number of variables of that block.

Sparse versions of CCA have also been proposed by [9, 14, 15]. However, they use the approximation described in Sect. 3.2.1, making sparse CCA equivalent to sPLS.

3.3 Preliminary Dimension Reduction Methods

3.3.1 Principal Component Based Dimension Reduction

Another approach to solve overfitting issues is the use of preliminary dimension reduction methods. We first used principal component analysis (PCA) to perform dimension reduction on each block of data (keeping as many components as necessary to explain 99 % of the block variance) before applying PLS (PCPLS) or CCA (PCKCCA). Regularization was not necessary anymore in that case, as the dimensionality had been dramatically reduced.

3.3.2 Univariate SNP Filtering

A second way to perform preliminary dimension reduction was to add to (s)PLS a first step of univariate filtering on the SNPs. We call this two-step method f(s)PLS. Similarly, we applied KCCA after a step of univariate filtering and called it fKCCA.

3.4 Comparison Study and Performance Evaluation

We compared the performances of the different methods on our simulated dataset. We first compared PLS and CCA, then investigated how their performance is improved by regularization with sparse PLS and L_2-regularized CCA. Finally we assessed the influence of a first dimension reduction step by PCA or filtering. Note that computations were always limited to the two first pairs of latent variables for computational time purposes. We also computed the performance of MULM to compare with the multivariate methods. Finally we assessed the performance of principal component regression (PCR) of the two first imaging principal components onto the genetic components explaining 99 % of the genetic variance.

3.4.1 Cross-Validation Procedure

We first evaluated the performances of the different methods by assessing the generalizability of the link they find between the blocks, using a five-fold cross-validation scheme. For each method, at each fold of the cross-validation (CV), the estimation of the model (weight vectors) was done on a training sample of 100 subjects and tested on a left-out sample of 400 subjects. Indeed, at each fold, the weights thus obtained were used to build the factorial scores of the "test" sample (the set of left out subjects) and the correlation coefficient between those factorial scores was computed. This yielded an average "test" correlation coefficient over folds, called the out-of-sample correlation coefficient, which reflects the link between the two blocks estimated on unseen subjects. Please note that at each fold, while the correlation coefficient obtained on the training samples is forced to be positive, the out-of-sample correlation coefficient may happen to be negative.

We performed a CV for MULM as well, where at each fold the two most significantly associated SNP/phenotype pairs on the training sample were extracted and tested by computing their correlation coefficient on the left-out sample.

3.4.2 Positive Predictive Value

Finally, since ground truth was known with simulated data, we could also compare the performances of the different methods by computing the Positive Predictive Value (PPV) when 50 SNPs are selected by each method. This is almost equivalent to the specificity of each method in our case where 50 SNPs are selected, since there are 56 causal SNPs in our simulated dataset. PPV curves were separately computed on 5 non-overlapping subsamples of 100 observations and averaged over these 5 subsamples. It should be noted that the informative SNPs that are not considered as causal are only slightly correlated to causal SNPs. Therefore they were removed to compute the PPV, since they could not really be identified as true or false effects.

4 Results

We investigated the performances of the different regularization and preliminary dimension reduction strategies combined with PLS or CCA, with a genetic dataset containing an increasing number of features varying between 200 (containing the informative SNPs) and 85,772 SNPs (the full dataset, mostly made of noise).

In Figs. 1 and 2 on the left column, we evaluated whether the link obtained between the imaging and SNP datasets could generalize to new unseen test subjects, using a cross-validation scheme. We computed the average out-of-sample correlation over the folds. On the right column, in order to evaluate the power of good detection of the causal SNPs, we calculated the PPV of each strategy, when 50 SNPs are selected, for each of the two first component pairs.

4.1 Influence of Regularization

We were first interested in comparing the performances of PLS and CCA when the number of SNPs p increases, and investigating the influence of L_1 regularization on PLS and of L_2 regularization on CCA.

Figure 1 shows regularization strategies starting with rKCCA with various L_2 regularization values (λ_2) that range from 0 (pure CCA) to 10,000 (which behaves like classical PLS). Then we pushed forward this penalization by experimenting sPLS with various L1 regularization values from 75 % of the SNPs with non null weights to an extreme penalization with only 10 % of selected SNPs.

On the left panel, we show the out-of-sample correlation coefficients obtained with the different methods for the two first component pairs, and it shows that in the lower dimensional space ($p = 200$) mostly made of informative features, the pure CCA, rKCCA without regularization ($\lambda_2 = 0$), has overfitted the "training" data on the first component pair ("training" corr. ≈ 1 and "test" corr. ≈ 0.20). Such a result highlights the limits of pure CCA to deal with situations where the number of training samples (100) is smaller than the number of dimensions ($p = 200$). However, with a suitable regularization in such a low-dimensional setting, rKCCA($\lambda_2 = 100$) performed better than all other methods, notably all (sparse) PLS. This results was expected since the evaluation criterion (correlation between factorial scores) is exactly the one which is maximized by CCA.

Nevertheless, the increase of space dimensionality (with irrelevant features) clearly highlighted the superiority of PLS and more notably sPLS over rKCCA in high-dimensional settings: the performance of rKCCA rapidly decreased while sPLS ($s_{\lambda_{1X}} = 0.1$) tolerated an increase of the dimensionality up to 1,000 features before its performance started to decrease. One may note that as expected theoretically, along with the increase of penalization (λ_2), rKCCA curves smoothly converged toward PLS.

On the second component pair, the results are less clearly interpretable. However (s)PLS curves were above the rKCCA ones.

The four graphs on the right panel of Fig. 1 demonstrate the superiority of sPLS methods to identify causal SNPs on the two first genetic components. Indeed, PPV curves show a smooth increase of the performance, when moving from unregularized CCA [rKCCA($\lambda_2 = 0$)] to strongly regularized PLS [sPLS($s_{\lambda_{1X}} = 0.10$)]. Moreover, while the out-of-sample correlation coefficient was not an appropriate measure to distinguish between the two causal patterns, PPV curves were computed for each pattern separately. One may note that the PPV on the first genetic component appears to be higher for the first pattern than for the second pattern (more visibly in low dimensions), while the opposite trend is observed on the second genetic component. That observation tends to show that the first causal pattern is captured by the first component pair, while the second pattern is captured by the second pair.

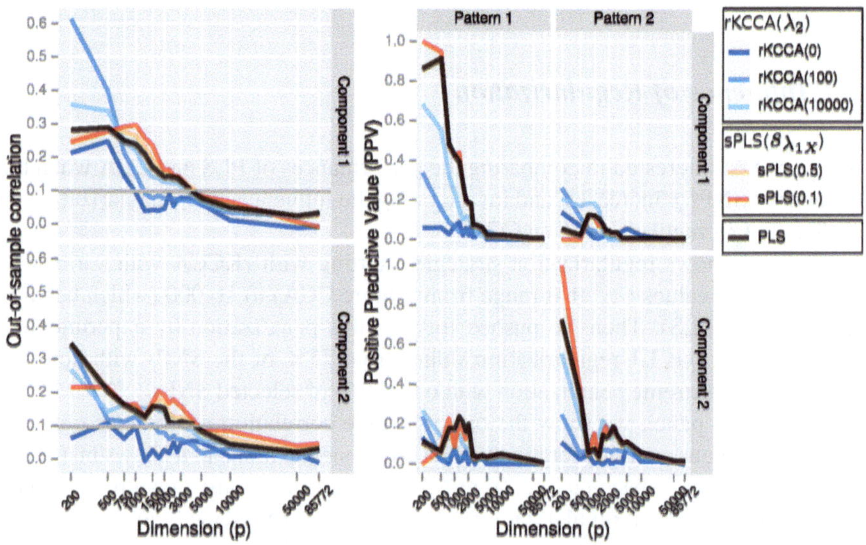

Fig. 1: Influence of regularization

4.2 Influence of the Dimension Reduction Step

We then investigated the influence of a first step of dimension reduction. Figure 2 presents different dimension reduction strategies: principal component (PC), filter (f), sparse (s) and combined filter + sparse (fs) methods. Here the parameter setting, 50 selected SNPs, was derived from the known ground truth (56 true causal SNPs). The 50 SNPs were either the 50 best ranked SNPs for (f) methods, the 50 non-null weights for sparse PLS or a combination of both for fsPLS: either 10 % of the 500 best ranked SNPs or 50 % of 100.

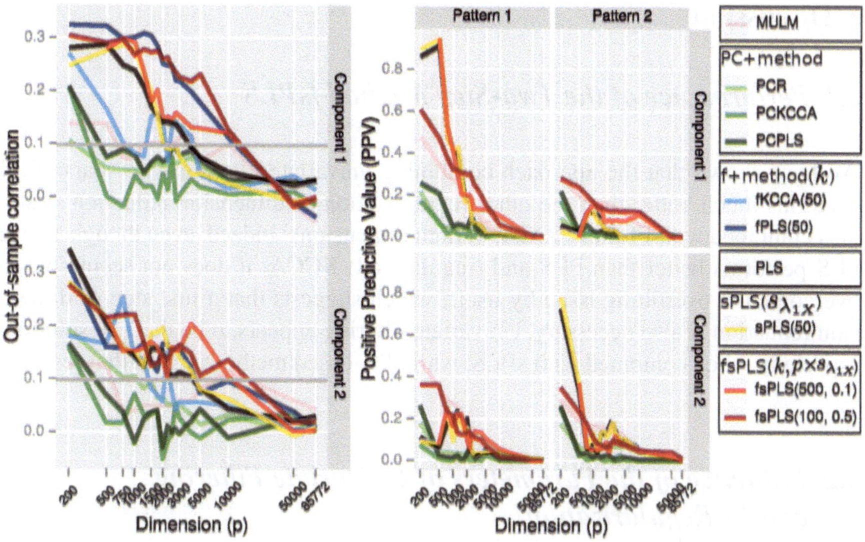

Fig. 2: Influence of dimension reduction. (f) Methods are superimposed with MULM for PPV

On the left panel, we show that all PC-based methods (green curves) failed to identify generalizable covariations when the number of irrelevant features increases.

Dimension reduction based on filtering failed with CCA but greatly improved the performance of classical PLS: fPLS($k = 50$) was the second best approach in our comparative study.

Moreover, as previously observed in Fig. 1, L_1 regularization limited the overfitting phenomenon (see sPLS($s_{\lambda_{1X}} * p = 50$) in Fig. 2) and delayed the decrease of PLS performance when dimensionality increased. Finally the best performance was obtained by combining filtering and L_1 regularization: fsPLS($k = 100$, $s_{\lambda_{1X}} = 0.50$), which kept 100 SNPs after filtering and selected 50 % of those SNPs by sPLS. Please note that the performance of fsPLS($k = 500$, $s_{\lambda_{1X}} = 0.10$) was lower and similar to that of sPLS(50) in low dimensions, but became more robust than sPLS and equivalent to fsPLS($k = 100$, $s_{\lambda_{1X}} = 0.50$) in higher dimensions. However, the purely univariate strategy based on MULM showed poor generalizability, which suggests that even though filtering is necessary to remove irrelevant features, it cannot capture the imaging/genetics link by itself and needs to be combined with a multivariate step.

Again, on the second component pair, the results are less clearly interpretable, but the curves of the strategies combining filtering and sparsity were above the others. The graphs on the right panel of Fig. 2 show similar results in terms of PPV.

5 Discussion

5.1 *Performance of the Two-Step Method fsPLS*

We have shown that the approach combining univariate filtering with sparse PLS performs much better than the other regularization or dimension reduction strategies combined with PLS or KCCA, in high dimensions. Indeed, even though sparse PLS performs better than PLS and (regularized) KCCA, it does not seem able to overcome the overfitting issue by itself, which suggests that a first step of dimension reduction is also necessary. Univariate filtering appears to be the best solution, especially when combined with sPLS, while PC-based methods fail in that respect.

5.2 *Influence of the Parameters of Univariate Filtering and L_1 Regularization*

We tried to assess the influence of the parameters of univariate filtering and sPLS selection on the generalizability of the link found by fsPLS between the two blocks of data, which explains why we repeated the cross-validation procedure and the PPV analysis for different pairs of parameters.

Our results show that when the dimensionality increases, fsPLS tends to extract the most generalizable neuroimaging/genetics link when considering 100 SNPs after filtering and 50 % of these SNPs selected by sPLS. Those results raise the question of the relative contribution of the univariate filtering and the sparsity constraint to select relevant features. A relatively large number of SNPs kept after filtering seems to be required, up to a trade-off between the numbers of true and false positives, to allow sPLS to extract a robust association between a multivariate pattern of SNPs and a multivariate neuroimaging pattern. However, univariate filtering appears to be a mandatory step to filter out the vast majority of irrelevant features, especially when the dimensionality increases.

Another reason to perform univariate filtering is that PLS and even sparse PLS are too sensitive to a large number of irrelevant features, as they try to explain the variance of each block while trying to find some link between the blocks. Indeed, let us remind the criterion that is maximized by PLS regression:

$$\max_{\|\mathbf{u}\|_2=\|\mathbf{v}\|_2=1} \operatorname{corr}(\mathbf{Xu},\mathbf{Yv})\sqrt{\operatorname{var}(\mathbf{Xu})}\sqrt{\operatorname{var}(\mathbf{Yv})}, \tag{6}$$

where the first term is the inter-block correlation between the two latent variables of each block and the two last terms the intra-block standard deviations of the latent variable of each block. In the case of very large blocks, the two terms of intra-block standard deviations weigh much more than the term of inter-block correlation, as discussed by [12]. Univariate filtering helps to solve this problem by reducing the number of SNPs and selecting the ones that are more correlated to the phenotypes.

5.3 Potential Limitations of fsPLS

Although common practice in genome wide association studies, univariate tests may not be the best filter and it could be interesting to consider multivariate filters that account for specific interactions between potential predictors [4]. For instance a limitation of univariate filtering may be that it filters out suppressor variables. Indeed such variables are useful to remove the non-specific variability of the relevant SNPs, improving their predictive power, while being themselves not correlated (and thus not detectable) with imaging phenotypes.

As for penalization, even though it is well-known that it plays an important role when trying to uncover specific relationships among high-dimensional data, the choice of the penalization is also important. For instance, an L_1, L_2 or L_1-L_2 (elastic net) penalization scheme does not give rise to the same results when data are correlated. Indeed in the case of correlated variables grouping into a few clusters, L_1 penalization tends to select one "representative" variable of each cluster, which facilitates the interpretation of the results but may lead to an unstable solution, whereas L_2 penalization and the elastic net criterion tend to emphasize the whole set of correlated predictors. In our case, we observed that the L_1 penalization combined with the implicit extreme L_2 penalization of PLS led to an elastic net behavior of sPLS. Indeed, SNPs are spatially correlated in blocks due to LD and sPLS tends to select LD blocks. One could investigate more sophisticated penalizations that take into account the correlation structure of the data.

6 Conclusion

To conclude, the originality of this work was to investigate a two-step method combining univariate filtering and sparse PLS, called fsPLS, and we showed that it performed much better than other regularization or dimension reduction strategies combined with PLS or KCCA, on high-dimensional simulated imaging genetics data. Even though univariate filtering may seem to contradict the very nature of multivariate methods such as PLS, it still allows sPLS to extract a multivariate genetic pattern (among the remaining SNPs) covarying with an imaging pattern and appears to be necessary to overcome the overfitting issue in very high dimensional settings. This suggests that if individual variability in the genome contains predictors of the observed variability in brain phenotypes, they can be detected by fsPLS even though they may not be detected by a univariate screening only, or by standard regularized multivariate methods.

References

[1] V. D. Calhoun, J. Liu, and T. Adali, "A review of group ICA for fMRI data and ICA for joint inference of imaging, genetic, and ERP data", *NeuroImage* **45** (1(Suppl 1)), pp. S163–S172, 2009.

[2] H. Chun, and S. Keleş, "Sparse partial least squares regression for simultaneous dimension reduction and variable selection", *Journal of the Royal Statistical Society - Series B*, **72** (1), pp. 3–25, 2010.

[3] P.I.W. de Bakker, R. Yelensky, I. Peer, S. B. Gabriel, M. J. Daly, and D. Altshuler, "Efficiency and power in genetic association studies", *Nature Genetics*, **37** (11), pp. 1217–1223, 2005.

[4] R. Díaz-Uriarte, and S. Alvarez de Andrés, "Gene selection and classification of microarray data using random forest", *BMC Bioinformatics*, **7** (3), 2006.

[5] I. Guyon, S. Gunn, M. Nikravesh, L. A. Zadeh, "Feature Extraction: Foundations And Applications", Springer-Verlag, 2006.

[6] H. Hotelling, "Relations between two sets of variates", *Biometrika* **28**, pp. 321–377, 1936.

[7] K.-A. Lê Cao, D. Rossouw, C. Robert-Granié, and P. Besse, "A sparse PLS for variable selection when integrating omics data", *Statistical Applications in Genetics and Molecular Biology* **7** (1), 2008.

[8] J. Li, and Y. Chen, "Generating samples for association studies based on HapMap data", *BMC Bioinformatics*, **9** (44), 2008.

[9] E. Parkhomenko, D. Tritchler, and J. Beyene, "Sparse canonical correlation analysis with application to genomic data integration", *Statistical Applications in Genetics and Molecular Biology* **8** (1), Article 1, 2009.

[10] C. Soneson, H. Lilljebjörn, T. Fioretos, and M. Fontes, "Integrative analysis of gene expression and copy number alterations using canonical correlation analysis", *BMC Bioinformatics*, **11** (191), 2010.

[11] J. Stein, X. Hua, S. Lee, A. Ho, A. Leow, A. Toga, A. Saykin, L. Shen, T. Foroud, N. Pankratz, M. Huentelman, D. Craig, J. Gerber, A. Allen, J. Corneveaux, B. DeChairo, S. Potkin, M. Weiner, and P. Thompson, "Voxelwise genome-wide association study (vGWAS)", *NeuroImage* **53**, pp. 1160–1174, 2010.

[12] A. Tenenhaus, and M. Tenenhaus, "Regularized Generalized Canonical Correlation Analysis", *Psychometrika*, **76** (2), pp. 257–284, 2011.

[13] M. Vounou, T. E. Nichols, and G. Montana, "Discovering genetic associations with high-dimensional neuroimaging phenotypes: A sparse reduced-rank approach", *NeuroImage*, **53**, pp. 1147–1159, 2010.

[14] S. Waaijenborg, P. Verselewel de Witt Hamer, and A. Zwinderman, "Quantifying the association between gene expressions and DNA-markers by penalized canonical correlation analysis", *Statistical Applications in Genetics and Molecular Biology* **7** (1), Article 3, 2008.

[15] D. M. Witten, and R. Tibshirani, "Extensions of sparse canonical correlation analysis, with applications to genomic data", *Statistical Applications in Genetics and Molecular Biology* **8** (1), Article 28, 2009.

[16] S. Wold, H. Martens, and H. Wold, "The multivariate calibration problem in chemistry solved by the PLS method", in *Proceedings Conference Matrix Pencils*, A. Ruhe and B. Kastrøm, eds., pp. 286–293, Springer-Verlag, 1983.

Revisiting PLS Resampling: Comparing Significance Versus Reliability Across Range of Simulations

Natasa Kovacevic, Hervé Abdi, Derek Beaton, for the Alzheimer's Disease Neuroimaging Initiative*, and Anthony R. McIntosh

Abstract PLS as a general multivariate method has been applied to many types of data with various covariance structures, signal strengths, numbers of observations and numbers of variables. We present a simulation framework that can cover a wide spectrum of applications by generating realistic data sets with predetermined effect sizes and distributions. In standard implementations of PLS, permutation tests are used to assess effect significance, with or without procrustes rotation for matching effect subspaces. This approach is dependent on signal amplitude (effect size) and, as such, is vulnerable to the presence of outliers with strong amplitudes. Moreover, our simulations show that in cases when the overall effect size is weak, the rate of false positives—and to a lesser extent—false negatives, is quite high. From the applications point of view, such as linking genotypes and phenotypes, it is often more important to detect reliable effects, even when they are very weak. Reliability in such cases is measured by the ability to observe the same effects supported by the same patterns of variables, no matter which sets of observations (subjects) are used.We implemented split-half reliability testing with thresholds based on null distributions and compared the results to the more familiar significance testing.

Key words: PLS, SVD, Significance, Simulations, Reliability

---

*Data used in preparation of this article were obtained from the Alzheimer's Disease Neuroimaging Initiative (ADNI) database (adni.loni.ucla.edu). As such, the investigators within the ADNI contributed to the design and implementation of ADNI and/or provided data but did not participate in analysis or writing of this report. A complete listing of ADNI investigators can be found at http://adni.loni.ucla.edu/wpcontent/uploads/how_to_apply/ADNI_Acknowledgement_List.pdf

N. Kovacevic (✉) • A.R. McIntosh
Rotman Research Institute, Baycrest Centre, 3560 Bathurst Street, Toronto, Canada
e-mail: nkovacevic@research.baycrest.org; rmcintosh@research.baycrest.org

H. Abdi • D. Beaton
The University of Texas at Dallas, School of Behavioral and Brain Sciences, MS:Gr.4.1 800 West Campbell Road, Richardson, TX 75080-3021, USA
e-mail: herve@utdallas.edu; derekbeaton@utdallas.edu

H. Abdi et al. (eds.), *New Perspectives in Partial Least Squares and Related Methods*, Springer Proceedings in Mathematics & Statistics 56, DOI 10.1007/978-1-4614-8283-3_10,

1 Introduction

Partial Least Squares (PLS) is a versatile multivariate method that has been applied to many data types in neuroimaging, Psychology, physiology, Genetics, and Chemometrics to name but a few [1–3]. In neuroimaging, the standard application of PLS is PLS-correlation (PLSC, see [1]) whose computational core consists in the singular value decomposition of the correlation matrix of the variables from two matrices. In this context, the singular vectors are called *saliences* and the associations between the two data tables are explored with latent variables (LVs) computed as the projection of each table on its corresponding saliences (see, e.g., [1], for details). More recently, researchers have been interested in investigating large scale data sets with weak signals, as found, for example, when relating genotypes and complex phenotypes (e.g., personality traits, body weight). For these problems, PLS methods offer an excellent framework. However, it is difficult to interpret and validate genotype-phenotype associations obtained by PLS because the ground truth is not known. For this reason it is important to generate simulated data and hone the ability of PLS to correctly detect weak but reliable signals.

More generally, PLS validation requires a simulation framework that can cover a wide range of applications that vary in covariance structure, signal strength, number of observations and variables. In this work, we present a set of simulations where significance testing—within the standard PLS implementation—shows a clear propensity to Type I errors, and much more so for certain data types. PLS methods seem to be more prone to this Type I error inflation when the cross-validation approach—used to derive the sampling distribution under the null hypothesis—involves Procrustes rotations to project the LVs from the permuted data onto the original LVs (as done in the current implementation of PLSC, see [2]).

It is well known that significance testing using permutation tests is essentially amplitude driven and vulnerable to the presence of outliers ([4]). We have found, in practice, another weakness of PLS when it is applied to a weak correlation structure between predictor and response variables, where one of these two data sets has a very weak covariance structure while the other set has a very strong covariance structure between variables (as is often the case in "brute force" approaches to genotype-phenotype associations). In such cases, the strength of correlations between response variables (e.g., highly redundant behavioral measures)—even though not necessarily related to the genes—can overpower the permutation tests and falsely identify genotype-phenotype associations. When this is the case, using completely random genetic data will produce similar results to the analysis performed on real genetic data (because the analysis is driven by the other set). Therefore, a more appropriate question then is: Can we detect associations (i.e., LVs) that reliably represent specific genotype-phenotype links, such that any set of subjects would produce similar LVs with simultaneous, better-than-chance similarity for both associated patterns (e.g., genotype and phenotype)?

Motivated by such examples, we designed a Monte-Carlo simulation framework, flexible enough to mimic many realistic scenarios, with the advantage that we could manipulate the ground truth. We also introduced a new split-half resampling framework for reliability testing, similar to [5], as an alternative to significance (null hypothesis) testing within the PLS approach. We then compared results obtained with classical PLSC analysis with and without Procrustes rotations to those obtained using split-half reliability testing.

Although the work presented here is general in nature and applies to different data types, our starting point was the PLSC methodology as implemented in the PLSC software package ([2]), which focuses on neuroimaging applications. We will therefore use terminology appropriate for neuroimaging applications, where predictor variables are typically some sort of brain imaging data, as in [2]. Typically, subjects are split across several groups and their data are collected under different experimental conditions. In this spirit, observations are condition specific subject data and predictor variables are voxels. Task-PLS refers to the data driven approach where stacked subjects voxel data are tested for group/condition membership patterns, called task effects. Seed-PLS refers to data driven analysis of the correlation matrix between entire brain data and a (typically small) subset of voxels, called seeds, where correlations are calculated across group and condition specific subject data. In this case, PLS also extracts group/condition patterns in correlations (see [1] for details).

2 Simulations

We used real data as a starting point for our simulations in order to create realistic scenarii while manipulating effect sizes and noise sizes and distributions. These data sets were chosen from brain imaging, behavior, and genetics to represent a wide range of data dimensions, specifically number of observations and number of predictor variables. These synthetic data sets were then tested with two main flavors of PLS, data driven task-PLS and seed-PLS.

2.1 Real Data Sources

We used three different types of real data: electro-encephalogram, behavior, and genetics.

2.1.1 Event Related Potentials (ERP) Data

The first set consists of electroencephalogram (EEG) data from a total of 48 subjects whose data were collected across 2 experimental conditions. In addition, subjects were divided into 3 age groups, with 16 subjects in each group: Young (mean age 22 ± 3 years), Middle (mean age 45 ± 6 years) and Older (mean age 66 ± 6 years). For the purposes of the present work, we used two visual perceptual matching tasks from the larger study that involved six conditions. Visual stimuli were presented simultaneously in a triangular array. In the perceptual matching task (PM), subjects indicated which of the three bottom stimuli matched the one on the top by pressing one of three buttons. In the delayed match to sample task (DMS), the instructions were the same as in the PM, except that the three bottom row stimuli were presented after a 2.5 s delay following the presentation of the top row stimulus.

EEG recordings from 76 electrodes were collected using BioSemi Active Two system with a bandwidth of 99.84 (0.16 100) Hz and sampling rate of 512 Hz. Data were recorded reference-free, but were converted to an average reference at Cz during the pre-processing. We utilized standard preprocessing steps for ERP data analysis. Continuous EEG recordings were bandpass filtered from 0.5 to 55 Hz. Data from trials with correct responses were "epoched" and base-lined into [−500 2,000] ms epochs with a [−500 0] ms pre-stimulus baseline. Artifact removal was performed using Independent Component Analysis (ICA). The data were averaged across trials for each condition separately. For our simulations we considered only [0 500] ms time window (257 time points) of the averaged data.

This represents a scenario with a small number of subjects (48), a large number of predictors (EEG channels × time points = 76 × 257 = 19,532), that are somewhat strongly correlated (see Fig. 1A) and a small number of group/condition dimensions (3 age groups × 4 conditions = 12).

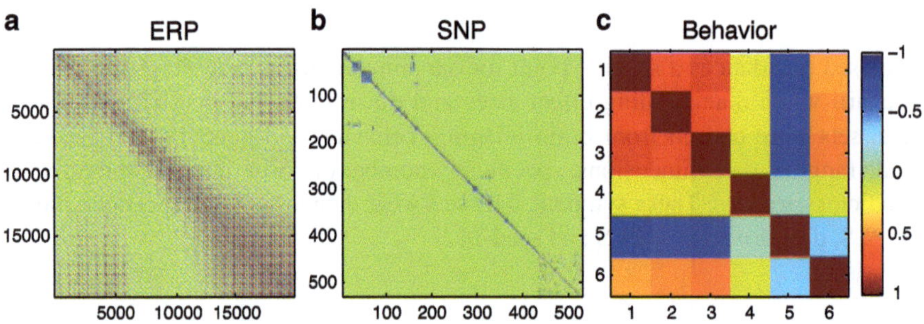

Fig. 1: Correlation matrices for three real data sets. Each matrix was derived from all available observations. Notice the wide variety in voxel space dimensionality and correlation strengths

2.1.2 Genes and Behavior: Genetic Data

Genetic and associated behavioral data used in the preparation of this article were obtained from the Alzheimer's Disease Neuroimaging Initiative (ADNI) database (adni.loni.ucla.edu). The ADNI was launched in 2003 by the National Institute on Aging (NIA), the National Institute of Biomedical Imaging and Bioengineering (NIBIB), the Food and Drug Administration (FDA), private pharmaceutical companies and non-profit organizations, as a $60 million, 5-year public–private partnership. The primary goal of ADNI has been to test whether serial magnetic resonance imaging (MRI), positron emission tomography (PET), other biological markers, and clinical and neuropsychological assessments can be combined to measure the progression of mild cognitive impairment (MCI) and early Alzheimer's disease (AD). Determination of sensitive and specific markers of very early AD progression is intended to aid researchers and clinicians to develop new treatments and monitor their effectiveness, as well as lessen the time and cost of clinical trials. The Principal Investigator of this initiative is Michael W. Weiner, MD, VA Medical Center and University of California at San Francisco. ADNI is the result of efforts of many co-investigators from a broad range of academic institutions and private corporations, and subjects have been recruited from over 50 sites across the U.S.A. and Canada. The initial goal of ADNI was to recruit 800 adults, aged from 55 to 90, to participate in the research, approximately 200 cognitively normal older individuals to be followed for 3 years, 400 people with MCI to be followed for 3 years and 200 people with early AD to be followed for 2 years. For up-to-date information, see www.adni-info.org.

In the ADNI database, the genetic information of each participant is a long list of pairs (one per chromosome) of DNA nucleotides (A, T, C, and G)—which could occur in $2^4 = 16$ different configurations—grouped in 23 chromosomes, amounting to roughly 600,000 genetic markers. However, after standard preprocessing with PLINK (pngu.mgh.harvard.edu/purcell/plink/, e.g., with a call rate of 90% and minor allele frequency of 5%, [6]) we were left with approximately 500,000 genomic locations that show enough variability in a population. These locations of variability are called single nucleotide polymorphisms (SNPs).

Because our goal was to understand the effects of resampling-based inference tests for PLS, we selected only some of the top reported, clinically relevant genetic markers, consisting of 178 SNP's. Because the work presented here is not concerned with data interpretation, we skip the details of clinical relevance. Each SNP has a major allele (e.g., A) which is the most frequent nucleotide (in a population) and a minor allele (e.g., T; rare in the population but required to be found in at least 5% of the population to be considered worth exploring). Thus, in practice only three variants for each location are used: the major homozygote (e.g., AA), the minor homozygote (e.g., TT), and the heterozygote (e.g., AT). Multivariate data sets of SNPs are most often re-coded through a process of counting the number of minor alleles. So, in our data: 0 represents the major allele homozygote (e.g., AA), 1 codes for the heterozygote (e.g., AT), and 2 represents the minor allele homozygote (e.g., TT). In most analyses, the SNPs are treated as quantitative data since most statistical methods used rely upon quantitative measures. Because the assumptions of a quantitative coding scheme seem unrealistic, we have decided to use a qualitative coding scheme and to consider that the values 0, 1, and 2 represent three different levels of a nominal variable and to code each possible variants with a 3 by 1 vector of binary variables (i.e., $AA = [100]$, $AT = [010]$, and $TT = [001]$).

The data were extracted from 756 subjects comprising three clinical groups and each clinical group was further split by sex. This produced a total of six, approximately equally populated, groups of subjects. This pattern of data represents a data analytic scenario with comparable numbers of observations (756) and predictors (SNPs $\times$ variants $= 178 \times 3 = 528$). Here, the predictor data are binary and weakly correlated (Fig. 1B). The number of group/condition dimensions is also small (six groups based on clinical diagnosis and sex).

2.1.3 Genes and Behavior: Behavioral Data

We extracted six behavioral measures from the same subjects as for the genetic data. Once again, as the interpretation of the behavioral data is not important for our simulations, we skip the details pertaining to the choice of behavioral measures. This represents a scenario with a large number of observations (756), small number of highly correlated predictors (six behavioral measures, see Fig. 1C), and a small number of group/condition dimensions.

2.2 *Simulation of Group/Condition Effects*

We start with a real data set stacked in a standard manner as a two-dimensional matrix $\mathbf{X}$ whose every row contains data for one subject (observation) in one condition ([2]). The rows are arranged such that observations are nested within condition blocks, which are in turn nested within group membership. From $\mathbf{X}$ we extracted two parameters: the covariance matrix $\mathbf{C}$ of the voxel space (covariance calculated across observations) and the group/condition specific mean signal $\mathbf{m}$ of the predictor variables across the real data observations. The mean signal $\mathbf{m}$ is further centered by subtracting the grand mean of all groups and conditions. To generate comparable simulated data with a controlled number of group/condition effects, we first decomposed $\mathbf{m}$ using a principal component analysis and then rebuilt the modified signal (denoted $\mathbf{m}_1$) using only the first K principal components. This allowed us to control the number of expected group/condition effects. In the simulations presented here, we chose $K = 3$ as the reasonable number of effects that can be expected with this type of data. To create a simulated voxel data set similar to the real data, we drew observations from a multivariate normal distribution with covariance $\mathbf{C}$ and mean $\mathbf{m}_1$. However, we wanted to test how well we can detect reliable task effects depending on the signal strength ($\mathbf{m}_1$ amplitude across voxels) and the noise distribution. For this reason, we used the signal amplitude as a scalar parameter denoted γ that was manipulated in order to vary the intensity of the signal as $\gamma\mathbf{m}_1$. In order to explore the effects of noise we removed the signal from a proportion (denoted *np*) of randomly selected voxels. To summarize, we designed a simulations scheme where we controlled:

1. The number of expected group/condition effects (set to 3 for all simulations)
2. The signal strength measured by γ, where $\gamma \in \{0, 0.5, 1, 3\}$
3. Percentage of noise-voxels (i.e., voxels for which $\mathbf{m}_1 = 0$) measured by *np*, where $np \in \{80, 40, 10, 0\%\}$.

2.3 Simulation of Correlation Effects

Once again we start with a real voxel data set $\mathbf{X}$, with voxel covariance matrix $\mathbf{C}$, as above. We create simulated versions of the data by drawing observations from a multivariate normal distribution with covariance $\mathbf{C}$ and zero mean vector. This produces a matrix $\mathbf{Y}$ with same dimensions as the real data $\mathbf{X}$. We then selected a small set of voxels as seeds (i.e., we extracted columns of $\mathbf{Y}$ corresponding to the selected voxels). Seed-PLS analyzes the correlation between $\mathbf{Y}$ and the seeds and searches for the group/condition effects within the correlation structure which is stacked by group and condition specificity in the same way as for task-PLS. In this case, the strength of the signal reflects the strength of the correlations between the columns of $\mathbf{Y}$ and the seeds. Note that the correlations are exactly 1 for the columns corresponding to the seeds across all groups and conditions. In this scenario, we manipulated the strength of the signal by permuting a random subset of rows of the seed matrix, while keeping the voxel data matrix $\mathbf{Y}$ unperturbed. The percentage of rows that were permuted, denoted *pp*, is inversely related to the strength of the correlations: if only few rows are permuted (e.g., $pp < 5\%$), the correlations change only slightly; if all rows are randomly permuted ($pp = 100\%$), all the correlations are destroyed. In the results presented here, we tested a range of *pp* values with $pp \in \{0\%, 30\%, 60\%, 100\%\}$.

3 Split-Half Reliability

The reliability of the latent variables is implemented in a split-half resampling framework similar to [5]. Here we give a brief description for the data driven PLS methods. The overview of the algorithm is shown in Fig. 2. We start by first decomposing the signal $\mathbf{D}$ (whether mean-centered group/condition mean signal in task-PLS or correlation signal between predictors and responses in seed-PLS) using the singular value decomposition (SVD). Specifically, assuming that $\mathbf{D}$ is in a group/condition by voxel format, then the SVD of $\mathbf{D}^T$ is obtained as:

$$\mathbf{D}^T = \mathbf{U}\mathbf{S}\mathbf{V}^T.$$

The columns of $\mathbf{U}$ store the left singular vectors (voxel patterns), the columns of $\mathbf{V}$ store the right singular vectors (group/condition effects) and $\mathbf{S}$ is the diagonal matrix of the singular values. In our framework, we will consider that a latent variable $\boldsymbol{\ell}_i$ comprises a matching set of right and left singular vectors (i-th column of $\mathbf{U}$, and $\mathbf{V}$ respectively) and singular value (i-th diagonal element of $\mathbf{S}$). In standard permutation tests, the significance of a given LV is focused on the amplitude of the singular value. However, in split-half reliability testing we are interested in the stability of the pairings between left and right singular vectors. Therefore, we randomly split every group of subjects and calculated the signals $\mathbf{D}_1$ and $\mathbf{D}_2$ by applying the same group

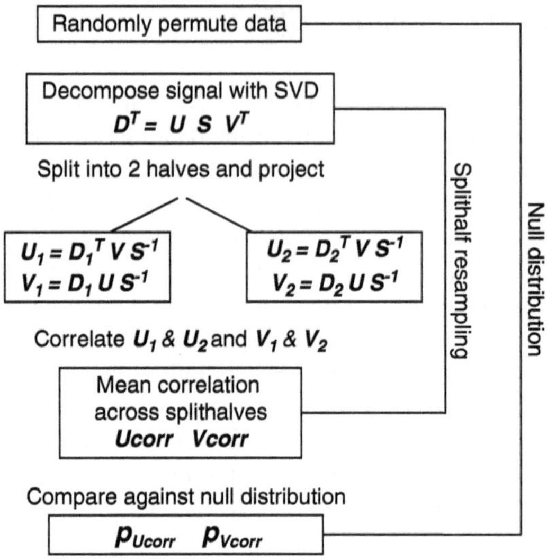

Fig. 2: Diagram of the algorithm for split-half reliability testing

and condition specific averaging/correlation procedure as originally performed on $\mathbf{D}$, working with only half of the subjects. We projected the original matrices $\mathbf{U}$ and $\mathbf{V}$ onto each half of $\mathbf{D}$ to obtain the corresponding half-sample matching pairings. Specifically, we computed:

$$\mathbf{U}_1 = \mathbf{D}_1^T \mathbf{V}\mathbf{S}^{-1} \quad \text{and} \quad \mathbf{U}_2 = \mathbf{D}_2^T \mathbf{V}\mathbf{S}^{-1}$$
$$\mathbf{V}_1 = \mathbf{D}_1 \mathbf{U}\mathbf{S}^{-1} \quad \text{and} \quad \mathbf{V}_2 = \mathbf{D}_2 \mathbf{U}\mathbf{S}^{-1}$$

The correlations between projected left and right split-half patterns (i.e., correlation between the matrices $\mathbf{U}_1$ and $\mathbf{U}_2$, and the matrices $\mathbf{V}_1$ and $\mathbf{V}_2$) are taken as measures of the correspondence between the voxel space and the $\mathbf{V}$ patterns, on one hand, and group/condition membership and the $\mathbf{U}$ patterns, on the other hand. By repeating this procedure many times, we obtain a robust estimate of split-half correlations for both left and right singular vectors. Note that this procedure uses the full sample to decompose the data structure into latent variables. This is particularly important for weak signals, where a half-sample may not reveal the signal. The purpose of the procedure is different from a standard split-half cross-validation, where each half-sample is independently analyzed. Instead, our focus is to evaluate the reliability of the associations—captured by the LV's—between voxel patterns and group/condition effects. In other words, our main question is: Given a group/condition effect, how reliable is the corresponding voxel pattern? Would the same group/condition effect links with a similar voxel pattern if we were to chose a different set of subjects? In an analogous way, given a voxel pattern (left singular vector), we want to estimate the reliability of the associated condition/group effect. For example, in the analysis of genotype/phenotype associations, the SVD decomposes the correlation matrix into latent variables, where each latent variable links

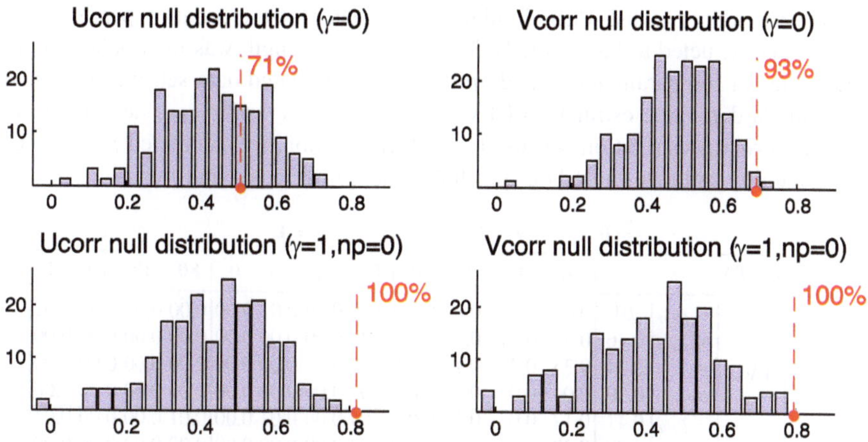

Fig. 3: Null-distribution for p_{Ucorr} and p_{Vcorr}. For illustration we chose the first LV from two ERP-type simulations of task effects. The *top row* corresponds to the "no signal" scenario with $\gamma = 0$. The *bottom row* corresponds to the simulation with the most realistic signal strength and distribution, with $\gamma = 1$ and $np = 0$. The *red dot* marks the split-half correlation of the original un-permuted data. The *red dotted line* and *red percent value* indicate the corresponding percentile of the null distribution. In both scenarios, the distributions are strongly skewed towards positive values, however the p_{Ucorr} and p_{Vcorr} percentile values suggest rejection of the null hypothesis for the realistic signal only

a particular weight from the SNPs with a particular weight of from the phenotype measures. In this case, our split-half procedure tests the reliability of this link.

It is important to notice that the distribution of the correlations between projected split-half patterns will be skewed even in a completely random data set. After all, the original SVD decomposition reflects the full sample, so it is not surprising that, on the average, the distribution of the values of the correlation between split halves is biased towards positive values (see Fig. 3). To deal with this systematic bias, we create a null distribution for the split-half correlations. This is done by randomly permuting observations (i.e., the rows of **X**) and repeating the split-half correlation estimation for each permuted data set. This allows us to estimate the probability of surpassing the correlations from the original un-permuted data set. We denote these probabilities by p_{Ucorr} and p_{Vcorr} and treat them as p-values that describe the stability of voxel patterns associated with **U** and group/condition patterns associated with **V**, respectively. In the present simulations, we performed 200 half-splits and 200 permutations to create the null distributions, and considered that a latent variable was reliable when both probabilities were smaller than 0.05 (i.e., $p_{\text{Ucorr}} < 0.05$ and $p_{\text{Vcorr}} < 0.05$).

Table 1: Results for task-PLS simulations. For each of the 3 data types, the signal was constructed to have exactly 3 LVs, and its strength was manipulated with the values of the parameters γ and np. For each simulated data set, we computed two standard p-value estimates of LV significance, p_{rot} and p_{nonrot} depending on weather Procrustes rotation was used or not. In addition, we calculated p-values of LV reliability estimates based on split-half resampling, p_{Ucorr} and p_{Vcorr}

			$\gamma=0$	$\gamma=0.5$				$\gamma=1$				$\gamma=3$			
		np(%)		**80**	**40**	**10**	**0**	**80**	**40**	**10**	**0**	**80**	**40**	**10**	**0**
ERP	LV1	p_{rot}	.01	0.03	0.01	0.01	0.01	0.01	0.00	0.01	0.00	0.00	0.01	0.01	0.01
		p_{nonrot}	0.46	0.35	0.27	0.20	0.17	0.20	0.00	0.00	0.00	0.00	0.00	0.00	0.00
		p_{Ucorr}	0.34	0.35	0.23	0.17	0.12	0.20	0.01	0.00	0.00	0.00	0.00	0.00	0.00
		p_{Vcorr}	0.10	0.04	0.04	0.01	0.04	0.00	0.00	0.00	0.00	0.00	0.00	0.00	0.00
	LV2	p_{rot}	0.41	0.17	0.03	0.01	0.01	0.04	0.00	0.01	0.00	0.01	0.00	0.00	0.01
		p_{nonrot}	0.78	0.29	0.03	0.01	0.01	0.03	0.00	0.00	0.00	0.00	0.00	0.00	0.00
		p_{Ucorr}	0.78	0.41	0.08	0.02	0.01	0.05	0.00	0.00	0.00	0.00	0.00	0.00	0.00
		p_{Vcorr}	0.18	0.04	0.02	0.03	0.01	0.00	0.00	0.00	0.00	0.00	0.00	0.00	0.00
	LV3	p_{rot}	0.72	0.62	0.41	0.34	0.31	0.35	0.11	0.06	0.04	0.01	0.01	0.00	0.01
		p_{nonrot}	0.50	0.34	0.10	0.02	0.04	0.03	0.00	0.00	0.00	0.00	0.00	0.00	0.00
		p_{Ucorr}	0.58	0.29	0.07	0.01	0.04	0.04	0.00	0.00	0.00	0.00	0.00	0.00	0.00
		p_{Vcorr}	0.97	0.90	0.78	0.10	0.07	0.15	0.00	0.00	0.00	0.00	0.00	0.00	0.00
SNP	LV1	p_{rot}	0.21	0.14	0.00	0.01	0.00	0.02	0.01	0.01	0.01	0.01	0.00	0.01	0.00
		p_{nonrot}	0.48	0.40	0.13	0.04	0.02	0.14	0.00	0.00	0.00	0.00	0.00	0.00	0.00
		p_{Ucorr}	0.65	0.56	0.12	0.04	0.02	0.24	0.00	0.00	0.00	0.00	0.00	0.00	0.00
		p_{Vcorr}	0.97	0.86	0.97	0.93	0.95	0.83	0.11	0.04	0.04	0.00	0.00	0.00	0.00
	LV2	p_{rot}	0.24	0.29	0.10	0.07	0.04	0.14	0.01	0.01	0.01	0.01	0.00	0.01	0.00
		p_{nonrot}	0.63	0.53	0.16	0.04	0.04	0.17	0.00	0.00	0.00	0.00	0.00	0.00	0.00
		p_{Ucorr}	0.34	0.33	0.23	0.15	0.09	0.16	0.00	0.00	0.00	0.00	0.00	0.00	0.00
		p_{Vcorr}	0.82	0.83	0.26	0.32	0.12	0.12	0.00	0.00	0.00	0.00	0.00	0.00	0.00
	LV3	p_{rot}	0.55	0.48	0.54	0.47	0.50	0.14	0.07	0.01	0.01	0.00	0.01	0.01	0.01
		p_{nonrot}	0.15	0.20	0.22	0.21	0.23	0.07	0.01	0.00	0.00	0.00	0.00	0.00	0.00
		p_{Ucorr}	0.47	0.40	0.54	0.41	0.51	0.08	0.01	0.00	0.00	0.00	0.00	0.00	0.00
		p_{Vcorr}	0.63	0.53	0.46	0.42	0.42	0.25	0.00	0.00	0.00	0.00	0.00	0.00	0.00
Behavior	LV1	p_{rot}	0.01	0.00	0.01	0.01	0.01	0.01	0.00	0.00	0.01	0.02	0.00	0.01	0.00
		p_{nonrot}	0.18	0.17	0.00	0.00	0.00	0.20	0.00	0.00	0.00	0.18	0.00	0.00	0.00
		p_{Ucorr}	0.28	0.28	0.00	0.00	0.00	0.26	0.00	0.00	0.00	0.19	0.00	0.00	0.00
		p_{Vcorr}	0.17	0.13	0.00	0.00	0.00	0.20	0.00	0.00	0.00	0.17	0.00	0.00	0.00
	LV2	p_{rot}	0.56	0.51	0.18	.04	0.26	0.43	0.12	0.05	0.17	0.07	0.09	0.21	0.10
		p_{nonrot}	0.26	0.21	0.00	0.00	0.00	0.15	0.00	0.00	0.00	0.00	0.00	0.01	0.00
		p_{Ucorr}	0.23	0.17	0.00	0.00	0.00	0.12	0.00	0.00	0.00	0.00	0.00	0.00	0.00
		p_{Vcorr}	0.20	0.16	0.00	0.14	0.01	0.07	0.00	0.07	0.00	0.00	0.00	0.00	0.00
	LV3	p_{rot}	0.91	0.92	0.71	0.87	0.74	0.74	0.78	0.87	0.72	0.63	0.72	0.42	0.51
		p_{nonrot}	0.51	0.51	0.07	0.34	0.10	0.27	0.17	0.28	0.07	0.01	0.00	0.00	0.00
		p_{Ucorr}	0.31	0.21	0.02	0.21	0.08	0.12	0.02	0.04	0.23	0.01	0.00	0.00	0.00
		p_{Vcorr}	0.47	0.42	0.07	0.17	0.01	0.27	0.04	0.08	0.01	0.13	0.00	0.00	0.00

Table 2: Results for seedPLS simulations. For each of the 3 datatypes correlation strengths were manipulated with parameter *pp*. For each simulated data set, we computed two standard *p*-value estimates of the LV significance, p_{rot} and p_{nonrot} depending on wether Procrustes rotation was used or not. In addition we calculated *p*-values of LV reliability estimates based on split-half resampling, p_{Ucorr} and p_{Vcorr}

		pp(%)	100	60	30	0
ERP	LV1	p_{rot}	0.00	0.01	0.01	0.00
		p_{nonrot}	0.15	0.00	0.00	0.00
		p_{Ucorr}	0.27	0.00	0.00	0.00
		p_{Vcorr}	0.34	0.02	0.01	0.00
	LV2	p_{rot}	0.03	0.04	0.12	0.04
		p_{nonrot}	0.17	0.24	0.86	0.14
		p_{Ucorr}	0.32	0.06	0.94	0.04
		p_{Vcorr}	0.52	0.30	0.55	0.06
	LV3	p_{rot}	0.15	0.23	0.45	0.14
		p_{nonrot}	0.58	0.60	0.88	0.14
		p_{Ucorr}	0.12	0.20	0.81	0.01
		p_{Vcorr}	0.85	0.16	0.20	0.21
SNP	LV1	p_{rot}	0.01	0.01	0.00	0.01
		p_{nonrot}	0.32	0.01	0.00	0.00
		p_{Ucorr}	0.07	0.00	0.00	0.00
		p_{Vcorr}	0.96	0.79	0.00	0.00
	LV2	p_{rot}	0.00	0.01	0.00	0.01
		p_{nonrot}	0.81	0.01	0.00	0.00
		p_{Ucorr}	0.12	0.01	0.00	0.00
		p_{Vcorr}	0.82	0.07	0.00	0.00
	LV3	p_{rot}	0.12	0.01	0.00	0.00
		p_{nonrot}	0.84	0.05	0.00	0.00
		p_{Ucorr}	0.99	0.32	0.00	0.00
		p_{Vcorr}	0.01	0.70	0.00	0.00
Behavior	LV1	p_{rot}	0.07	0.00	0.01	0.01
		p_{nonrot}	0.85	0.00	0.00	0.00
		p_{Ucorr}	0.91	0.00	0.00	0.00
		p_{Vcorr}	0.62	0.22	0.00	0.00
	LV2	p_{rot}	0.28	0.00	0.00	0.01
		p_{nonrot}	0.20	0.00	0.00	0.00
		p_{Ucorr}	0.32	0.00	0.00	0.00
		p_{Vcorr}	0.10	0.00	0.00	0.00
	LV3	p_{rot}	0.74	0.36	0.78	0.90
		p_{nonrot}	0.69	0.02	0.78	0.86
		p_{Ucorr}	0.41	0.05	0.66	0.41
		p_{Vcorr}	0.82	0.02	0.46	0.26

4 Results and Discussion

Each of the three real data sets were used to generate simulations for the two flavors of PLS. In the case of task-PLS, simulations were designed to have exactly three significant LVs, however the strength of the signal captured by these LVs was varied from no signal ($\gamma = 0$) to weak signal (e.g, $\gamma = 0.5, np = 40\%$) and strong signal ($\gamma = 3, np = 0\%$). Simulations for seedPLS were simpler, where partial permutations of the seed data resulted in a reduction of the initial correlations, going from no reduction ($pp = 0\%$) to more reduction ($pp = 30\%, 60\%$) and full reduction ($pp = 100\%$). For each simulation, we calculated two standard p-value estimates of the LV significance, p_{rot} and p_{nonrot} depending on weather Procrustes rotation was used or not. In addition, we calculated p-values of LV reliability estimates based on split-half resampling, p_{Ucorr} and p_{Vcorr}. The results are presented in Tables 1 and 2.

Acknowledgements Data collection and sharing for this project was funded by the Alzheimer's Disease Neuroimaging Initiative (ADNI) (National Institutes of Health Grant U01 AG024904). ADNI is funded by the National Institute on Aging, the National Institute of Biomedical Imaging and Bioengineering, and through generous contributions from the following: Abbott; Alzheimers Association; Alzheimers Drug Discovery Foundation; Amorfix Life Sciences Ltd.; AstraZeneca; Bayer HealthCare; BioClinica, Inc.; Biogen Idec Inc.; Bristol-Myers Squibb Company; Eisai Inc.; Elan Pharmaceuticals Inc.; Eli Lilly and Company; F. Hoffmann-La Roche Ltd and its affiliated company Genentech, Inc.; GE Healthcare; Innogenetics, N.V.; IXICO Ltd.; Janssen Alzheimer Immunotherapy Research & Development, LLC.; Johnson & Johnson Pharmaceutical Research & Development LLC.; Medpace, Inc.; Merck & Co., Inc.; Meso Scale Diagnostics, LLC.; Novartis Pharmaceuticals Corporation; Pfizer Inc.; Servier; Synarc Inc.; and Takeda Pharmaceutical Company. The Canadian Institutes of Health Research is providing funds to support ADNI clinical sites in Canada. Private sector contributions are facilitated by the Foundation for the National Institutes of Health (www.fnih.org). The grantee organization is the Northern California Institute for Research and Education, and the study is coordinated by the Alzheimer's Disease Cooperative Study at the University of California, San Diego. ADNI data are disseminated by the Laboratory for Neuro Imaging at the University of California, Los Angeles. This research was also supported by NIH grants P30 AG010129 and K01 AG030514.

References

[1] A. Krishnan, L. J. Williams, A. R. McIntosh, and H. Abdi, "Partial Least Squares (PLS) methods for neuroimaging: A tutorial and review,"*NeuroImage* **56**, pp. 455–475, 2011.

[2] A. R. McIntosh and N. Lobaugh, "Partial least squares analysis of neuroimaging data: Applications and advances,"*NeuroImage* **23**, pp. 250–263, 2004.

[3] H. Abdi, "Partial least squares regression and projection on latent structure regression (PLS Regression)," *WIREs Computational Statistics* **2**, pp. 97–106, 2010.

[4] R. Wilcox, "Introduction to Robust Estimation and Hypothesis Testing," *Academic Press*, New York, 2012.

[5] S. C. Strother, J. Anderson, L. K. Hansen, U. Kjems, R. Kustra, J. Sidtis, S. Frutiger, S. Muley, S. LaConte, and D. Rottenberg, "The quantitative evaluation of functional neuroimaging experiments: The NPAIRS data analysis framework,"*NeuroImage* **15**, pp. 747–771, 2002.

[6] S. Purcell, B. Neale, K. Todd-Brown, L. Thomas,MAR Ferreira, D. Bender, J. Maller, P. Sklar, PIW de Bakker, M. J. Daly, and P. C. Sham, ' 'PLINK: a toolset for whole-genome association and population-based linkage analysis," *American Journal of Human Genetics* **81**, 559–575.

The Stability of Behavioral PLS Results in Ill-Posed Neuroimaging Problems

Nathan Churchill, Robyn Spring, Hervé Abdi, Natasa Kovacevic, Anthony R. McIntosh, and Stephen Strother

Abstract Behavioral Partial-Least Squares (PLS) is often used to analyze ill-posed functional Magnetic Resonance Imaging (*f*MRI) datasets, for which the number of variables are far larger than the number of observations. This procedure generates a latent variable (LV) brain map, showing brain regions that are most correlated with behavioral measures. The strength of the behavioral relationship is measured by the correlation between behavior and LV scores in the data. For standard behavioral PLS, bootstrap resampling is used to evaluate the reliability of the brain LV and its behavioral correlations. However, the bootstrap may provide biased measures of the generalizability of results across independent datasets. We used split-half resampling to obtain unbiased measures of brain-LV reproducibility and behavioral prediction of the PLS model, for independent data. We show that bootstrapped PLS gives biased measures of behavioral correlations, whereas split-half resampling identifies highly stable activation peaks across single resampling splits. The ill-posed PLS solution can also be improved by regularization; we consistently improve the prediction accuracy and spatial reproducibility of behavioral estimates by (1) projecting *f*MRI data onto an optimized PCA basis, and (2) optimizing data preprocessing on an individual subject basis. These results show that significant improvements in generalizability and brain pattern stability are obtained with split-half versus bootstrapped resampling of PLS results, and that model performance can be further improved by regularizing the input data.

N. Churchill (✉) • R. Spring • N. Kovacevic • A.R. McIntosh • S. Strother
Rotman Research Institute, Baycrest Center 3560 Bathurst Street, Toronto, Canada
e-mail: nchurchill@research.baycrest.org; rspring@research.baycrest.org; nkovacevic@research.baycrest.org; rmcintosh@research.baycrest.org; sstrother@research.baycrest.org

H. Abdi
The University of Texas at Dallas, School of Behavioral and Brain Sciences, MS:Gr.4.1 800 West Campbell Road, Richardson, TX 75080-3021, USA
e-mail: herve@utdallas.edu

H. Abdi et al. (eds.), *New Perspectives in Partial Least Squares and Related Methods*, Springer Proceedings in Mathematics & Statistics 56, DOI 10.1007/978-1-4614-8283-3_11,

Key words: *f*MRI, Behavioral PLS, Bootstrap, Split-half resampling, Prediction, Reproducibility, PCA

1 Introduction

A central goal of functional magnetic resonance imaging (*f*MRI) studies of the human brain is to identify networks of brain areas that are tightly linked to measures of behavior [1, 2]. This problem is typically highly ill-posed and ill-conditioned, with the number of variables P being very large (i.e. more than 20,000 voxels in brain images), and the number of data samples N being quite small (i.e. typically less than one hundred subjects, with behavioral measures and brain images), a configuration of data known as the "$P \gg N$" problem. To address this issue, two general approaches have emerged in the neuroimaging literature to measure behavioral relations with *f*MRI. The first approach defines a priori a small number of brain regions expected to relate to the behavior of interest. This provides a much better conditioned problem, because the number of brain regions is now roughly of the same order as the number of observations ($P \approx N$). The second approach uses most of the available voxels, and attempts to find the brain locations that best reflect the behavioral distribution in a data-driven multivariate analysis. This method attempts to control the ill-conditioned nature of the problem, by using resampling and regularization with dimensionality reduction techniques. A leading approach of this second type is behavioral PLS, as provided in the open-source MATLAB™"PLS package" developed by McIntosh and et al. [3].

The closely related problem of building discriminant or so called "mind reading" approaches has also been developed and explored in the neuroimaging community [4–7]. When defined as a data-driven multivariate problem with large P, mind reading is also ill-conditioned. Resampling techniques have been developed to control for instability and optimize the reliability of the voxels most closely associated with the discriminant function [6, 9, 10]. These approaches use cross-validation forms of bootstrap resampling [11] or split-half resampling [6]. Split-half resampling is particularly interesting, because it has been shown theoretically to provide finite sample control of the error rate of false discoveries in general linear regression methods when applied to ill-posed problems, provided certain exchangeability conditions are met [12].

Behavioral PLS and linear discriminant analysis belong to the same linear multivariate class of techniques, as both are special cases of the generalized singular value decomposition or generalized eigen-decomposition problem [20]. Specifically, let $\mathbf{Y}$ be a $N \times K$ matrix of K behavioral measures or categorical class labels for N subjects, and $\mathbf{X}$ be a $N \times P$ matrix of brain images, where $P \gg N$. The eigen-solution of expression:

$$(\mathbf{Y}^\mathsf{T}\mathbf{Y})^{-1/2}\mathbf{Y}^\mathsf{T}\mathbf{X}(\mathbf{X}^\mathsf{T}\mathbf{X})^{-1/2} \tag{1}$$

reflects the linear discriminant solution for categorical class labels in $\mathbf{Y}$ [13]. When $P > N$, $(\mathbf{X}^\mathsf{T}\mathbf{X})$ will be singular and therefore $(\mathbf{X}^\mathsf{T}\mathbf{X})^{-1/2}$ cannot be computed without some form of regularization. When $\mathbf{X}$ and $\mathbf{Y}$ are centered and normalized (i.e. each column of these matrices has a mean of zero and a norm of 1), and $(\mathbf{X}^\mathsf{T}\mathbf{X}) = (\mathbf{Y}^\mathsf{T}\mathbf{Y}) = \mathbf{I}$ (i.e. $\mathbf{X}$ and $\mathbf{Y}$ are orthogonal matrices), then Eq. 1 corresponds to the general partial least squares correlation approach defined in Krishnan et al. [3, 14], for which behavioral PLS with $\mathbf{Y}$ containing subject behavioral scores is a special case. Given the similar bivariate form of PLS and linear discriminants, the goal of this study was to use the split-half techniques developed in the discriminant neuroimaging literature to test the stability of solutions from behavioral PLS, which uses standard bootstrap resampling methods as implemented in the neuroimaging PLS package [3] (code located at: www.rotman-baycrest.on.ca/pls/source/).

2 Methods and Results

2.1 Functional Magnetic Resonance Imaging (fMRI) Data Set

Twenty young normal subjects (20–33 years, 9 male) were scanned with *f*MRI while performing a forced-choice, memory recognition task of previously encoded line drawings [15], in an experiment similar to that of Grady et al. [16]. We used a 3 Tesla *f*MRI scanner to acquire axial, interleaved, multi-slice echo planar images of the whole brain ($3.1 \times 3.1 \times 5$ mm voxels, TE/TR=30/2000 ms). Alternating scanning task and control blocks of 24 s were presented 4 times, for a total task scanning time per subject of 192 s. During the 24 s task blocks, every 3 s subjects saw a previously encoded figure side-by-side with two other figures (semantic and perceptual foils) on a projection screen, and were asked to touch the location of the original figure on an *f*MRI-compatible response tablet [17]. Control blocks involved touching a fixation cross presented at random intervals of 1–3 s.

The resulting 4D *f*MRI time series were preprocessed using standard tools from the AFNI package, including rigid-body correction of head motion (`3dvolreg`), physiological noise correction with RETROICOR (`3dretroicor`), temporal detrending using Legendre polynomials and regressing out estimated rigid-body motion parameters (`3dDetrend`, see [8] for an overview of preprocessing choices in *f*MRI). For the majority of results (see Sects. 2.2 and 2.3), we preprocessed the data using a framework that optimizes the specific processing steps independently for each subject, as described in [18, 19], within the split-half NPAIRS resampling framework [6]. In Sect. 2.4, we provide more details of pipeline optimization, and demonstrate the importance of optimizing preprocessing steps on an individual subject basis in the PLS framework.

We performed a two-class linear discriminant analysis separately for each dataset (Class 1: Recognition scans; Class 2: Control scans), which produced an optimal Z-scored statistical parametric map [$SPM(Z)$] per subject. For each subject, the Z-score value of each voxel reflects the extent to which this voxel's brain location contributes to the discrimination of recognition vs. control scans, for that subject.

2.2 Split-Half Behavioral PLS

The 20 subjects' $SPM(Z)$s were stacked to form a $20 \times 37{,}284$ matrix $\mathbf{X}$ as described in Eq. 1, and a 20×1 $\mathbf{y}$ vector was formed from the differences of the mean (Recognition – Control) block reaction times per subject (in milli-seconds). After centering and normalizing $\mathbf{X}$ and $\mathbf{y}$, a standard behavioral PLS was run, as outlined in [3], with 1,000 bootstrap replications. The resulting distribution is reported in Fig. 1 (left) under "Bootstrapped PLS." For each bootstrap sample, a latent variable (LV) brain map was also calculated. At each voxel, the mean was divided by the standard error on the mean (SE), computed over all bootstrap measures; this is reported as a bootstrap ratio brain map SPM_{boot} (horizontal axes of Fig. 3).

The behavioral PLS procedure was modified to include split-half resampling as follows. After centering and normalizing $\mathbf{X}$ and $\mathbf{y}$, subjects were randomly assigned 1,000 times to split-half matrices $\mathbf{X}_1$ and $\mathbf{X}_2$, and behavioral vectors $\mathbf{y}_1$ and $\mathbf{y}_2$. For each split-half matrix/vector pair, we obtained the projected brain pattern LV defined by $\mathbf{e}_i = \mathbf{y}_i^{\mathsf{T}}\mathbf{X}_i$ that explained the most behavioral image variance for $i = 1,2$. The correlation $r_{(i,\mathrm{train})} = \rho(\mathbf{y}_i, \mathbf{X}_i\mathbf{e}_i^{\mathsf{T}})$ reflects the correlation between behavior and expression of the latent brain pattern $\mathbf{e}_i$, for each split-half training set. The distribution of the 2,000 split-half $r_{(i,\mathrm{train})}$ values is plotted in Fig. 1 (middle). We also obtained an independent test measure of the behavioral prediction power of each $\mathbf{e}_i$ by calculating $r_{(i,\mathrm{test})} = \rho(\mathbf{y}_{j \neq i}, \mathbf{X}_{j \neq i}\mathbf{e}_i^{\mathsf{T}})$ for i and $j = 1,2$. The distribution of these 2,000 $r_{(i,\mathrm{test})}$ values is plotted in Fig. 1(right). The test $r_{(i,\mathrm{test})}$ behavioral correlations are consistently lower than both training and bootstrap estimates. The reproducibility of the two split-half brain patterns may also be measured as the correlation of all paired voxel values $r_{\mathrm{spatial}} = \rho(\mathbf{e}_1, \mathbf{e}_2)$; this measures the stability of the latent brain pattern across independent datasets. The overall reproducibility of this pattern is also relatively low but consistently greater than zero, with median r_{spatial} of 0.025 (ranging from 0.014 to 0.043; plotted in Fig. 4).

Figure 2 plots the median latent variable (LV) score of each subject, as training-data ($\mathbf{X}_i\mathbf{e}_i^{\mathsf{T}}$ scores, Fig. 2a) and as test-data ($\mathbf{X}_{j \neq i}\mathbf{e}_i^{\mathsf{T}}$ scores, Fig. 2b); we plotted the median LV scores vs. behavior over the 1,000 resamples. The median training scores show a consistently stronger linear trend than for test. In addition, there is a subject (red circle) whose brain-behavior relation cannot be predicted by the other subjects' data in the test space (it is a significant outlier by Cooks D test, with statistic $d = 0.90$ exceeding the outlier threshold $4/N$ [21]). By comparison, in the training space, this subject is not a significant outlier.

The split-half brain patterns $\mathbf{e}_1$ and $\mathbf{e}_2$ can also be used to estimate a behavioral SPM that is robust to subject heterogeneity. As described in [6], this is done by normalizing each $\mathbf{e}_i$ to mean zero and variance one, and then projecting the pairwise voxel values onto the line of identity (the first component of a principal component analysis (PCA) on the scatterplot of $\mathbf{e}_1$ vs. $\mathbf{e}_2$ voxel values), which gives a signal-axis estimate: $\mathbf{e}_{\mathrm{signal}} = (\mathbf{e}_1 + \mathbf{e}_2)/\sqrt{2}$. The orthogonal, minor-axis projection (second component of a PCA on the scatter-plot), forms the noise axis. This measures uncorrelated, non-reproducible signal at each voxel, giving noise vector: $\mathbf{e}_{\mathrm{noise}} = (\mathbf{e}_1 - \mathbf{e}_2)/\sqrt{2}$. This is used to estimate a reproducible Z-scored map

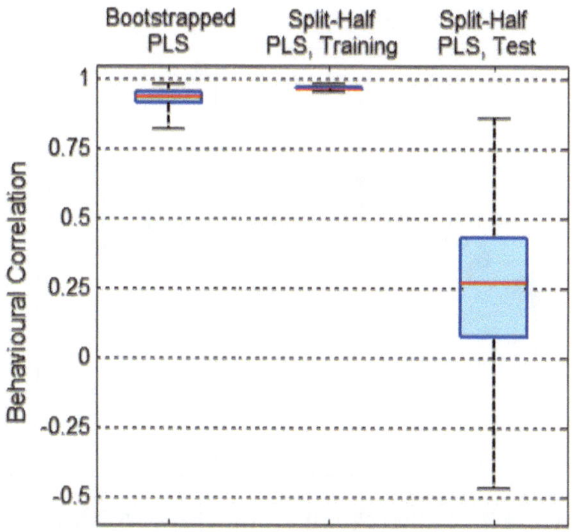

Fig. 1: Behavioral correlation distributions for standard bootstrapped behavioral PLS *(left)*, and split-half training *(middle)* and test *(right)* distributions. Distributions are plotted as box-whisker plots with min.– max. whisker values, a 25th–75th percentile box and the median (*red bar*); results shown for 1,000 bootstrap or split-half resampling iterations

$rSPM(Z)_{\text{split}} = \mathbf{e}_{\text{signal}}/SD(\mathbf{e}_{\text{noise}})$, where $SD(\mathbf{e}_{\text{noise}})$ provides a single spatially global noise estimator. The average of the 1,000 $rSPM(Z)_{\text{split}}$ voxel values are plotted on the vertical axis in Fig. 3a, against SPM_{boot} values. The $rSPM(Z)_{\text{split}}$ shows generally lower signal values than SPM_{boot}, with a nonlinear relationship. However, this difference is partly a function of the global versus local noise estimators. We can instead estimate the mean $\mathbf{e}_{\text{signal}}$ value at each voxel, and normalize by the SD on $\mathbf{e}_{\text{noise}}$ for each voxel (each computed across 1,000 resamples), generating voxel-wise estimates of noise in the same manner as SPM_{boot}. This $rSPM(Z)$ is plotted against SPM_{boot} in Fig. 3b, demonstrating a strong linear trend, albeit with increased scatter for high-signal voxels. This scatter is primarily due to differences in the local noise estimates: the mean bootstrap LV and $\mathbf{e}_{signal}$ patterns are highly consistent (correlation equal to 0.99), whereas the local noise estimates are more variable between the two methods (plotted in Fig. 3c; correlation equal to 0.86).

2.3 Behavioral PLS *on a Principal Component Subspace*

For standard behavioral PLS, we project the behavioral vector $\mathbf{y}$ directly onto $\mathbf{X}$ (the subject *SPM*s) to identify the latent basis vector $\mathbf{e} = \mathbf{y}^{\text{T}}\mathbf{X}$. However, taking our cue from the literature on split-half discriminant analysis in *f*MRI (see, e.g.

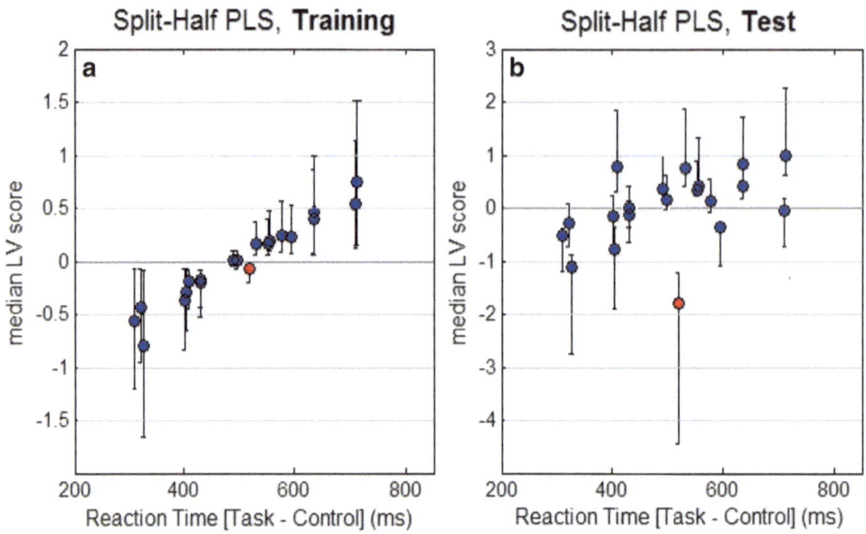

Fig. 2: Median subject behavioral LV scores are plotted against difference in reaction time between (Task-Control) experimental conditions. The error bars give upper and lower quartile ranges on LV scores, for each subject (for 1,000 split-half resamples). Results are shown for **(a)** the training-space LV scores, and **(b)** the test-space LV scores. The subject represented by a *red* dot is a significant outlier in test-space, based on Cook's *D* statistic (see text for details)

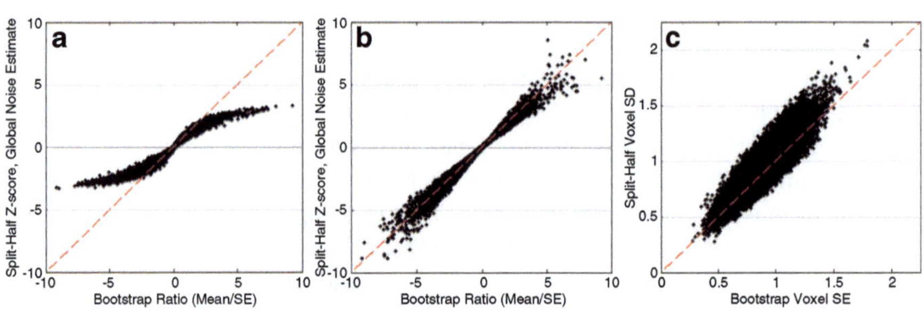

Fig. 3: Scatter plot of pairs of voxel *SPM* values: we compare standard bootstrapped behavioral PLS analysis producing (mean voxel salience)/(standard error), to split-half signal/noise estimates. This includes **(a)** standard NPAIRS estimation of voxel signal, normalized by global noise standard deviation (*Z*-scored) for each resample, and **(b)** voxel signal, normalized by standard error (bootstrap ratios) or standard deviation (split-half *Z*-scores) estimated at each voxel. **(c)** plot of voxels' standard error (bootstrap) against standard deviation (split-half). Results are computed over 1,000 split-half/bootstrap resamples

[7, 10, 18, 19, 22]), we can regularize and de-noise the data space in which the analysis is performed, by first applying PCA to $\mathbf{X}$, and then running a PLS analysis on a reduced PCA subspace.

The singular value decomposition [20] produces $\mathbf{X} = \mathbf{USV}^{\mathsf{T}}$, where $\mathbf{U}$ is a set of orthonormal subject-weight vectors, $\mathbf{S}$ is a diagonal matrix of singular values, and $\mathbf{V}$ is a set of orthonormal image basis vectors. We represent $\mathbf{X}$ in a reduced k-dimensional PCA space ($k \leq N$), by projecting onto the subset of 1 to k image bases, $\mathbf{V}^{(k)} = [\mathbf{V}_1\mathbf{V}_2 \ldots \mathbf{V}_k]$, giving $\mathbf{Q}^{(k)} = \mathbf{XV}^{(k)}$. We performed PLS analysis on $\mathbf{Q}^{(k)}$, by normalizing and centering subject scores of each PC-basis, and then obtaining the projection $\mathbf{w}_i = \mathbf{y}_i^{\mathsf{T}}\mathbf{Q}_i$ that explained the most behavior variance in the new PC basis.

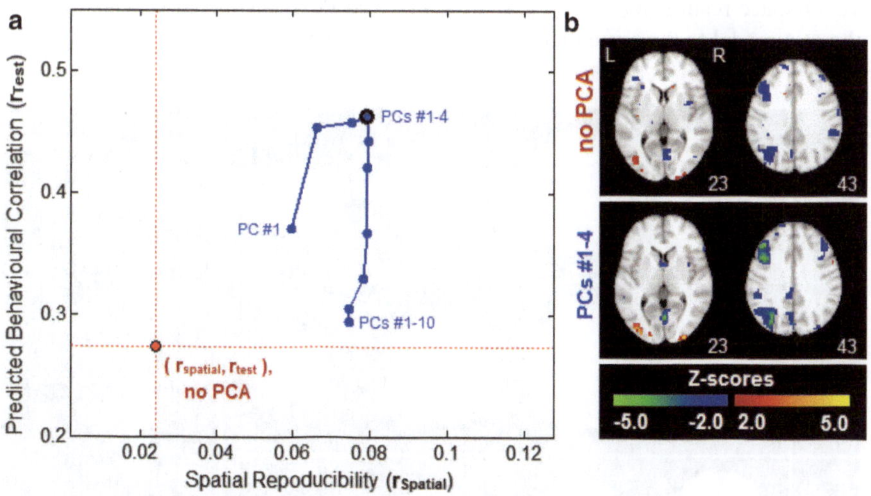

Fig. 4: **(a)** Plot of median predicted behavioral correlation r_{test} and spatial reproducibility r_{spatial} of the LV brain map, for PLS performed on a PCA subspace of the subject data (*blue*). These subspaces include the 1 to k Principal Components (PCs), where we vary ($1 \leq k \leq 10$). The (r_{spatial}, r_{test}) values are plotted for each k (subspace size) as points on the curve; a subspace of PCs 1–4 simultaneously optimized (r_{spatial}, r_{test}), circled in *black*. We also plot the median (r_{test},r_{spatial}) point, estimated directly from matrix $\mathbf{X}$ for reference (*red circle*). **(b)** Plots of split-half Z-scored *SPM*s with global noise estimation, for no PCA estimation (*red*), and an optimized PCA dimensionality $k = 4$ (*blue*). Positive Z-scores indicate positive correlation with the behavioral measure of reaction time, and negative Z-scores indicate negative correlation. Voxel values are computed as the mean over 1,000 split-half resamples, with spatially global noise estimation from each split-half pair

The predicted behavioral correlation is measured by projecting the test data onto the training PC-space, and then onto $\mathbf{w}_i$, giving behavioral correlations $r_{(i,\text{test})} = \rho(\mathbf{y}_{j \neq i}, \mathbf{w}_i(\mathbf{X}_{j \neq i}\mathbf{V}_i)$. We also obtained eigen-images by projecting back onto the voxel space (i.e. $\mathbf{e}_i = \mathbf{w}_i\mathbf{V}_i^{(k)}$), to compute the $rSPM(Z)_{\text{split}}$ and reproducibility,

$r_{spatial}$. The resulting median behavioral prediction $r_{(test)}$ and reproducibility $r_{(spatial)}$ are plotted in Fig. 4a, as a function of the number of PC bases k. From this curve, we identify the PC subspace $k = 4$, that maximizes both r_{test} and $r_{spatial}$. Note that the median r_{test} and $r_{spatial}$ are consistently higher when performed on a PCA basis than PLS performed directly on $\mathbf{X}$, for all subspace sizes $k = 1 \ldots 10$. The predicted behavioral correlation is generally higher for the $k = 4$ PC subspace than PLS performed directly on $\mathbf{X}$ (median $\Delta r_{test} = 0.17$; increased for 891 of the 1,000 resamples), as is spatial reproducibility ($\Delta r_{spatial} = 0.05$; increased for all 1,000 resamples). Figure 4b depicts slices from the mean $rSPM(Z)s$ of PLS performed directly on $\mathbf{X}$ (top) and in an optimized PC subspace (bottom). The PCA optimization tends to increase mean Z-scores in the same areas of activation previously identified by voxel-space results, indicating that the optimized PC basis increases sensitivity of the PLS model to the same underlying set of brain regions.

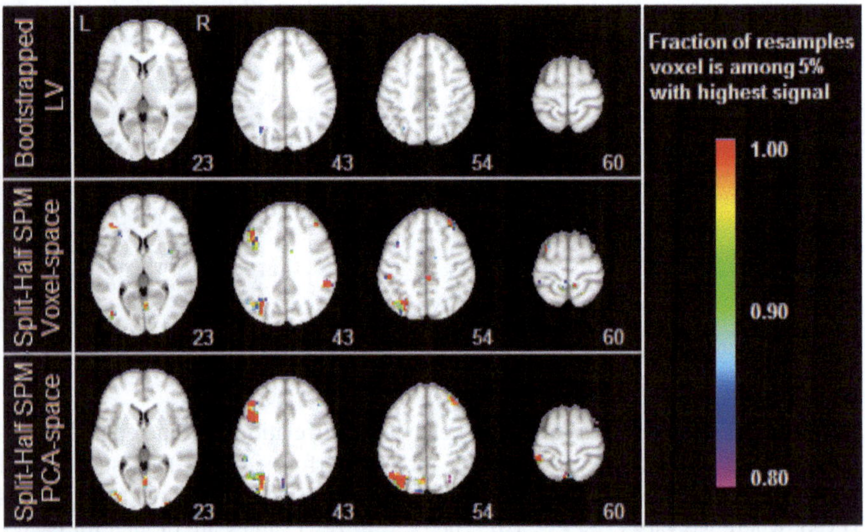

Fig. 5: Plot showing the reliability of peak voxel values. *(top)* peak LV values are shown across standard PLS bootstrap replications. Peak voxels of the split-half reproducible $rSPM(Z)s$, with global noise estimated at each split, are shown across resamples for *(middle)* voxel-space estimation, and *(bottom)* estimation on an optimized PCA subspace. For each of the 1,000 bootstrap/split-half resamples, we identified the top 5% highest-signal voxels (LV values for bootstrap estimation; Z-scores for split-half estimation). This plot measures the fraction of resamples where each voxel is part of the top 5%

In Fig. 5, we depict the stability of bootstrap and split-half resampling estimates. We compared the reliability of peak voxels across bootstrap LVs (top), relative to split-half $rSPM(Z)_{split}$ estimates with global noise estimation; the split-half model estimates a Z-scored SPM from each resampling split. Results are shown for

$rSPM(Z)_{\text{split}}$ estimated directly from data matrix $\mathbf{X}$ (middle), and $rSPM(Z)_{\text{split}}$ estimated from the optimized PCA subspace of $\mathbf{X}$, $k = 4$ PCs (bottom). We measured peak signal as the top 5% of voxel signal values, for each resample (bootstrap-estimated LV scores or split-half-estimated Z-scores). At each voxel, we measured the fraction of resamples where it was a peak voxel (i.e. among the top 5%). For bootstrap LVs, only 2 of 37,284 voxels (less than .001%) were active in more than 95% of resamples, compared to split-half Z-scored estimates of 324 voxels (0.87%; PLS computed on $\mathbf{X}$) and 343 voxels (0.92%; PLS on an optimized PCA basis). This demonstrates that although $rSPM(Z)_{\text{split}}$ with global noise estimation produces lower mean signal values than SPM_{boot} (Fig. 3a), the location of peak $rSPM(Z)_{\text{split}}$ values are highly stable across resampling splits. We can therefore identify reliable SPM peaks with relatively few resampling iterations.

2.4 Behavioral PLS *and Optimized Preprocessing*

For results in Sects. 2.2 and 2.3, we preprocessed the *f*MRI data to correct for noise and artifact, as outlined in [18, 19]. For this procedure, we included/excluded every combination of the preprocessing steps: (1) motion correction, (2) physiological correction, (3) regressing head-motion covariates and (4) temporal detrending with Legendre polynomial of orders 0–5, evaluating $2^3 \times 6 = 48$ different combinations of preprocessing steps ("pipelines").

For each pipeline, we performed an analysis in the NPAIRS split-half framework [6], and measured spatial reproducibility and prediction accuracy (posterior probability of correctly classifying independent scan volumes). We selected the pipeline that minimized the Euclidean distance from perfect prediction and reproducibility:

$$D = \sqrt{(1 - \text{reproducibility})^2 + (1 - \text{prediction})^2}, \tag{2}$$

independently for each subject. This may be compared to the standard approach in *f*MRI literature, which is to apply a single fixed pipeline to all subjects. We compared the current "individually optimized" results with the optimal "fixed pipeline," of motion correction and 3rd-order detrending; this was the set of steps that, applied to all subjects, minimized the D metric across subjects (details in [18]).

Figure 6 compares fixed pipeline results (*red*) to individually optimized data (*blue*), for PLS on a PCA subspace. Figure 6a show, for both pipelines, median behavioral prediction $r_{(\text{test})}$ and reproducibility $r_{(\text{spatial})}$ plotted as a function of PCA dimensionality. Data with fixed preprocessing (*red*) optimized $r_{(\text{test})}$ and $r_{(\text{spatial})}$ at PC #1, a lower dimensionality than individually optimized preprocessing (*blue*), at PCs #1–4. For the optimized PC bases (circled in *black*), individual pipeline optimization improves over fixed pipelines with median $\Delta r_{(\text{test})} = 0.11$ (increased for 898 out of the 1,000 resamples), and $\Delta r_{(\text{spatial})} = 0.06$ (increased for 810 out of the 1,000 resamples). Figure 6b shows sample slices from the mean Z-scored SPMs, in the optimized PC subspaces. Individual subject pipeline optimization generally produces higher peak Z-scores, and sparser, less noisy SPMs, than fixed preprocessing.

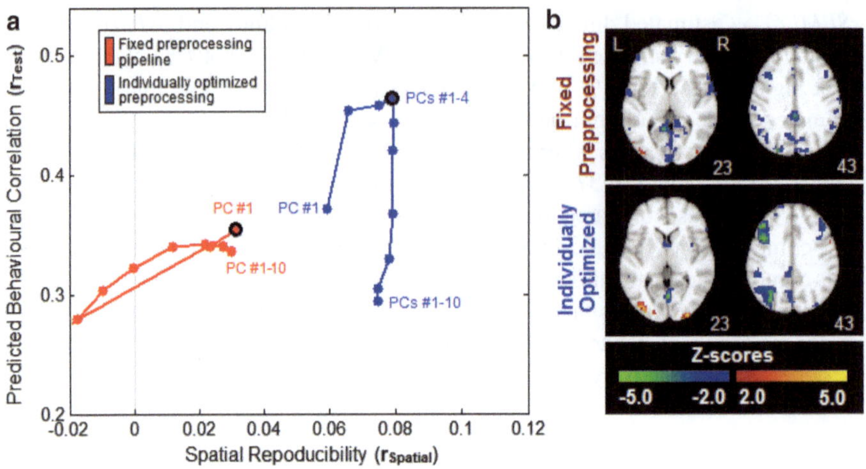

Fig. 6: (**a**) Plot of median predicted behavioral correlation r_{test} and spatial reproducibility r_{spatial} of the LV brain map for PLS, when performed on a PCA subspace of the subject data. Results are plotted for data preprocessed with a fixed set of steps (*red*; all subjects have the same preprocessing applied), and with preprocessing individually optimized for each subject (*b*lue; this is the pipeline used for all previous results). For both datasets, subspaces include the 1 to k principal components (PCs), where we vary $(1 < k < 10)$. The $(r_{\text{spatial}}, r_{\text{test}})$ values are plotted for each k (subspace size) as points on the curve; we circled in *black* the PC-space that optimized $(r_{\text{spatial}}, r_{\text{test}})$ for each pipeline set. (**b**) Plots of split-half Z-scored *SPM*s with global noise estimation under the optimal PC subspace, for the optimal fixed pipeline (*red*; PC #1), and individually optimized pipelines (*blue*; PCs #1–4). Positive Z-scores indicate areas of positive correlation with the behavioural measure (reaction time), and negative Z-scores indicate negative correlation. Voxel values are computed as the mean over 1,000 split-half resamples

3 Discussion and Conclusions

The results presented in Fig. 1 indicate that bootstrapping behavioral PLS values may result in a large upward bias in estimated behavioral correlation values (Fig. 1, left) that is similar to the prediction biases encountered from training sets (Fig. 1, middle) in training-test frameworks such as split-half resampling. Based on our prior experience with such prediction models, this upward bias is caused by over-fitting a low-dimensional categorical or behavioral vector in the high dimensional space spanned by the brain images, without appropriate model regularization. Therefore, the measured correlations from bootstrapped behavioral PLS apply only to the data set used for their estimation and cannot be generalized. In contrast, the much lower split-half test estimates of behavioral correlation in Fig. 1 (right) are generalizable but are potentially biased downwards, being based on relatively small training/test groups of only 10 subjects.

Non-generalizable training bias is also reflected in the plots of median LV scores vs. behavioral measures, in Fig. 2. If the scores are computed from the training-space estimates, we obtain a stronger linear trend and less variability across splits, compared to independent test data projected onto the training basis. As shown in Fig. 2, plotting the test-space scores may also reveal potential prediction outliers that are not evident in the training plots.

The Fig. 3a plot also shows that bootstrapped peak *SPM* signals are consistently higher than standard split-half global *Z*-scoring. However, Fig. 3b shows that on this is primarily a function of the different noise estimators, as the voxel-wise, split-half noise estimation *SPM* is highly correlated with the bootstrap estimated *SPM*. Both of the scatter-plots show a strong monotonic relation between SPM_{boot} and the $rSPM(Z)s$, indicating that regardless of the estimation procedure, approximately the same spatial locations drive both bootstrap and split-half analyses. Even for voxel-wise noise estimation, the difference between split-half and bootstrap *SPMs* is primarily driven by the local noise estimates (plotted in Fig. 3c), whereas mean signal values are highly similar.

Figure 4 shows that the original **X** data space can be better regularized and stabilized, by projecting data onto a PC subspace prior to analysis. By adapting the number of PC dimensions, we trace out a behavioral correlation vs. reproducibility curve as a function of the number of PCs, similar to the prediction vs. reproducibility curves observed in discriminant models [10, 22]. These results highlight, again, the ill-posed nature of the PLS data-analysis problem, and the importance of regularizing *f*MRI data. We also note that even a full-dimensionality PC-space model (e.g. PCs 1–10 included in each split-half) outperforms estimation directly on the matrix **X**. The PCA projects data onto the bases of maximum variance, prior to standard PLS normalization (giving zero mean and unit variance to scores of each PC basis). The superior performance of PCs 1–10 over no PC basis (Fig. 4) indicates that the variance normalization in voxel space may significantly limit the predictive generalizability of behavioral PLS results for some analyses.

Figure 5 demonstrates the advantages of split-half resampling with global noise estimation. For each split, we generate a single Z-scored $rSPM(Z)$, for which peak voxels tend to be highly consistent across $rSPM(Z)s$ of individual resampling splits. This allows us to measure voxel *Z*-scores on a little as one resampling split. The stability of the peak activations also allows us to identify reliable brain regions from a single split, which is not available to voxel-wise bootstrap estimation. The cross-validation framework is therefore particularly useful when only limited *f*MRI data is available, and has been previously used to optimize preprocessing in brief task runs of less than 3 min in length (e.g. [18, 19]).

The results of Fig. 6 compared data with preprocessing choices optimized on an individual subject basis, relative to the standard *f*MRI approach of using a single fixed pipeline. Results indicate that optimizing preprocessing choices on an individual subject basis can significantly improve predicted test correlation and the spatial reproducibility of LV maps in behavioral PLS. Note that pipeline optimization was performed independently of any behavioral measures, as we chose preprocessing steps to optimize *SPM* reproducibility and prediction accuracy of the linear discriminant analysis model. These results demonstrate that improved preprocessing may help to better detect brain-behavior relationships in *f*MRI data.

References

[1] D. Wilkinson, and P. Halligan, "The relevance of behavioral measures for functional-imaging studies of cognition," *Nature Review Neuroscience* **5**, pp. 67–73, 2004.

[2] A. R. McIntosh, "Mapping cognition to the brain through neural interactions," *Memory* **7**, pp. 523–548, 1999.

[3] A. Krishnan, L. J. Williams, A. R. McIntosh, and H. Abdi, "Partial Least Squares (PLS) methods for neuroimaging: A tutorial and review," *Neuroimage* **56**, pp. 455–475, 2011.

[4] N. Morch, L. K. Hansen, S. C. Strother, C. Svarer, D. .A. Rottenberg, B. Lautrup, R. Savoy, and O. B. Paulson, "Nonlinear versus linear models in functional neuroimaging: Learning curves and generalization crossover," *Information Processing in Medical Imaging*, J. Duncan and G. Gindi, eds.; Springer-Verlag, New York, pp. 259–270, 1997.

[5] A. J. O'Toole, F. Jiang, H. Abdi, N. Pénard, J. P. Dunlop, and M. A. Parent, "Theoretical, statistical, and practical perspectives on pattern-based classification approaches to the analysis of functional neuroimaging data," *Journal of Cognitive Neuroscience* **19**, pp. 1735–1752, 2007.

[6] S. C. Strother, J. Anderson, L. K. Hansen, U. Kjems, R. Kustra, J. Sidtis, S. Frutiger, S. Muley, S. LaConte, and D. Rottenberg, "The quantitative evaluation of functional neuroimaging experiments: the NPAIRS data analysis framework," *Neuroimage* **15**, pp. 747–771, 2002.

[7] S. C. Strother, S. LaConte, L. K. Hansen, J. Anderson, J. Zhang, S. Pulapura, and D. Rottenberg, "Optimizing the *f*MRI data-processing pipeline using prediction and reproducibility performance metrics: I. A preliminary group analysis," *Neuroimage* **23 Suppl 1**, pp. S196–S207, 2004.

[8] S. C. Strother, "Evaluating *f*MRI preprocessing pipelines," *IEEE Engineering in Medicine and Biology Magazine* **25**, pp. 27–41, 2006

[9] H. Abdi, J. P. Dunlop, and L. J. Williams, "How to compute reliability estimates and display confidence and tolerance intervals for pattern classifiers using the Bootstrap and 3-way multidimensional scaling (DISTATIS)," *Neuroimage* **45**, pp. 89–95, 2009.

[10] S. Strother, A. Oder, R. Spring, and C. Grady, "The NPAIRS Computational statistics framework for data analysis in neuroimaging," *presented at the 19th International Conference on Computational Statistics, Paris, France*, 2010.

[11] R. Kustra, and S. C. Strother, "Penalized discriminant analysis of [15O]-water PET brain images with prediction error selection of smoothness and regularization hyperparameters," *IEEE Transactions in Medical Imaging* **20**, pp. 376–387, 2001.

[12] N. Meinshausen, and P. Bühlmann, "Stability selection," *Journal of the Royal Statistical Society: Series B (Statistical Methodology)* **72**, pp. 417–473, 2010.

[13] K. V. Mardia, J. T. Kent, and J. M. Bibby, *Multivariate Analysis*, Academic Press, London, 1979.

[14] H. Abdi, "Partial least squares regression and projection on latent structure regression (PLS Regression)," *WIREs Computational Statistics* **2**, pp. 97–106, 2010.

[15] J. G. Snodgrass, and M. Vanderwart, "A standardized set of 260 pictures: norms for name agreement, image agreement, familiarity, and visual complexity," *Journal of Experimental Psychology: Human Learning* **6**, pp. 174–215, 1980.

[16] C. Grady, M. Springer, D. Hongwanishkul, A. R. McIntosh, and G. Winocur, "Age-related changes in brain activity across the adult fifespan: A failure of inhibition?," *Journal of Cognitive Neuroscience* **18**, pp. 227–241, 2006.

[17] F. Tam, N. W. Churchill, S. C. Strother, and S. J. Graham, "A new tablet for writing and drawing during functional MRI," *Human Brain Mapping* **32**, pp. 240–248, 2011.

[18] N. W. Churchill, A. Oder, H. Abdi, F. Tam, W. Lee, C. Thomas, J. E. Ween, S. J. Graham, and S. C. Strother, "Optimizing preprocessing and analysis pipelines for single-subject *f*MRI: I. Standard temporal motion and physiological noise correction methods," *Human Brain Mapping* **33**, pp. 609–627, 2012.

[19] N. W. Churchill, G. Yourganov, A. Oder, F. Tam, S. J. Graham, and S. C. Strother, "Optimizing preprocessing and analysis pipelines for single-subject *f*MRI: 2. Interactions with ICA, PCA, task contrast and inter-subject heterogeneity," *PLoS One* **7**, (e31147), 2012.
[20] H. Abdi, "Singular value decomposition (SVD) and generalized singular value decomposition (GSVD)," in *Encyclopedia of Measurement and Statistics*, N. Salkind, ed., pp. 907–912, Sage, Thousand Oaks, 2007.
[21] K. A. Bollen, and R. W. Jackman, "Regression diagnostics: An expository treatment of outliers and influential cases," in J. Fox, and J.S. Long, (eds.), *Modern Methods of Data Analysis*, pp. 257–291. Sage, Newbury Park, 2012.
[22] P. M. Rasmussen, L. K. Hansen, K. H. Madsen, N. W. Churchill, and S. C. Strother, "Pattern reproducibility, interpretability, and sparsity in classification models in neuroimaging," *Pattern Recognition* **45**, pp. 2085–2100, 2012.

Part IV
Multiblock Data Modeling

Two-Step PLS Path Modeling Mode B: Nonlinear and Interaction Effects Between Formative Constructs

Alba Martínez-Ruiz and Tomas Aluja-Banet

Abstract A two-step PLS path modeling MODE B procedure is implemented to estimate nonlinear and interaction effects among formative constructs. The procedure preserves the convergence properties of PLS MODE B with centroid scheme (Wold's algorithm) and offers a way to build proper indices for linear, nonlinear and interaction terms, all of which are unobservable, and to estimate the relationships between them. A Monte Carlo simulation study is carried out in order to provide empirical evidence of its performance. Linear, nonlinear and interaction effects are underestimated. Accuracy and precision increase by increasing the sample size. Significant nonlinear and interaction effects and an increase in the predictability of models are detected with medium or large sample sizes. The procedure is well-suited to estimate nonlinear and interaction effects in structural equation models with formative constructs and few indicators.

Key words: Partial least squares path modeling, MODE B, Nonlinear effects, Interaction effects, Monte Carlo simulation

1 Introduction

Wold's PLS-MODE-B algorithm with centroid scheme is a consistent method for building a sequence of unobservable variables—also called constructs—for structural equation models (SEMs) with formative blocks of variables. Overlapping

A. Martínez-Ruiz (✉)
Universidad Católica de la Santísima Concepcíon, Concepción, Chile
e-mail: amartine@ucsc.cl

T. Aluja-Banet
Universitat Politecnica de Catalunya, Campus Nord UPC. C5204.
Jordi Girona 13, 08034 Barcelona, Spain
e-mail: tomas.aluja@upc.edu

H. Abdi et al. (eds.), *New Perspectives in Partial Least Squares and Related Methods*, Springer Proceedings in Mathematics & Statistics 56, DOI 10.1007/978-1-4614-8283-3_12,

successive iterations, Wold's algorithm uses the latest estimates of the constructs available at each iteration to compute the unobservable indices [4, 14, 15]. Once constructs are obtained, the PLS path modeling procedure estimates linear relationships among unobservables using multiple regression. Mathes (1993) [11] has shown that "the PLS estimation of MODE B with centroid weighting scheme is a critical point of the function sum of absolute correlations of the adjacent latent variables in the structural systems" (p. 235). Hanafi (2007) [4] has proven for Wold's implementation that PLS MODE B with centroid weighting scheme is monotonically convergent.

In a PLS framework, Jakobowicz (2007) [8] and Henseler (2012) [6] have addressed nonlinearities. Jakobowicz has studied PLS-MODE-A SEMs and Henseler PLS-MODE-C SEMs.[1] The main idea of Jakobowicz is to transform nonlinear variables into linear variables using monotonic B-spline transformations and alternating least squares (ALS). All exogenous construct of an endogenous unobservable variable are transformed such that the square multiple correlation coefficient of this endogenous construct is maximized. The ALS procedure is used to estimate the parameters of the monotonic B-spline function and the path coefficients for the inner model. On the other hand, Henseler compares the hybrid, product indicator, two-stage, and orthogonal approaches for estimating the quadratic effect of an exogenous formative construct on an endogenous construct. The author recommends the hybrid approach—the original solution of Wold [15]—that computes the quadratic construct in the iterative stage of the PLS algorithm. Both, Jakobowicz's and Henseler's approaches, require the modification of the PLS algorithm. Moreover, and as Jakobowicz has pointed out, in his approach the inner relationships are obtained maximizing the explained variance of the endogenous construct which is supposed to have nonlinear relationships with other exogenous constructs, but not considering in the same manner other endogenous unobservable variables included in the same model.

In this paper, we are interested in estimating the effect of a nonlinear or interaction term in a endogenous construct, and so far, there is no evidence for assessing whether PLS-MODE-B is well-suited for (1) building nonlinear and interaction terms from formative constructs and for (2) estimating the structural relationships between them and an endogenous construct. We have implemented a two-step PLS path modeling with MODE B (TsPLS) procedure to estimate nonlinear and interaction effects in SEMs with formative outer models. The procedure considers the score estimation of linear terms using the PLS-MODE-B algorithm with centroid scheme (step one). Hence, the TsPLS procedure preserves the convergence properties of Wold's algorithm and offers a way to build proper indices for linear, nonlinear and interaction terms, all of which are unobservable. Next, scores of nonlinear and interaction terms are directly computed from linear terms (step two). Finally, the dependent construct is regressed on the linear, nonlinear and interaction unobservable terms.

The TsPLS procedure may be proper to estimate nonlinear and interaction effects in SEMs with formative constructs in information systems [1], marketing and

[1] "The PLS algorithm is called PLS-MODE-C if each of Modes A and B is chosen at least one in the model" [15, p. 10].

business research [5, 9], among others. We describe the TsPLS procedure in Sect. 2. To assess how well the TsPLS procedure detects the presence or absence of nonlinear and/or interaction effects between formative constructs, a Monte Carlo simulation study is designed and implemented (Sect. 3). Section 4 reports the results and we conclude the paper with some final remarks in Sect. 5.

2 Two-Step PLS Path Modeling Mode B (TsPLS) Procedure

For considering nonlinearities in a PLS SEM, we have replaced the linear model that relates the unobservable variables by a linear polynomial model. Thus, quadratic and cross-product terms of the constructs are introduced in the relationship between exogenous and endogenous unobservable variables. For instance, if ξ_1, ξ_2, and ξ_3 are exogenous formative constructs and η is an endogenous formative construct, the following nonlinear and interaction terms may be related to η: ξ_2^2 and $\xi_1\xi_3$. Equation (1) describes the structural relationship between the dependent construct and the linear and nonlinear unobservable terms

$$\eta = \beta_{j0} + \sum_j \beta_j \xi_j + \sum_j \alpha_j \xi_j^2 + \sum_j \sum_l \gamma_{jl} \xi_j \xi_l + \zeta \tag{1}$$

where β_j, α_j, and γ_j are path coefficients and ζ is the disturbance term. There is no linear relationship between predictor and residual, that is, $E(\eta/\xi_j) = \sum_j \beta_j \xi_j + \sum_j \alpha_j \xi_j^2 + \sum_j \sum_l \gamma_{jl} \xi_j \xi_l$. Thus, this condition implies that $E(\zeta/\forall\xi_j) = 0$, $E(\zeta/\forall\xi_j^2) = 0$, $E(\zeta/\forall\xi_l\xi_j) = 0$, $cov(\zeta,\xi_j) = 0$, $cov(\zeta,\xi_j^2) = 0$ and $cov(\zeta,\xi_l\xi_j) = 0$.

Each linear construct ξ_j is formed by a set of manifest variables as a linear function of them plus a residual (Eq. (2)). The weights π_{ji} determine the extent to which an indicator contributes to the formation of a construct. Each block of manifest variables may be multidimensional. The residual δ has a zero mean, and it is uncorrelated with the manifest variables x_{ji}. Since each construct is formed by a linear combination of the manifest variables, the sign of each weight π_{ji} should be the same sign as the correlation between x_{ji} and ξ_j [14, p. 165].

$$\xi_j = \sum_i \pi_{ji} x_{ji} + \delta_j. \tag{2}$$

Based on PLS path modeling with MODE B, a two-step score construction procedure is implemented to estimate the polynomial model (Eq. (1)). Standardized estimates Y_j of the unobservable variables are computed as usual in the iterative stage of the Wold's procedure (step one). The Wold's PLS-MODE-B algorithm starts choosing an arbitrary weight vector—outer weights—to first relate each construct with their own manifest variables. Usually this vector is a vector of ones. Each standardized unobservable variable Y_j—zero mean, unit variance—is computed as an exact linear combination of its own centered manifest variables (external estimation). An auxiliary unobservable variable Z_j is introduced as a counterpart to the variable Y_j. Each Z_j—at iteration s—is computed as a weighting sum of standardized unobservable variables computed in the iteration s and $s+1$. There are three different inner weighting schemes: the centroid, the factorial and the path weighting schemes. The simplest scheme is the centroid scheme where the inner weights are equal to the signs of the correlations between Y_j and its adjacent Y_i's. Once the auxiliary variables are estimated, the outer weights are recomputed. MODE B is considered for recomputing the outer weights when outer models are formative. The vector w_j of weights w_{ji} is the vector of the regression coefficients in the multiple regression of Z_j on the manifest variables related to the same unobservable variable Z_j. The first stage is iterated until convergence. Once scores are obtained, quadratic (nonlinear) and cross-product (interaction) terms are directly computed from the value of the standardized estimates of constructs (step two). The TsPLS procedure ends when the endogenous variable is regressed on the linear, nonlinear and interaction terms. Finally, to estimate structural relationships, we have followed the recommendation already made for multiple regression. Recall that in multiple regression, standardized coefficients of interaction effects are affected by changes in the means of the variables or the correlations between predictor and moderator variables [7]. Therefore, independent and dependent variables, which are both linear and unobservable, are standardized; and nonlinear and interaction unobservable terms are not standardized. If the regression coefficients are significant, this procedure ensures the interpretability of the coefficients.

3 Monte Carlo Simulation Study

We have performed a simulation study to examine the performance of the TsPLS procedure to estimate linear, nonlinear and interaction effects in SEMs with formative measurement models [3, 12]. The aims were:

- To examine the performance of the TsPLS procedure to estimate linear, nonlinear and interaction effects in SEMs with formative measurement models.
- To examine the performance of the TsPLS procedure when few indicators are considered per unobserved variable.
- To examine the performance of the TsPLS procedure when considering different sample sizes.
- To inspect the conditions under which the TsPLS procedure detects a significant nonlinear or interaction effect, by assuming that they actually exist (statistical power).
- To inspect the conditions under which the TsPLS procedure detects an increase in the predictability of a model.

The underlying true model considered a simple structure with three formative exogenous constructs (ξ_1, ξ_2 and ξ_3) related to one formative endogenous construct (η). We have investigated the nonlinear effect of the second construct (ξ_2), and the moderating effect of the first construct (ξ_1) on the third unobservable variable

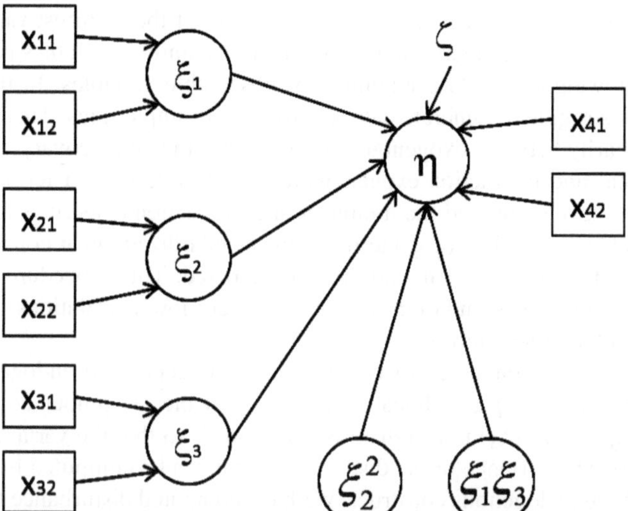

Fig. 1: Structural and measurement models of the simulated setups; these measurement models consider two indicators per construct; a structural equation model with linear (ξ_1, ξ_2 and ξ_3), nonlinear (ξ_2^2), and interaction ($\xi_1\xi_3$) effects

(ξ_3) (see Fig. 1). The experimental design considered models with two, four, six and eight indicators per unobservable variable and three sample sizes (100, 250, 500), a total of 12 different specifications. We have generated 500 random data sets for each of the 3×4 cells of the two-factor design in 5 steps as follows [10].

1. Manifest variables of exogenous constructs. Based on the PLS models, we have generated standardized manifest variables x_{ji} as random normal data for each formative exogenous construct ξ_j ($x_{ji} \sim N(0,1)$). There is no multicollinearity between indicators, but they may covary.
2. Formative and linear exogenous terms. To compute the formative exogenous constructs, we have assumed that all variables forming a construct are considered. Thus, variance of disturbance terms of the formative relationships, δ_j, are constrained to zero. The variance of ξ_j is described by Eq. (3) where $V(.)$ is the variance operator.

$$V(\xi_j) = \pi_j'\mathbf{X}_j'\mathbf{X}_j\pi_j \,. \tag{3}$$

We have chosen a set of permissible weights π_j for each block of variables, so that the variance of formative constructs is one.[2] Then, we have computed

[2] If $\mathbf{X}$ is a set of p variables, the variance of a linear combination $Y = \mathbf{X}b$ may be computed as $S_Y^2 = b'S_{\mathbf{XX}}b$. Thus, if Y and $\mathbf{X}$ are standardized variables, to derive a set of permissible weights (or b in this example) for relationships among variables is straightforward. These weights or true values are those that the PLS-MODE-B algorithm attempts to recover. Recall that in PLS-MODE-B, the outer weights are equal to the regression coefficients that are obtained once the scores of unobservable

each formative construct ξ_j as a weighted sum of the manifest variables. It is relevant to note that increasing the number of manifest variables per construct involves weighting a larger number of observable variables. In this case the routines struggle to generate data sets for small sample sizes. There is no multicollinearity between exogenous constructs, but they may covary.

3. Nonlinear and interaction exogenous terms. Once the linear terms are generated, we have computed the nonlinear and interaction constructs as the power (ξ_j^2) and cross-product ($\xi_j \xi_l$) terms of the standardized linear constructs. Nonlinear and interaction terms are not standardized. This procedure ensures the interpretability of simulation results. There are low correlations between the linear and nonlinear terms.
4. Formative endogenous construct. We have calculated the endogenous unobservable variable η as a linear combination of the exogenous constructs (ξ_1, ξ_2, ξ_3, ξ_2^2, and $\xi_1 \xi_3$) plus a disturbance term ζ, so that the variance of η is 1 (Eq. (4)). The variance of the disturbance term can be computed from the variance of the endogenous construct. We have computed disturbance terms of the inner relationships as random normal data with a zero mean and the corresponding standard deviation. They are distributed independently of the unobservable variables.

$$V(\eta) = \beta' \xi' \xi \beta + V(\zeta). \tag{4}$$

where ξ is a five-dimensional vector containing all the linear and nonlinear exogenous constructs and β a vector with the corresponding path coefficients β_j, α_j and γ_{jl}. It is worth noting that, for small sample sizes, it may happen that the standard deviation of disturbance terms is undefined for a given set of true values of path coefficients.

5. Manifest variables of endogenous construct. Once the formative endogenous construct is computed, we have created the manifest variables for the formative endogenous outer model. We have generated $i-1$ standardized manifest variables as random normal data $x_{ji} \sim N(0,1)$. The ith observable variable is calculated as the difference between the endogenous construct and the weighted manifest variables, that is as a linear combination of normal variables (Eq. (5)).

$$x_{ji} = \frac{1}{\pi_{ji}}(\eta - \sum_i \pi_{ji} x_{ji}). \tag{5}$$

To set the true parameters, we have taken into account different combinations of permissible values in order to show whether they are recovered by the TsPLS procedure. Table 1 shows the true values of weights, linear, nonlinear and interaction effects. As can be seen, we have considered different permissible values of weights in weight vectors. However, we also studied the case where weights in weight vectors are all equal. For instance, we set a true weight of 0.6 for models with two

---

variables are computed. PLS path modeling belongs to the family of fixed-point methods, "fixed-points are found iteratively by means of a sequence of regressions starting with an arbitrary choice for $\hat{w}$" (see [2], p. 78).

indicators per construct as well as values of 0.4, 0.35, and 0.3 for models with four, six and eight indicators per construct, respectively. Even though, these values may not be considered permissible weights, obtained results were quite similar to those shown in Sect. 4. We have implemented the TsPLS procedure in R-project [13]. The R-package dgmb implements the data generation procedure and offers a Graphical User Interface (GUI) to fix the simulation parameters and generate data [10].

To achieve the objectives of this research, we have examined the accuracy and precision with which the TsPLS procedure retrieves the true values. Accuracy is reported in terms of the mean bias ($\frac{1}{t}\sum_{i=1}^{t} E[\theta_i] - \theta$, where t is the number of runs) and mean relative bias (MRB $= 100 * \frac{1}{t}\sum_{i=1}^{t} \frac{\theta - E[\theta_i]}{\theta}$, [1]) of the estimates. Precision is reported in terms of the mean square error of the estimates (MSE $= Bias^2 + Variance$). Furthermore, we have examined the relative frequency with which the TsPLS procedure detects a significant nonlinear or interaction effect, assuming that this actually exists. Statistical power is determined considering an $\alpha = 0.05$ and the corresponding sample size and effect. Finally, we have inspected the conditions under which the procedure detects an increase in the predictability of a model. The increase in the predictability is reported in terms of the p-values of F-statistics in two cases. First, we have analyzed whether there is a significant increase in predictive power of a linear model with the addition of a nonlinear term. A second case considers the addition of an interaction term to the nonlinear model.

Table 1: True values for weights, and linear, nonlinear and interaction effects; a model with three formative exogenous constructs and one formative endogenous construct; cases for two, four, six and eight manifest variables (MVs) in each outer model

Coefficient	2 MVs	4 MVs	6 MVs	8 MVs
Weights of exogenous	(0.8,0.4)	(0.2,0.3,0.5,0.7)	(0.5,0.3,0.4, 0.3,0.5,0.1)	(0.3,0.3,0.4,0.3, 0.4,0.3,0.2,0.3)
constructs (π_j)	(0.4,0.8)	(0.2,0.4,0.6,0.5)	(0.2,0.4,0.6, 0.4,0.2,0.3)	(0.3,0.3,0.4,0.3, 0.2,0.3,0.4,0.2)
	(0.1,0.9)	(0.3,0.5,0.7,0.2)	(0.3,0.6,0.2, 0.3,0.4,0.2)	(0.4,0.5,0.4,0.3, 0.2,0.1,0.3,0.2)
Linear effects (β_j)	(0.5,0.4,0.3)	(0.5,0.4,0.3)	(0.5,0.4,0.3)	(0.5,0.4,0.3)
Nonlinear effects (α_j)	0.3	0.3	0.3	0.3
Interaction effects (γ_{jl})	0.3	0.3	0.3	0.3
Weights of endogenous	(0.4,0.8)	(0.2,0.3,0.5,0.5)	(0.5,0.3,0.4, 0.3,0.5,0.1)	(0.3,0.3,0.4,0.3, 0.4,0.3,0.2,0.3)
construct (π_j)				

4 Results

Figure 2 shows the mean relative biases of linear, nonlinear and interaction effects for models with two, four, six and eight indicators and different sample sizes. True linear effects of 0.5 and 0.4 are illustrated in Fig. 2a, b, respectively. True nonlinear and interaction effects are 0.3 (Fig. 2c, d, respectively). The TsPLS procedure underestimates the true values of all inner relationships. All estimates significantly improve by increasing the number of observations. The biases of linear effects are much smaller than the biases of nonlinear and interaction effects. For models with two indicators per construct, the largest MRBs of linear effects range from 13.24% (N = 100) to 9.90% ($N = 500$) whereas in models with eight indicators the largest MRBs range from 17.87% ($N = 100$) to 12.10% ($N = 500$). The largest MRBs of interaction effects range from 32.38% ($N = 100$) to 19.34% ($N = 500$) in models with two indicators per construct whereas in models with eight indicators per construct the largest MRBs range from 68.41% ($N = 100$) to 30.47% ($N = 500$). In the case of nonlinear effects, the largest MRBs range from 29.33% ($N = 100$) to 20.48% ($N = 500$) in models with two indicators per construct whereas in models with eight indicators per construct, the largest MRBs range from 67.35% ($N = 100$) to 28.33% ($N = 500$). The estimates of nonlinear and interaction effects may not be accurate when models are estimated with small sample sizes ($\leq N = 100$), but the true values are underestimated.

Figure 3 shows the mean square errors of linear, nonlinear and interaction effects by increasing the sample size and the number of indicators per construct. By increasing the sample size, the MSE clearly approaches zero. Interestingly, the TsPLS procedure performed better for the nonlinear effects than for interaction effects.

Figure 4a, b show the relative frequency (statistical power) with which the TsPLS procedure detects a significant nonlinear effect and a significant interaction effect, respectively. The estimated model included three linear effects, a nonlinear effect and an interactive effect on the endogenous construct. As seen in the figures, the statistical power increases by increasing the sample size. When outer models have two, four, six and eight indicators per construct, the statistical power is $\geq$0.8 when the sample size is $\geq$250. In addition, the TsPLS procedure also detects significant nonlinear and interaction effects with a relative frequency $\geq$0.8 when outer models have two indicators per construct and $N = 100$. The procedure most frequently detected a significant nonlinear effect rather than an interaction effect.

Figure 5a shows the p-values of the F-statistics to test the effect of the nonlinear term in the model. Here, the compared models are (1) a nonlinear model with three linear effects and a nonlinear effect on the endogenous construct, and (2) a linear model with three linear effects on the endogenous construct. A statistically significant F indicates that the nonlinear effect exists. Thus, the TsPLS procedure detects significant nonlinear effects (p-value $<$0.01) with samples sizes $\geq$100 in models with two, four and six indicators per construct. When models consider eight indicators per construct, significant nonlinear effects are also captured with a sample size of 100 and a p-value $<$0.05.

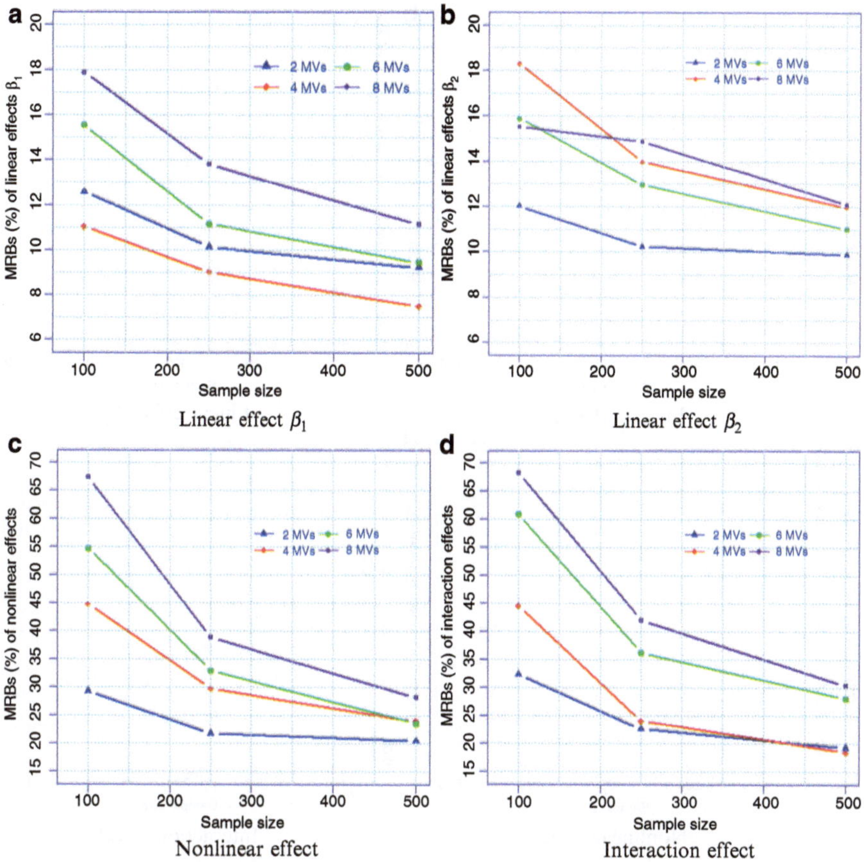

Fig. 2: Mean relative biases of linear, nonlinear and interaction effects. SEMs with two, four, six and eight indicators per construct and different sample sizes. (**a**) Linear effect β_1. (**b**) Linear effect β_2. (**c**) Nonlinear effect. (**d**) Interaction effect

Figure 5b shows the p-values of the F-statistics to account for an increase in the explained variance in the dependent variable with the addition of an interactive effect to the nonlinear model. Results are similar to the previous case and the procedure detects a significant interactive effect (p-value <0.01) with samples sizes ≥ 250. For $N = 100$, the procedure detects a significant interactive effects with p-value <0.05 in models with two and four indicators per construct.

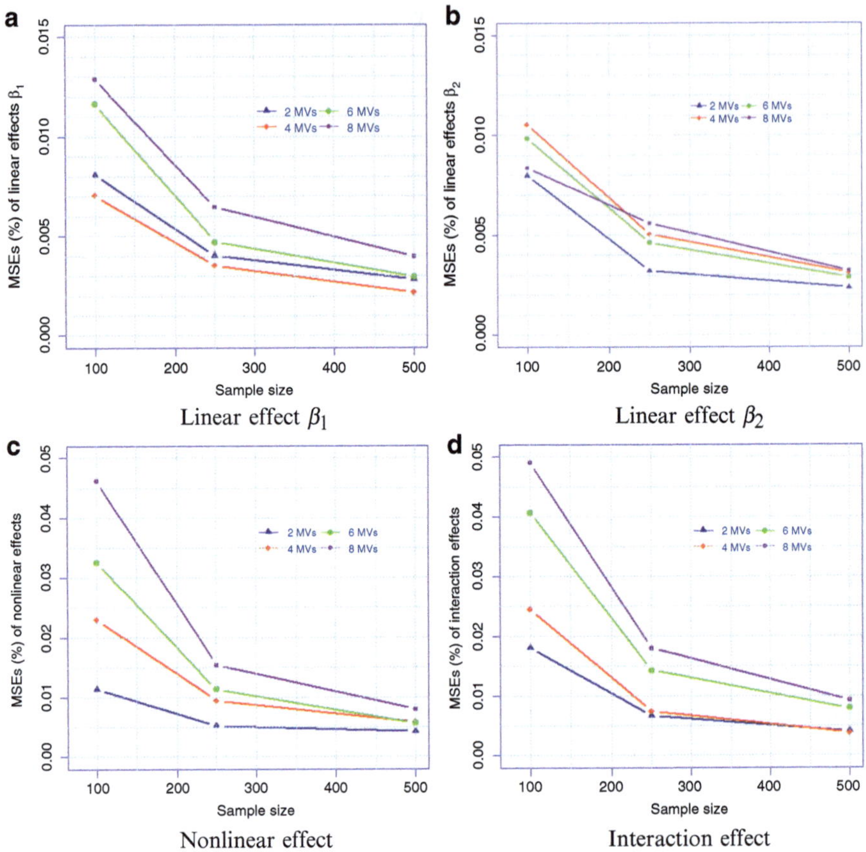

Fig. 3: Mean square errors of linear, nonlinear and interaction effects. SEMs with two, four, six and eight indicators per construct and different sample sizes. (**a**) Linear effect β_1. (**b**) Linear effect β_2. (**c**) Nonlinear effect. (**d**) Interaction effect

5 Final Remarks

Results are quite conclusive. The TsPLS procedure preserves the convergence properties of Wold's algorithm for PLS path modeling with MODE B. Hence, the TsPLS estimates are consistent; that is, by increasing the sample size, bias and error approach zero. Through this extensive Monte Carlo simulation study, we have shown that, under the conditions considered here, the TsPLS procedure underestimates the linear, nonlinear and interaction effects between exogenous and endogenous formative constructs.

When the sample size is $N = 100$, the procedure struggles to recover the true value of the nonlinear and interaction effects, especially when the formative outer models have six and eight indicators per construct. When the measurement models

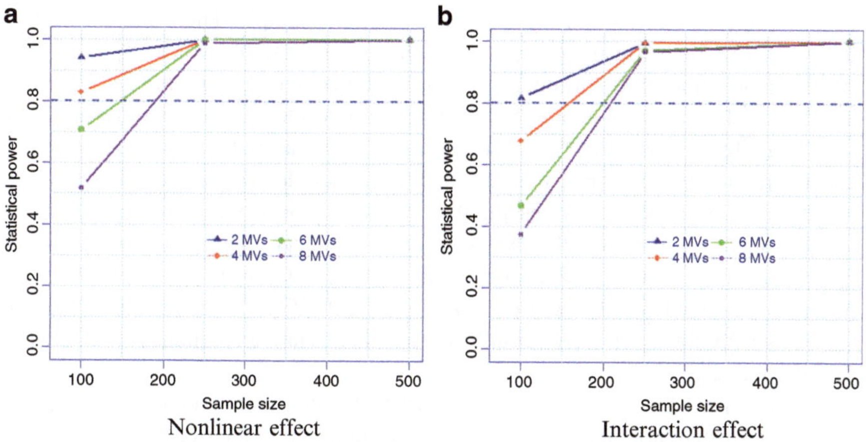

Fig. 4: Relative frequency with which the TsPLS procedure detects a significant nonlinear or interaction effect (statistical power). A nonlinear-interactive model with three linear effects, a nonlinear effect and an interactive effect on the dependent variable. The *dashed horizontal line* indicates a statistical power of 0.8. (**a**) Nonlinear effect. (**b**) Interaction effect

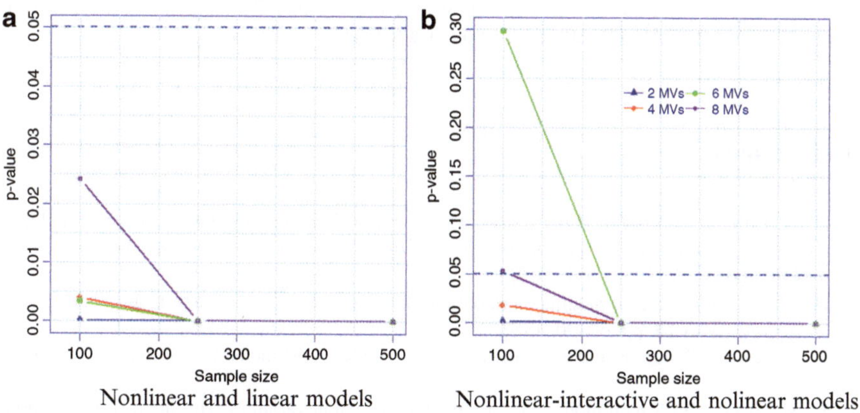

Fig. 5: *p*-values for *F*-statistics for the difference between the R-squares of the dependent construct in two cases. (**a**) A nonlinear effect is added to the linear model, and (**b**) an interaction effect is added to the nonlinear model. The *dashed horizontal line* indicates the *p*-value < 0.05

include two or four indicators per construct, accuracy and precision significantly improve. Thus, the TsPLS procedure is suited to estimate SEMs with formative blocks of variables and few indicators per construct. However, it is important to note that if a construct is well-formed by few indicators, this is a property of the model, not the method. The set of manifest variables of a formative construct should be a census of the variables that form the construct, not a sample. Therefore, a tendency to define formative constructs with few indicators should be avoided if there is no strong theory supporting this decision. For medium and large sample sizes ($N = 250$ and $N = 500$), biases are about 15% for linear effects and 30% for nonlinear and interaction effects. Thus, the TsPLS procedure is more suited to estimating nonlinear and interaction effects when medium and large sample sizes are available. These results are confirmed when assessing the predictive power of the model and the relative frequency with which the TsPLS procedure detects a significant or interaction effect (statistical power). However, a significant increase of the predictability of the model is observed whether sample sizes are ≥ 100. The TsPLS procedure detects significant nonlinear or interaction effects with a relative frequency ≥ 0.80 when sample sizes are ≥ 250.

A main drawback of the TsPLS procedure is that nonlinear and interaction unobservable terms are not taken into account when computing the linear effects. This is because, the latter are calculated within the Wold's algorithm (limited-information approach). Therefore, to consider other approaches is a task for future research. Finally, the data generation procedure described here allow us to study other simulation set ups in future research.

Acknowledgements We thank reviewer's comments and suggestions. This research was supported by a grant from the Universidad Católica de la Santísima Concepción and the Comisión Nacional de Investigación Científica y Tecnológica (CONICYT), Chile.

References

[1] W. Chin, B. Marcolin, and P. Newsted, "A partial least squares latent variable modeling approach for measuring interaction effects," *Information Systems Research* **14**(2), pp. 189–217, 2003.
[2] T. Dijkstra, "Some comments on maximum likelihood and partial least squares," *Journal of Econometrics* **22**, pp. 67–90, 1983.
[3] J. Gentle, "Random number generation and Monte Carlo methods," Springer, New York, 2003.
[4] M. Hanafi, "PLS path modeling: Computation of latent variables with the estimation mode B," *Computational Statistics* **22**, pp. 275–292, 2007.
[5] J. Henseler, C. Ringle, and R. Sinkovics, "The use of partial least squares path modeling in international marketing," *Advances in International Marketing* **20**, pp. 277–319, 2009.
[6] J. Henseler, G. Fassott, T. Dijkstra, and B. Wilson, "Analysing quadratic effects of formative constructs by means of variance-based structural equation modelling," *European Journal of Information Systems* **21**, pp. 99–112, 2012.
[7] J. Jaccard and R. Turrisi, "Interaction effects in multiple regression," Sage Publications, Thousand Oaks, 2003.
[8] E. Jakobowicz, "Contributions aux modèles d'équations structurelles à variables latentes," PhD Thesis Conservatoire National des Arts et Métiers, Paris, 2007.
[9] C. Jarvis, S. Mackenzie, and P. Podsakoff, "A critical review of construct indicators and measurement model misspecification in marketing and consumer research," *Journal of Consumer Research* **30**, pp. 199–218, 2003.

[10] A. Martinez-Ruiz and C. Martinez-Araneda, "A graphical user interface for generating data for structural equation models with formative constructs," in *Proceedings: 7th International Conference on Partial Least Squares and Related Methods*, (Houston,USA), May 2012.
[11] H. Mathes, "Global optimization criteria of the PLS-algorithm in recursive path models with latent variables," in *Statistical Modelling and Latent Variables*, K. Haagen, D. J. Bartholomew and M. Deistler, eds., (Amsterdam, The Netherlands), 1993.
[12] P. Paxton, P. Curran, K. Bollen, J. Kirby, and F. Chen, "Monte Carlo experiments: Design and implementation," *Structural Equation Modeling: A Multidisciplinary Journal* **8**(2), pp. 287–312, 2001.
[13] R. D. C. Team, "R: A language and environment for statistical computing," R Foundation for Statistical Computing, Vienna, Austria, 2010.
[14] M. Tenenhaus, V. Esposito-Vinzi, Y. M. Chatelin, and C. Lauro, "PLS path modeling," *Computational Statistics and Data Analysis* **48**, pp. 159–205, 2005.
[15] H. Wold, "Soft modeling: The basic design and some extensions," in *Systems Under Indirect Observation, Part II*, K. G. Jöreskog and H. Wold, eds., (Amsterdam, The Netherlands), 2010.

[illegible]

A Comparison of PLS and ML Bootstrapping Techniques in SEM: A Monte Carlo Study

Pratyush N. Sharma and Kevin H. Kim

Abstract Structural Equation Modeling (SEM) techniques have been extensively used in business and social science research to model complex relationships. The two most widely used estimation methods in SEM are the Maximum Likelihood (ML) and Partial Least Square (PLS). Both the estimation methods rely on Bootstrap re-sampling to a large extent. While PLS relies completely on Bootstrapping to obtain standard errors for hypothesis testing, ML relies on Bootstrapping under conditions in violation of the distributional assumptions. Even though Bootstrapping has several advantages, it may fail under certain conditions. In this Monte Carlo study, we compare the accuracy and efficiency of ML and PLS based Bootstrapping in SEM, while recovering the true estimates under various conditions of sample size and distributional assumptions. Our results suggest that researchers might benefit by using PLS based bootstrapping with smaller sample sizes. However, at larger sample sizes the use of ML based bootstrapping is recommended.

Key words: Bootstrapping, Partial least squares, Maximum likelihood, Structural equation modeling

1 Introduction

Structural Equation Modeling techniques, such as the covariance based SEM (CBSEM) and the Partial Least Squares based SEM (PLS), have gained enormous popularity as the key multivariate analysis methods in empirical research in the

P.N. Sharma (✉)
Joseph M. Katz Graduate School of Business, University of Pittsburgh, Pittsburgh, PA, USA
e-mail: pns9@pitt.edu

K.H. Kim
School of Education and Joseph M. Katz Graduate School of Business,
University of Pittsburgh, Pittsburgh, PA, USA
e-mail: khkim@pitt.edu

H. Abdi et al. (eds.), *New Perspectives in Partial Least Squares and Related Methods*, Springer Proceedings in Mathematics & Statistics 56, DOI 10.1007/978-1-4614-8283-3_13,

past few years. These techniques have been applied in diverse disciplines such as management information systems (MIS) [19], marketing [18] and psychology [10]. While there are many similarities in the two techniques, there are major differences among them especially in the estimation approaches they utilize. CBSEM focuses on estimating a set of model parameters so that the theoretical covariance matrix implied by the system of structural equations is as close as possible to the sample covariance matrix [18]. One of the most common estimation methods in CBSEM is the Maximum Likelihood (ML), which assumes multivariate normality and large sample theory. However, since researchers often work with relatively small samples from non-normal populations, bootstrap re-sampling offers a viable alternative [16]. Unlike CBSEM, PLS does not work with latent variables rather it works with block variables, and estimates model parameters to maximize the variance explained for all endogenous constructs through a series of ordinary least squares regression [18]. Thus, Partial Least Squares (PLS) based Structural Equation Models do not assume normality, and hence employ bootstrapping to obtain standard errors for hypothesis testing. Instead they assume that the sample distribution is a reasonable representation of the intended population distribution [7, 8].

Bootstrapping is a nonparametric approach to statistical inference that does not make any distributional assumptions of the parameters like traditional methods. Bootstrapping draws conclusions about the characteristics of a population strictly from the sample at hand, rather than making unrealistic assumptions about the population. That is, given the absence of information about the population, the sample is assumed to be the best estimate of the population. Hence, bootstrapping has advantages in situations where there is weak or no statistical theory about the distribution of a parameter, or when the underlying distributional assumptions needed for valid parametric inference are violated [14].

Bootstrapping estimates the empirical sampling distribution of a parameter by re-sampling from a sample with replacement. Although each re-sample has the same number of elements as the original sample, the replacement method ensures that each of these re-samples is likely to be slightly and randomly different than the original sample [15]. If the sample is a good approximation of the population then bootstrapping will provide a good approximation of the sampling distribution of the parameter. This necessitates a sufficiently large and unbiased sample. Unsurprisingly, researchers have cautioned against blind faith in bootstrapping and advocated investigation of bootstrapping, especially under conditions of insufficient sample size [9, 21].

In Structural Equation Models, Bootstrapping allows for the possibility to conduct significance testing of a statistic (θ) such as a path or a factor loading. Such significance tests analyze the probability of observing a statistic of that size or larger when the null hypothesis $H_0 : \theta = 0$, is true. However, Bollen and Stine (1992) have argued that while such a naïve bootstrap procedure works well in many cases, it can fail if the sample that is used to generate bootstrap samples doesn't represent the population. Under the naïve bootstrapping, the mean of the bootstrap population (i.e. the average of the observed sample) is unlikely to be equal to zero. In such cases, the bootstrap samples are drawn from a population for which the null hypothesis does not hold, regardless of whether H_0 holds for the unknown population from the original sample was drawn. Hence the bootstrap values of the test statistic are likely to reject H_0 too often [2]. This is most likely for misspecified models, or when the true population model is unknown. As a remedy, Bollen and Stine proposed a simple transformation of the data that seeks to make the null hypothesis true under the bootstrap re-sampling by centering the data around the sample mean. Re-sampling from the centered values forces the mean of the bootstrap population to be zero so that H_0 holds, resulting in fewer Type-1 errors [2].

Given the reliance of CBSEM and PLS on bootstrapping under most conditions, we argue that researchers need a better understanding of bootstrapping behavior especially under the limiting conditions of sample size, distributional assumptions and model misspecifications. We also seek to contribute to the ongoing debate in the MIS and marketing literatures about the use of PLS under conditions of insufficient sample sizes and distributional assumption violations [7, 11, 12, 18, 19]. While there are a few studies comparing CBSEM and PLS under various sets of design factors [1, 6, 18], none of them have focused on bootstrapping behaviors of the estimation methods used. Our goal is to provide researchers with some additional guidelines based on bootstrapping behavior while choosing among CBSEM and PLS. We conducted a Monte Carlo study to evaluate the efficiency and accuracy in model parameter recovery by naïve bootstrapping in PLS, and ML and Bollen-Stine bootstrapping in CBSEM. Specifically, our research question is: In terms of the efficiency and accuracy of model parameter recovery, how does naïve bootstrapping in PLS compare to ML and Bollen-Stine bootstrapping in SEM across various conditions of sample size and distributional assumptions? We analyzed this question using a mixed ANOVA design.

2 Method

We conducted a Monte Carlo study to analyze the behavior of bootstrapping techniques under the most commonly used estimation methods in SEM. The latent variable model used in this study involved two exogenous variables and one endogenous variable each with three reflective indicators with no cross-loadings, no model misspecifications and no interaction effects. The factor loadings (lambdas) for the measurement model were set to 0.6 and the path loadings (betas) for the structural model were set at 0.3 (Fig. 1).

Data were generated using Fleishman and Vale-Maurelli's method [4, 20] for this underlying model under five conditions of sample size (50, 100, 150, 200 and 500) and four distributions (normal, χ^2 with $df = 3$, t-distributed with $df = 5$, and uniform). One hundred dataset replications were performed for each of the 20 conditions, with 250 bootstrap replications for each dataset. Standardized parameter estimates from PLS, ML, and ML Bollen-Stine bootstraps were compared. All simulations were run on the R computing environment [17] using the sem [5] and

Fig. 1: A 3-factor theoretical model

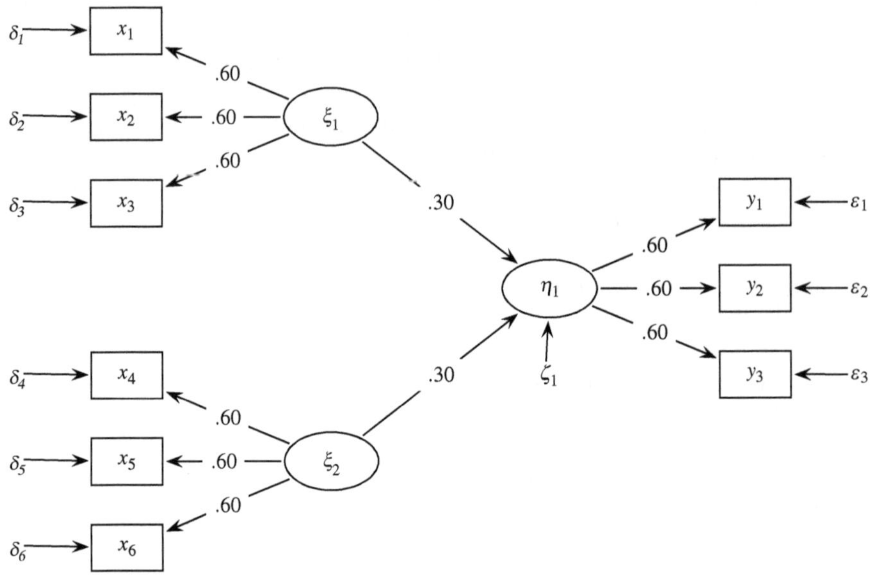

sempls packages [13]. PLS parameters were estimated using path weighting scheme for inner weight computation and Mode A (reflective) outer weight computation. To assess the accuracy in the parameter recovery, we performed a $3 \times 5 \times 4$ mixed design ANOVA on average bias and root mean square deviation (RMSD). To assess efficiency, we conducted mixed ANOVA to analyze the averages of the standard deviations of the sampling distribution of parameter estimates (i.e., standard errors).

3 Results

In order to analyze if there were any differences among the techniques in terms of achieving proper solutions, we checked the instances for non-convergence of solutions (i.e., the model didn't coverage within 500 iterations) and the valence of the variance estimates. Table 1 presents the frequency pattern of non-convergence of ML based CBSEM, PLS, and the three bootstrap techniques. We found that for ML based CBSEM all non-convergences occurred at sample size 50, however PLS always converged. Bootstrap techniques produced higher non-convergences and the non-convergence rates decreased as sample size increased. Bollen-Stine Bootstrap had the lowest number of non-convergence issues and almost all such non-convergences occurred at sample size 50. Surprisingly, PLS Bootstrap also suffered from non-convergences at larger sample sizes of 200. ML based bootstrap fared the worst among the three techniques, however, all non-convergences occurred at smaller sample sizes 50 and 100. Overall, non-convergence rates up to sample size 100 were below 5%, while the non-convergence rates at sample size 150 or above were very low (less than 1%).

Table 1: The number of non-convergence frequencies of CBSEM, PLS, and bootstrap

				Bootstrap		
Sample size	Distribution	ML-SEM	PLS	ML	Bollen-Stine	PLS
50	Normal	7		12	11	6
	χ^2	5		15	12	10
	t-distributed	5		16	7	2
	Uniform	5		12	6	12
100	Normal			4		3
	χ^2			2		3
	t-distributed			1		3
	Uniform				1	
150	Normal			1		
	χ^2					1
	t-distributed					
	Uniform					1
200	Normal					1
	χ^2					2
	t-distributed					1
	Uniform					2
500	Normal					
	χ^2					
	t-distributed					
	Uniform					
Total				63	37	47

Table 2: Mean bias and RMSD of measurement and structural model by sample size and estimation method. Since ML and Bollen-Stine SEM bootstrap values were similar, we only present ML bootstrap values

Sample size	Method	Measurement model				Structural model			
		Bias	SE	RMSD	SE	Bias	SE	RMSD	SE
50	ML	0.008	0.002	0.163	0.002	0.048	0.012	0.210	0.007
	PLS	0.115	0.003	0.214	0.003	−0.032	0.003	0.121	0.004
100	ML	0.000	0.001	0.103	0.002	0.009	0.010	0.147	0.006
	PLS	0.134	0.002	0.184	0.002	−0.060	0.003	0.097	0.003
150	ML	0.000	0.001	0.083	0.002	−0.008	0.009	0.111	0.006
	PLS	0.143	0.002	0.174	0.002	−0.076	0.003	0.100	0.003
200	ML	0.000	0.001	0.071	0.002	0.005	0.009	0.097	0.006
	PLS	0.148	0.002	0.167	0.002	−0.080	0.003	0.097	0.003
500	ML	−0.001	0.001	0.044	0.002	0.000	0.009	0.064	0.006
	PLS	0.153	0.002	0.160	0.002	−0.087	0.003	0.094	0.003

Next, we analyzed the accuracy and efficiency of the bootstrap techniques. Since in this study there were no model misspecifications, we expected that ML and Bollen-Stine bootstrapping in SEM would result in similar accuracy and efficiency of the parameter recovery. The ANOVA results showed that the bias, RMSD and the averages of the standard deviations of the parameter estimates for both ML and Bollen-Stine bootstraps in CBSEM were similar, confirming our expectations. However, these estimates differed significantly when compared to naïve bootstrapping in PLS (Table 2). In terms of measurement model accuracy, we found that the mean bias and RMSD values for the retrieved factor loading estimates in naïve PLS bootstrapping were larger than both ML and Bollen-Stine SEM bootstraps. Surprisingly, we found that as sample size increased PLS bias increased, suggesting that naïve PLS bootstrapping overestimated the factor loadings at larger sample sizes. However, we also found that in general naïve PLS bootstrap had larger bias but smaller RMSD than both the ML and Bollen-Stine bootstraps for the structural model estimates (i.e., regression coefficients among latent variables). The RMSD values for structural model suggested that the naïve PLS bootstrap outperformed ML and Bollen-Stine SEM bootstraps up to a sample size of 200, after which the situation was reversed. The effect of distributional conditions on bootstrap accuracy and efficiency was not significant.

In terms of the measurement model efficiency, we found that the mean of standard errors of ML and Bollen-Stine SEM bootstraps were smaller than the naïve PLS bootstrap up to a sample size of 200 (Table 3). However, at a sample size of 500, naïve PLS bootstrap had similar efficiency as ML and Bollen-Stine bootstraps. For the structural model efficiency, we found that the naïve PLS bootstrap outperformed ML and Bollen-Stine SEM bootstraps at all levels of sample size.

Table 3: Mean standard errors of measurement and structural models by sample size and estimation methods

Sample size	Method	Measurement model		Structural model	
		M	*SE*	*M*	*SE*
50	ML	0.181	0.002	0.276	0.006
	PLS	0.215	0.005	0.146	0.003
100	ML	0.107	0.002	0.183	0.006
	PLS	0.127	0.005	0.090	0.003
150	ML	0.085	0.002	0.146	0.006
	PLS	0.099	0.005	0.076	0.003
200	ML	0.074	0.002	0.123	0.006
	PLS	0.076	0.005	0.067	0.003
500	ML	0.045	0.002	0.079	0.006
	PLS	0.045	0.005	0.044	0.003

4 Discussion

In this investigation of bootstrap methods we wished to ascertain if there were differences between naïve PLS bootstrap and ML or Bollen-Stine SEM bootstraps under conditions free from model misspecifications. We found that, in general, the naïve PLS bootstrap was more accurate and efficient than ML and Bollen-Stine SEM bootstraps for estimating structural model parameters. However the reverse was true for measurement model estimates. As per our expectations, we found that the ML and Bollen-Stine bootstraps in SEM had similar accuracy and efficiency in recovering the parameter estimates. We leave it for future research to perform additional analysis by incorporating model misspecifications, different number of indicators, and compare the three bootstrap methods mentioned above. It would be interesting to investigate the effect of the complex interplay of model misspecifications, sample size, and the estimation methods (PLS vs. ML) on bootstrap accuracy and efficiency. Specifically, the research question that can be addressed in the next phase is: In terms of the efficiency and accuracy of model parameter recovery, how does naïve bootstrapping in PLS compare to ML and Bollen-Stine bootstrapping in SEM across various conditions of measurement and structural model misspecifications, and sample size?

Like all research, our study was limited in several ways. First, all our variables were generated on a continuous scale, while in practice researchers most often work with categorical and nominal data. Second, all our constructs were reflective in nature rather than formative. Reflective constructs aim to identify measures that are inter-correlated, have unidimensionality and have strong internal consistency. Formative constructs on the other hand aim to explain unobservable variance, reduce multicollinearity and consider the indicators as predictors of the construct [3]. While this distinction is important to capture the congruence between the theoretical definition of the construct and its measurement, it was out of scope of our study.

5 Conclusion

PLS outperformed ML in smaller sample sizes; not only did it always converge but it also led to smaller bias and RMSD than ML. However, as sample size increased, the difference between the techniques disappeared. At larger sample sizes, ML produced smaller bias and RMSD than PLS. PLS was more accurate at reproducing structural parameters than ML but it was less efficient at the measurement parameters. Hence, at large sample sizes, ML is preferred over PLS but at small sample sizes, a researcher might benefit from using PLS over ML.

References

[1] Areskoug, B. (1982). The first canonical correlation: theoretical PLS analysis and simulation experiments. In K. G. Joreskog & H. Wolds (Eds.), Systems under direct observation: causality, structure, and prediction. Amsterdam: North Holland.
[2] Bollen, K. A., & Stine, R. A. (1992). Bootstrapping goodness of fit measures in structural equation modeling. Sociological methods and research, 21, 205–229.
[3] Diamanatopoulos, A., & Siguaw, J. A. (2006). formative versus reflective indicators in organizational measure development: a comparison and empirical illustration. British Journal of Management, 17, 263–282.
[4] Fleishman, A. I. (1978). A method for simulating non-normal distributions. Psychometrika, 43, 521–532.
[5] Fox, J. (2006). Structural equation modeling with the sem package in R. Structural Equation Modeling, 13, 465–486.
[6] Goodhue, D., Lewis, W., & Thompson, R. (2006). PLS small sample size and statistical power in MIS research. Paper presented at the 39th Hawaii International Conference on System Sciences, Maui, HI.
[7] Hair, J. F., Ringle, C. M., & Sarstedt, M. (2011). PLS: indeed a silver bullet. Journal of Marketing Theory and Practice, 19, 139–151.
[8] Hair, J. F., Ringle, C. M., & Sarstedt, M. (2011). PLS-SEM: indeed a silver bullet. Journal of Marketing Theory and Practice, 19, 139–151.
[9] Ichikawa, M., & Konishi, S. (1995). Application of bootstrap methods in factor analysis. Psychometrika, 60, 77–93.
[10] MacCallum, R. C., & Austin, J. T. (2000). Application of structural equation modeling in psychological research. Annual Review of Psychology, 51, 201–226.
[11] Marcoulides, G. A., Chin, W., & Saunders, C. (2009). Foreward: a critical look at partial least squares modeling. MIS Quarterly, 33, 171–175.
[12] Marcoulides, G. A., & Saunders, C. (2006). PLS: a silver bullet? A commentary on sample size issues in PLS modeling. MIS Quarterly, 30, 3–10.
[13] Monecke, A. semPLS: an R package for structural equation models using partial least squares.
[14] Mooney, C. (1996). Bootstrap statistical inference: examples and evaluations for political science. American Journal of Political Science, 40, 570–602.
[15] Mooney, C., & Duval, R. (1993). Bootstrapping: a nonparametric approach to statistical inference. Newbury, CA: Sage.
[16] Nevitt, J., & Hancock, G. R. (2001). Performance of bootstrapping approaches to model test statistics and parameter standard error estimation in structural equation modeling. Structural Equation Modeling, 8, 353–377.
[17] R Development Core Team: a language and environment for statistical computing. http://www.r-project.org
[18] Reinartz, W. J., Heinlein, M., & Henseler, J. (2009). An emprical comparison of the efficacy of covariance based and variance based SEM. International Journal of Research in Marketing, 26, 332–344.
[19] Ringle, C. M., Sarstedt, M., & Straub, D. (2012). A critical look at the use of PLS-SEM in MIS Quarterly. MIS Quarterly, 36, 3–14.
[20] Vale, C. D., & Maurelli, V. A. (1983). Simulating multivariate non-normal distributions. Psychometrika, 48, 465–471.
[21] Yung, Y. F., & Bentler, P. M. (1994). Bootstrap corrected ADF test statistics in covariance structure analysis. British Journal of Mathematical and Statistical Psychology, 47, 63–84.

Multiblock and Path Modeling with OnPLS

Tommy Löfstedt, Mohamed Hanafi, and Johan Trygg

Abstract OnPLS was recently proposed as a general extension of O2PLS for applications in multiblock and path model analysis. OnPLS is very similar to O2PLS in the case with two matrices, but generalizes symmetrically to cases with more than two matrices without giving preference to any matrix.

OnPLS extracts a minimal number of globally joint components that exhibit maximal covariance and correlation. A number of locally joint components are also extracted. These are shared between some matrices, but not between all. These components are also maximally covarying with maximal correlation. The variation that remains after the joint and locally joint variation has been extracted is unique to a particular matrix. This unique variation is orthogonal to all other matrices and captures phenomena specific in its matrix.

The method's utility has been demonstrated by its application to synthetic datasets with very good results in terms of its ability to decompose the matrices. It has been shown that OnPLS affords a reduced number of globally joint components and increased intercorrelations of scores, and that it greatly facilitates interpretation of the models. Preliminary results in the application on real data has also given positive results. The results are similar to previous results using other multiblock and path model methods, but afford an increased interpretability because of the locally joint and unique components.

Key words: OnPLS, Principal component analysis, Multi-block analysis

---

T. Löfstedt (✉) • J. Trygg
Department of Chemistry, Computational Life Science Cluster (CLiC),
Umeå University, Umeå SE-90187, Sweden
e-mail: tommy.lofstedt@chem.umu.se; johan.trygg@chem.umu.se

M. Hanafi
Sensometrics and Chemometrics Laboratory, Rue de la Géraudière,
BP 82 225 Nantes, 44322, France
e-mail: mohamed.hanafi@oniris-nantes.fr

H. Abdi et al. (eds.), *New Perspectives in Partial Least Squares and Related Methods*,
Springer Proceedings in Mathematics & Statistics 56, DOI 10.1007/978-1-4614-8283-3_14,

1 Introduction

OnPLS was recently proposed as a general extension of O2PLS [13, 15] for applications in multiblock and path model analysis [5, 7]. OnPLS is very similar to O2PLS in the case with two matrices, but generalizes symmetrically to cases with more than two matrices without giving preference to any matrix.

OnPLS extracts a minimal number of globally joint components that exhibit maximal covariance and correlation. A number of locally joint components are also extracted. These are shared between some matrices, but not between all. These components are also maximally covarying with maximal correlation. The variation that remains after the joint and locally joint variation has been extracted is unique to a particular matrix. This unique variation is orthogonal to all other matrices and captures phenomena specific in its matrix.

The method's utility has been demonstrated by its application to synthetic datasets with very good results in terms of the ability to decompose the matrices. It has been shown that OnPLS affords a reduced number of globally joint components and increased intercorrelations of scores, and that it greatly facilitates interpretation of the models. Preliminary results in the application on real data has also given positive results. The results are similar to previous results using other multiblock and path model methods, but afford an increased interpretability because of the locally joint and unique components.

2 Method and Theory

Latent variable methods often produce scores vectors $\mathbf{v}t$ as linear combinations of the columns of a matrix $\mathbf{X}$ using loading weight vectors $\mathbf{v}w$ such that $\mathbf{v}t = \mathbf{X}\mathbf{v}w$. The rows of the matrix represent observations (patients, samples etc.) and the columns are the measured variables on these samples. The score vectors are usually maximizing some criterion, e.g., maximal correlation or covariation between scores from different matrices or maximal variance of a score vector within a matrix, under some constraints on the weights or the scores. For example, in PLS regression there are two matrices, $\mathbf{X}$ and $\mathbf{Y}$, and the objective is to maximize the covariance between their score vectors as

$$(M-1)\mathrm{Cov}(\mathbf{v}t, \mathbf{v}u) = \mathbf{v}t^{\mathrm{T}}\mathbf{v}u = \mathbf{v}w^{\mathrm{T}}\mathbf{X}^{\mathrm{T}}\mathbf{Y}\mathbf{v}c, \tag{1}$$

assuming the matrices $\mathbf{X}$ and $\mathbf{Y}$ have been mean centred, and where M is the number of rows in the matrices.

In the context of latent variable methods in general, but of PLS regression in particular, assume now that we can decompose $\mathbf{X}$ as

$$\mathbf{X} = \mathbf{X}_1 + \mathbf{X}_2 \tag{2}$$

such that $\mathbf{Y}^{\mathrm{T}}\mathbf{X}_1 = \mathbf{Y}^{\mathrm{T}}\mathbf{X}$ is maximal according to PLS regression and $\mathbf{Y}^{\mathrm{T}}\mathbf{X}_2 = \mathbf{0}$. Then when we calculate the score vectors $\mathbf{v}t$ we get

$$\mathbf{v}t = \mathbf{X}\mathbf{v}w = (\mathbf{X}_1 + \mathbf{X}_2)\mathbf{v}w = \mathbf{X}_1\mathbf{v}w + \mathbf{X}_2\mathbf{v}w \tag{3}$$

in which $\mathbf{X}_2\mathbf{v}w$ need not be zero while at the same time $\mathbf{Y}^{\mathrm{T}}\mathbf{X}_2\mathbf{v}w = \mathbf{v}0$. The scores thus contain variation that won't change the relation to $\mathbf{Y}$ but will surely affect the interpretation of the score vector $\mathbf{v}t$. PLS is still able to model these matrices and create a very good regression model, but the model will require more components and thus be more difficult to interpret [14, 16].

Methods that find and extract variation with this property were originally called orthogonal signal correction methods (or filters) [17], but have moved beyond being just pre-processing filters to being an integral part of the model building and interpretation [13–15].

The objective in OnPLS was originally to generalize this to several matrices in order to separate variation that is joint in *all* matrices from variation that is not joint in *all* matrices (i.e., from the variation that was unique in a particular matrix or related to at most all but one of the other matrices). But OnPLS has recently been extended beyond this to decompose a set of matrices ($\mathbf{X}_i$, $M \times N_i$, $i = 1, \ldots, n$) in several parts such that each part represents the variation that is joint (in terms of covariation) between the different subsets of matrix combinations [6]. This idea is illustrated in Fig. 1.

This decomposition is thus made for the variation that is joint with *all* other matrices, the variation that is locally joint between *some but not all* matrices, i.e., for subsets of matrices, and the variation that is *unique* in a matrix, i.e., that is not shared with any other matrices. When simplifying the locally joint part, the model of each matrix looks like

$$\mathbf{X}_i = \underbrace{\mathbf{T}_{G,i}\mathbf{P}_{G,i}^{\mathrm{T}}}_{\substack{\mathbf{X}_{G,i} \\ \text{globally joint}}} + \underbrace{\mathbf{T}_{L,i}\mathbf{P}_{L,i}^{\mathrm{T}}}_{\substack{\mathbf{X}_{L,i} \\ \text{locally joint}}} + \underbrace{\mathbf{T}_{U,i}\mathbf{P}_{U,i}^{\mathrm{T}}}_{\substack{\mathbf{X}_{U,i} \\ \text{unique}}} + \mathbf{E}_i, \tag{4}$$

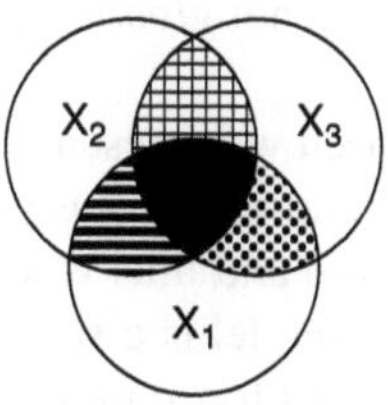

Fig. 1: The method described in this paper aims to divide each matrix in several parts. One globally joint part, i.e. the part that each matrix shares with *all* other matrices (the *black area* in the centre); several locally joint parts that contain the variation that each matrix shares with *some* of the other matrices (the areas with *lines*, *dots* and *squares*); and one part with variation unique in a particular matrix (the *open areas*)

for $i = 1, \dots, n$ (i.e., for n matrices). Note, that there are $2^{n-1} - 2$ possible locally joint models in each matrix, which is why, for simplicity, we lump them all together in $\mathbf{X}_{L,i}$.

This decomposition is sought for each matrix such that the covariation between the matrices $\mathbf{X}_{G,i}$ is maximal, such that the covariation between the relevant locally joint matrices $\mathbf{X}_{L,i}$ is maximal, such that the variation in $\mathbf{X}_{U,i}$ is maximal while covarying minimally with the other matrices, and also such that the different parts are mutually orthogonal.

Globally joint variation, captured in a score vector $\mathbf{v}t_{G,i}$, is understood as the variation in $\mathbf{X}_i$ that is joint with *all* other matrices, i.e., $\mathbf{X}_{G,j}^{\mathrm{T}}\mathbf{v}t_{G,i} \neq \mathbf{v}0$, for *all* $j \neq i$. Locally joint variation, on the other hand, captured in a score vector $\mathbf{v}t_{L,i}$ is understood as variation in $\mathbf{X}_i$ that is joint with *some*, but not all other matrices, i.e., such that $\mathbf{X}_{G,j}^{\mathrm{T}}\mathbf{v}t_{L,i} = \mathbf{v}0$, for all j, but such that possibly $\mathbf{X}_{L,j}^{\mathrm{T}}\mathbf{v}t_{L,i} \neq \mathbf{v}0$ for any but not all j. Unique variation is variation captured in a score vector $\mathbf{v}t_{U,i}$ such that $\mathbf{X}_j^{\mathrm{T}}\mathbf{v}t_{U,i} = \mathbf{v}0$ for all $j \neq i$.

The set analogy in Fig. 1 clearly illustrates that each matrix can be split in 2^{n-1} parts. I.e. in the $n = 3$ case we would split $\mathbf{X}_1$ into four parts as

$$\mathbf{X}_1 = \underbrace{(\mathbf{X}_1 \cap \mathbf{X}_2 \cap \mathbf{X}_3)}_{\text{globally joint}} + \underbrace{((\mathbf{X}_1 \cap \mathbf{X}_2) \setminus \mathbf{X}_3) + ((\mathbf{X}_1 \cap \mathbf{X}_3) \setminus \mathbf{X}_2)}_{\text{locally joint}} + \underbrace{(\mathbf{X}_1 \cap \overline{(\mathbf{X}_2 \cup \mathbf{X}_3)})}_{\text{unique}},$$

where $\cup$ is the set union operator, $\cap$ is the set intersection operator, $\setminus$ is the set difference operator and $\overline{\mathbf{S}}$ is the set complement. For matrix $\mathbf{X}_1$, $\mathbf{X}_1 \cap \mathbf{X}_2 \cap \mathbf{X}_3$ would thus be the globally joint part (shared with *all* matrices), $(\mathbf{X}_1 \cap \mathbf{X}_2) \setminus \mathbf{X}_3$ and $(\mathbf{X}_1 \cap \mathbf{X}_3) \setminus \mathbf{X}_2$ would be the locally joint parts (shared with *some* but not all) and $\mathbf{X}_1 \cap \overline{(\mathbf{X}_2 \cup \mathbf{X}_3)}$ would be the unique part (shared with *no* other matrices).

The joint models are either multiblock models where all matrices are assumed to be related to all other matrices and the objective is to investigate the nature of these relationships, or the joint models are path models in which a more or less complicated set of paths are assumed to exist between matrices.

The multiblock OnPLS approach was first presented in [7] such that it splits each matrix in two parts, one that relates all matrices to each other (the joint part) and one that contains all variation that is not *globally* joint, i.e., that contains the locally joint and the unique parts.

The OnPLS path model approach was presented in [5] as a generalization of the multiblock model in [7]. The generalization was such that some of the connections between matrices could be removed and thus would not contribute to the model of all other matrices; the multiblock model is a special case of this, with all matrices connected. This approach thus allows for more general relationships to be analyzed since path modeling connects a number of data sets and allows for analysis of the path along which information is considered to flow from one matrix to another. These paths can for instance represent a known time sequence, an assumed causality order, or some other chosen organizational scheme [2, 12].

In order to describe these relationships, an $n \times n$ adjacency matrix $\mathbf{C}$ was introduced in [5] that has elements $c_{i,j} = 1$ if the matrices $\mathbf{X}_i$ and $\mathbf{X}_j$ are connected and 0 otherwise. The matrix **C** thus describes a graph, such as the one in Fig. 2a. If all elements (except for the diagonal) of **C** are 1, then the problem seeks a multiblock solution as described in [7], and illustrated in Fig. 2b.

2.1 Joint Variation

The objective in OnPLS is thus to split each matrix in parts belonging to the different intersections of the Venn diagram in Fig. 1. The approach that we suggest is described below and begins with trying to find the black colored center intersection of Fig. 1.

The key to finding the globally joint variation is in the singular value decomposition of the cross-product matrix of each pair of connected matrices

$$\mathbf{C}_{i,j}\Sigma_{i,j}\hat{\hat{\mathbf{W}}}_{i,j}^{\mathrm{T}} = c_{i,j}\mathbf{X}_j^{\mathrm{T}}\mathbf{X}_i, \tag{5}$$

for $j \neq i$ and $c_{i,j} = 1$. The columns of $\hat{\hat{\mathbf{W}}}_{i,j}$ corresponding to non-zero (or "sufficiently large," when considering noise levels and so on) singular values of $\Sigma_{i,j}$ represents the variation in $\mathbf{X}_i$ that is joint with $\mathbf{X}_j$. Note that if $c_{i,j} = 0$, then the matrices $\mathbf{X}_i$ and $\mathbf{X}_j$ are not connected. In that case we have $\hat{\hat{\mathbf{W}}}_{i,j} = \mathbf{0}$ and the weight matrix will not contribute to the solution.

All globally and locally joint variation is captured in this decomposition, but the unique variation is excluded. This weight matrix, $\hat{\hat{\mathbf{W}}}_{i,j}$, is thus a representation of all variation in $\mathbf{X}_i$ that is joint with $\mathbf{X}_j$. If we want to maximize

$$\mathbf{X}_j^{\mathrm{T}}\mathbf{v}t_i = \mathbf{X}_j^{\mathrm{T}}\mathbf{X}_i\mathbf{v}w_i \tag{6}$$

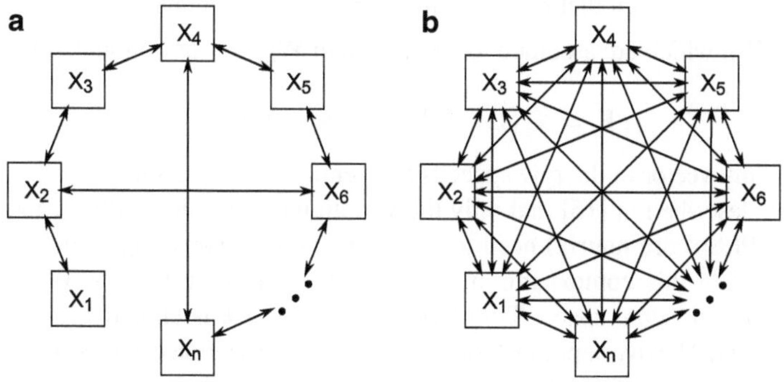

Fig. 2: OnPLS can build two conceptually different types of models. (**a**) Those in which the matrix **C** contains zeros outside of the diagonal, which conceptualizes path models. (**b**) And those in which the matrix **C** only have ones outside of the diagonal

for a score vector $\mathbf{v}t_i$ of $\mathbf{X}_i$, then we are looking for the right singular vector corresponding to the largest singular value of $\mathbf{X}_j^{\mathrm{T}}\mathbf{X}_i$. Analogously, if we want to minimize Eq. 6 to find score vectors of $\mathbf{X}_i$ orthogonal to $\mathbf{X}_j$ then we want the right singular vector corresponding to the smallest singular value of $\mathbf{X}_j^{\mathrm{T}}\mathbf{X}_i$ (those we exclude here).

The assumption now is that the globally joint variation for $\mathbf{X}_i$ is represented in all weight matrices, $\hat{\hat{\mathbf{W}}}_{i,j}$, and that it can be extracted from them. The method we have used, and which works very well in practice, is to take the singular value decomposition of an augmented matrix with all these bases such that

$$\hat{\mathbf{W}}_i \boldsymbol{\Sigma}_i \mathbf{V}_i^{\mathrm{T}} = \left[\hat{\hat{\mathbf{W}}}_{i,1}|\ldots|\hat{\hat{\mathbf{W}}}_{i,i-1}|\hat{\hat{\mathbf{W}}}_{i,i+1}|\ldots|\hat{\hat{\mathbf{W}}}_{i,n}\right], \tag{7}$$

in which $\hat{\mathbf{W}}_i$ is a recollection of the common structures in all covariation bases. Directions common in all $\hat{\hat{\mathbf{W}}}_{i,j}$ must end up in the first singular vectors of $\hat{\mathbf{W}}_i$ because of the nature of the singular value decomposition. This is basically the same as using SUM-PCA [9] to find the common structures in the weight matrices. Many other methods could be used here, e.g., generalised PCA, canonical correlation etc., but in our experience this approach works fast and yields very good results.

Thus, in the $n = 3$ case for $\mathbf{X}_1$, the augmented matrix in Eq. 7 would contain all variation covered by the filled (globally joint), the striped (locally joint between $\mathbf{X}_1$ and $\mathbf{X}_2$) and the dotted (locally joint between $\mathbf{X}_1$ and $\mathbf{X}_3$) fields in Fig. 1 and the objective is to extract from this the black field (the globally joint field in the centre of Fig. 1). This is what is attempted to be found in $\hat{\mathbf{W}}_1$.

Thus, $\hat{\mathbf{W}}_i$ is a representation of all variation in $\mathbf{X}_i$ that is shared with *all* other matrices. Any vector $\mathbf{v}w_{LU,i}$ in the row space of $\mathbf{X}_i$ orthogonal to the columns of $\hat{\mathbf{W}}_i$ will yield a score vector $\mathbf{v}t_{LU,i} = \mathbf{X}_i\mathbf{v}w_{LU,i}$ orthogonal to the globally joint space since

$$\mathbf{X}_j^{\mathrm{T}}\mathbf{v}t_{LU,i} = \mathbf{X}_j^{\mathrm{T}}\mathbf{X}_i\mathbf{v}w_{LU,i} = \mathbf{C}_{i,j}\Sigma_{i,j}\hat{\hat{\mathbf{W}}}_{i,j}^{\mathrm{T}}\mathbf{v}w_{LU,i}, \tag{8}$$

where $j \neq i$ for j such that $c_{i,j} = 1$, since $\mathbf{v}w_{LU,i}$ is orthogonal to the globally joint weights, $\hat{\mathbf{W}}_i$, which are assumed to be present in $\hat{\hat{\mathbf{W}}}_{i,j}$. By orthogonalizing $\mathbf{X}_i$ to $\hat{\mathbf{W}}_i$ we get

$$\mathbf{E}_i = \mathbf{X}_i\left(\mathbf{I} - \hat{\mathbf{W}}_i\hat{\mathbf{W}}_i^{\mathrm{T}}\right) = \mathbf{X}_i - \mathbf{X}_i\hat{\mathbf{W}}_i\hat{\mathbf{W}}_i^{\mathrm{T}}, \tag{9}$$

in which any vector in the row space is a potential $\mathbf{v}w_{LU,i}$ vector.

It was described in [15] that only the variation overlapping with the joint scores (e.g., the PLS score matrix) need to be extracted, not *everything* orthogonal to the joint space. In this context this means that by projecting $\mathbf{X}_i$ onto $\hat{\mathbf{W}}_i$ we obtain a potential joint score space, $\mathbf{T}_i = \mathbf{X}_i\hat{\mathbf{W}}_i$, but this space is tainted by variation *not* globally joint. The overlapping "non-globally" joint variation can therefore be found by maximizing

$$\left\|\mathbf{T}_i^{\mathrm{T}}\mathbf{v}t_{LU,i}\right\|^2 = \left\|\mathbf{T}_i^{\mathrm{T}}\mathbf{E}_i\mathbf{v}w_{LU,i}\right\|^2 = \mathbf{v}w_{LU,i}^{\mathrm{T}}\mathbf{E}_i^{\mathrm{T}}\mathbf{T}_i\mathbf{T}_i^{\mathrm{T}}\mathbf{E}_i\mathbf{v}w_{LU,i}. \tag{10}$$

This overlap is thus given by the eigenvector corresponding to the largest eigenvalue, λ_i, of

$$\mathbf{E}_i^{\mathrm{T}}\mathbf{T}_i\mathbf{T}_i^{\mathrm{T}}\mathbf{E}_i\mathbf{v}w_{LU,i} = \lambda_i\mathbf{v}w_{LU,i}. \tag{11}$$

When the "non-globally joint" weight vector $\mathbf{v}w_{LU,i}$ and the corresponding score vector $\mathbf{v}t_{LU,i}$ have been found, the corresponding "non-globally joint" loading vector is calculated as $\mathbf{v}p_{LU,i} = \mathbf{X}_i^{\mathrm{T}}\mathbf{v}t_{LU,i}/(\mathbf{v}t_{LU,i}^{\mathrm{T}}\mathbf{v}t_{LU,i})$ and the matrix is deflated by

$$\mathbf{X}_i \leftarrow \mathbf{X}_i - \mathbf{v}t_{LU,i}\mathbf{v}p_{LU,i}^{\mathrm{T}} = \left(\mathbf{I} - \frac{\mathbf{v}t_{LU,i}\mathbf{v}t_{LU,i}^{\mathrm{T}}}{\mathbf{v}t_{LU,i}^{\mathrm{T}}\mathbf{v}t_{LU,i}}\right)\mathbf{X}_i. \tag{12}$$

This updated matrix is used to find the next "non-globally joint" component starting over from Eq. 9 and onwards.

A number of such score vectors are extracted from its corresponding matrix. The matrix that is left does not have any more "non-globally joint variation", and can therefore be used successfully in a standard multiblock or path model to find the globally joint model. Note also that the globally joint variation and the non-globally joint variation will be orthogonal by construction, since the deflation procedure orthogonalizes against $\mathbf{v}t_{LU,i}$ in Eq. 12.

An optimisation criterion analogous to the criterion used in PLS regression is used here for building the multiblock and path models. The objective is to maximize the covariation of all connected matrices. This is achieved by maximizing the function

$$f_{\mathbf{C}}(\mathbf{w}_1,\ldots,\mathbf{w}_n) = \sum_{i=1}^{n}\sum_{j=1,j\neq i}^{n} c_{i,j}\mathbf{v}w_i^T\mathbf{X}_i^T\mathbf{X}_j\mathbf{v}w_j = \sum_{i=1}^{n}\sum_{j=1,j\neq i}^{n} c_{i,j}\mathbf{v}t_i^T\mathbf{v}t_j, \tag{13}$$

using the matrix $\mathbf{C}$ as explained above and subjecting the weight vectors to the constraint that $\|\mathbf{v}w_i\| = 1$, for $i = 1,\ldots,n$. It is straight-forward to use the method of Lagrange multipliers to solve this problem. Doing that we end up with the system

$$\lambda_i\mathbf{v}w_i = \sum_{j=1,j\neq i}^{n} c_{i,j}\mathbf{X}_i^{\mathrm{T}}\mathbf{X}_j\mathbf{v}w_j \tag{14}$$

for all $i = 1,\ldots,n$. These equations are thus the conditions required to maximize the objective function, $f_{\mathbf{C}}$, in Eq. 13.

Two monotonic convergent procedures were recently proposed [5] for the computation of the weights $\mathbf{v}w_i$, for $i = 1,\ldots,n$, in the maximization of Eq. 13. The first procedure is based on Jacobi iteration and was initially proposed by Horst [4] and its monotonic convergence was proven by ten Berge [10]. This procedure has been further extended to more general classes of problems by Hanafi and Kiers [3]. The second procedure is based on Gauss-Siedel iteration and was reported and its monotonic convergence proven by ten Berge [10]. The principle of this procedure has been used by Hanafi [2] and Tenenhaus and Tenenhaus [11] in a PLS path modelling context.

The two procedures are iteratively building a monotonically convergent sequence of weights, $\mathbf{v}w_i^{(s)}$, $s = 1, 2, \ldots$, using two different iteration schemes. New and simplified proofs of the monotonic convergence of these procedures were given in [5] for the maximisation of $f_{\mathbf{C}}$.

The criterion in Eq. 13 is equivalent to MAXDIFF [3] in the special case when all $c_{i,j} = 1$, for $i \neq j$, and in the path model case it ends up being equivalent to PLS path modeling using Horst's inner weighting scheme (where the inner weighting scheme is the identity) and what has been called "New Mode A" [11].

Note that any method could be used in this step instead of the proposed approach, as long as it is sufficiently similar to the one used in the filtering. Their score and loading vectors should be found using sufficiently similar approaches in order to compare and contrast the globally joint, locally joint and unique models. This should, however, not be a big problem in general.

To mimic the model of $\mathbf{X}$ in PLS regression and in order to allow for more complex models to be built, OnPLS allows more than one joint component to be extracted from the matrices. The proposed method of deflation is just as in PLS regression. Once the weights $\mathbf{v}w_i$ have been found, we calculate the scores

$$\mathbf{v}t_i = \mathbf{X}_i \mathbf{v}w_i \tag{15}$$

and the loadings

$$\mathbf{v}p_i = \frac{\mathbf{X}_i^{\mathrm{T}} \mathbf{v}t_i}{\mathbf{v}t_i^{\mathrm{T}} \mathbf{v}t_i} \tag{16}$$

and deflate each matrix by

$$\mathbf{X}_i \leftarrow \mathbf{X}_i - \mathbf{v}t_i \mathbf{v}p_i^{\mathrm{T}} = \mathbf{X}_i \left(\mathbf{I} - \mathbf{v}w_i \mathbf{v}p_i^{\mathrm{T}}\right) = \left(\mathbf{I} - \frac{\mathbf{v}t_i \mathbf{v}t_i^{\mathrm{T}}}{\mathbf{v}t_i^{\mathrm{T}} \mathbf{v}t_i}\right) \mathbf{X}_i. \tag{17}$$

Note that when all "non-globally joint" variation has been extracted we have $\mathbf{v}p_i \approx \mathbf{v}w_i$ and in that case this deflation approach is very similar to the deflation approach in MAXDIFF. But to make the contrast with MAXDIFF clear the proposed approach was named nPLS in [7].

2.2 Locally Joint Variation

The variation extracted above is globally joint. This means that the variation that is left is either unique to a particular matrix or shared with *some*, but not all other matrices.

The locally joint variation can be found by performing OnPLS recursively on models using a subset of the n matrices with at least two connected matrices. Removing one component at the time from the strongest submodel until there are no more submodels will extract locally joint variation with maximal variance.

There are $2^{n-1}-2$ possible locally joint models that can be built for each matrix. One approach would be to build all of these for one component after first extracting the variation that is not "global" from the perspective of this submodel, select the one that yields the highest value for Eq. 13. Deflate this component from the matrices it belongs to and start over again until there are no more significant components. While this is a possible approach that would extract the maximal locally joint variation, it is hardly feasible with large n because of the exponential growth in the number of submodels when n increases.

Another approach would be to assume that all n matrices are included, build a model of them and then remove those whose components are insignificant or "weak" according to some criteria. At least one must be insignificant since it would otherwise be part of the globally joint model. Then the model is rebuilt while including only those matrices that had significant components. This would be a greedy approach for finding the locally joint model with maximum value of Eq. 13 one component at the time. While this approach should extract the same number of locally joint components, it should also terminate more quickly since it would automatically disregard submodels where there is no locally joint variation.

2.3 Unique Variation

If the components of the locally joint variation are extracted just as the components for the globally joint variation (as in Eq. 17), then the variation that is left is by construction orthogonal to all variation already extracted. But this also means that it is orthogonal to all other matrices since it was not caught by the globally joint model and not by one of the locally joint models either. I.e. this variation must be unique in the particular matrix. A PCA model of the variation that is left will therefore separate the unique systematic variation from noise, while also extracting the components in order or decreasing variance.

3 Results

Several different synthetic data sets have been used in order to evaluate OnPLS. They were all created in the following way: A set of loading vectors were created using either a Gaussian loading profile or a unit pulse loading profile (rectangular loading profile), and the loadings were correlating to different degrees. All score vectors were mutually orthogonal random vectors. The score vectors were given arbitrary lengths (i.e., different norms). Each matrix was then created as a sum of products of the matrices' loading profiles and the scores. The matrices were created as:

$$\mathbf{X}_i = \underbrace{\sum_{j=1}^{A_G} \mathbf{v}t_j\mathbf{v}p_{i,j}^{\mathrm{T}}}_{\mathbf{X}_{G,i}} + \underbrace{\sum_{j=A_G+1}^{A_G+A_{L,i}} \mathbf{v}t_j\mathbf{v}p_{i,j}^{\mathrm{T}}}_{\mathbf{X}_{L,i}} + \underbrace{\sum_{j=1}^{A_{U,i}} \mathbf{v}t_{i,j}\mathbf{v}p_{i,A_G+A_{L,i}+j}^{\mathrm{T}}}_{\mathbf{X}_{U,i}} + \mathbf{R}_i, \tag{18}$$

where $\mathbf{R}_i$ is normally distributed random noise (about 1 % was added to each matrix). The matrices thus share globally joint score vectors, and a different number of locally joint score vectors, but have their own unique loadings.

Example 1 in [7] illustrates how "non-globally joint" variation distorts both the joint components and the unique components, and how extracting this variation greatly improves interpretability of both the globally joint model and the model of the unique variation. The loading vectors are clearly distorted if this type of filtering is not applied, as seen in Fig. 3b, but when the "non-globally joint" variation is removed the extracted model is the true underlying joint model as seen in Fig. 3c.

By performing the decomposition as described above the correlation of the joint score vectors increases and converges to the true maximum. This is seen in Fig. 4 (the results of Example 3 in [7]). In this example there were six matrices sharing two joint components and having a different number of locally joint components (zero through eight). The variation that is captured by the joint score vectors in the OnPLS model is fine-tuned towards the globally joint part and contains almost no locally joint variation (see [7]).

A simulation study shows that the variation found for the different parts is close to the true variation put in the matrices. Several hundred models were built and for each of them the modified RV coefficient [8] (a correlation coefficient for matrices) was calculated between the true and the extracted components for each part. The results showed that the modified RV coefficients were between 0.8–0.9 on average for the globally joint model, the locally joint model and the unique model.

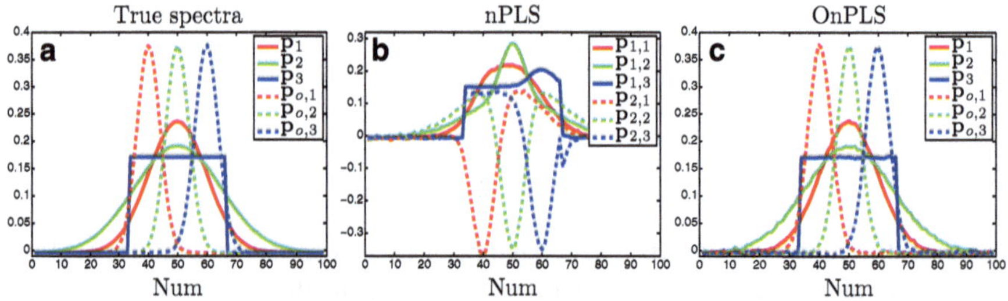

Fig. 3: The true joint and unique components that make up the data sets in this example are displayed in (**a**). A two-component nPLS model was created for each matrix without extracting any locally joint or unique components; the result is displayed in (**b**). An OnPLS model was also created for each matrix; the extracted joint and unique OnPLS loadings are displayed in (**c**) (Reprinted from [7] with kind permission from John Wiley & Sons)

An analysis of infra red data used in the monitoring of protein structure changes during cheese ripening was performed in [5]. The results are similar to previous multiblock and path model analyses of these data with the added benefit of being able to interpret the "non-globally joint" components.

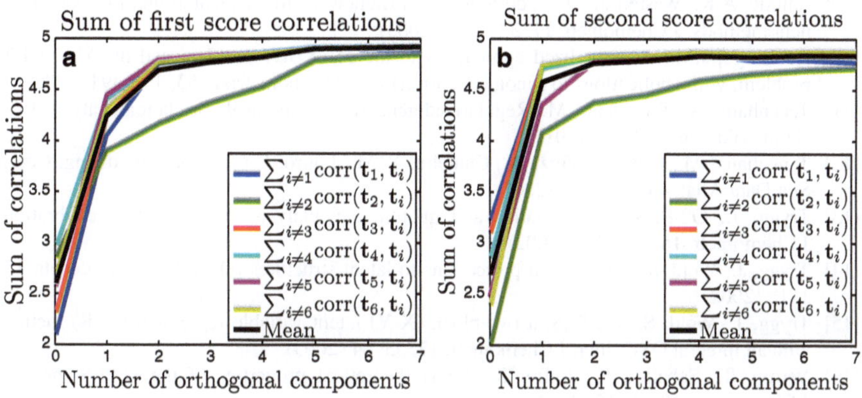

Fig. 4: When an increasing number of non-globally joint components is extracted the sum of correlations of the joint score vectors of the OnPLS model increases. The sum of all joint first score vectors is shown in (**a**) and the sum of all joint second score vectors is shown in (**b**) (Reprinted from [7] with kind permission from John Wiley & Sons)

In [1] the authors built two O2PLS models in series to analyze three data sets. The score matrices **T** and **U** of the first O2PLS model were put in a single matrix and then used with the third matrix in a second O2PLS model. This approach was used to find the globally joint variation in the analysis of metabolite, protein and transcript data of hybrid aspen. These data were also analyzed using OnPLS [6], which gave very similar results for the globally joint variation with the added benefit of transparent analysis of locally joint and unique variation.

References

[1] Bylesjö, M., Nilsson, R., Srivastava, V., Grönlund, A., Johansson, A.I., Jansson, S., Karlsson, J., Moritz, T., Wingsle, G., Trygg, J.: Integrated analysis of transcript, protein and metabolite data to study lignin biosynthesis in hybrid aspen. J Proteome Res. **8**, 199–210 (2009)
[2] Hanafi, M.: PLS path modelling: Computation of latent variables with the estimation mode B. Comp Stat. **22**, 275–292 (2007)
[3] Hanafi, M., Kiers, H.A.L.: Analysis of *k* sets of data, with differential emphasis on agreement between and within sets. Comp Stat Data Anal. **51**, 1491–1508 (2006)
[4] Horst, P.: Relations among *m* sets of measures. Psychometrika. **26**, 129–149 (1961)
[5] Löfstedt, T., Hanafi, M., Mazerolles, G., Trygg, J.: OnPLS path modelling. Revised in Chemometr Intell Lab. July 2012

[6] Löfstedt, T., Hoffman, D., Trygg, J.: Global, local and unique decompositions in OnPLS for multiblock data analysis. Submitted to Anal Chim Acta. July 2012.
[7] Löfstedt, T., Trygg, J.: OnPLS—a novel multiblock method for the modelling of predictive and orthogonal variation. J Chemometr. **25**, 441–455 (2011)
[8] Smilde, A.K., Kiers, H.A.L., Bijlsma, S., Rubingh, C.M., van Erk, M.J.: Matrix correlations for high-dimensional data: the modified RV-coefficient. Bioinformatics. **25**, 401–405 (2009)
[9] Smilde, A.K., Westerhuis, J.A., de Jong, S.: A framework for sequential multiblock component methods. J Chemometr. **17**, 323–337 (2003)
[10] ten Berge, J.M.F.: Generalized approaches to the MAXBET problem and the MAXDIFF problem, with applications to canonical correlations. Psychometrika. **53**, 487–494 (1988)
[11] Tenenhaus, A., Tenenhaus, M.: Regularized generalized canonical correlation analysis. Psychometrika. **76**, 257–284 (2011)
[12] Tenenhaus, M., Esposito Vinzi, V., Chatelin, Y.-M., Lauro, C.: PLS path modeling. Comp Stat Data Anal. **48**, 159–205 (2005)
[13] Trygg, J.: O2-PLS for qualitative and quantitative analysis in multivariate calibration. J Chemometr. **16**, 283–293 (2002)
[14] Trygg, J., Wold, S.: Orthogonal projections to latent structures (O-PLS). J Chemometr. **15**, 1–18 (2002)
[15] Trygg, J., Wold, S.: O2-PLS, a two-block (X-Y) latent variable regression (LVR) method with an integral OSC filter. J Chemometr. **17**, 53–64 (2003)
[16] Verron, T., Sabatier, R., Joffre, R.: Some theoretical properties of the O-PLS method. J Chemometr. **18**, 62–68 (2004)
[17] Wold, S., Antti, H., Lindgren, F., Öhman, J.: Orthogonal signal correction of near-infrared spectra. Chemometr Intell Lab. **44**, 175–185 (1998)

Testing the Differential Impact of Structural Paths in PLS Analysis: A Bootstrapping Approach

Wynne W. Chin, Yong Jin Kim, and Gunhee Lee

Abstract Researchers are often interested in examining the relative impact of PLS structural paths. As such, this paper focuses on how one assesses the impact of various antecedent constructs to a particular endogenous construct. In a survey of recent papers in the field of Information Systems employing structural equation modeling, discussion of differences or conversely equivalency of paths were typically made without any statistical tests. In a few cases, a traditional *t*-test was used. This paper begins with a didactic presentation of how such *t*-tests are estimated followed with introducing an alternative bootstrapping approach. Results from both empirical and simulated data show different conclusions are made between these two approaches. In particular, we show that under data conditions of high kurtosis, bootstrapping is less likely to commit a Type I error of stating substantial differences among paths when none exist.

Key words: PLS-PM, Path model, Endogenous contstruct, *t*-test

W.W. Chin (✉)
Department of Decision and Information Systems, C. T. Bauer College of Business, University of Houston, Houston, TX 77204-6021, USA
e-mail: wchin@uh.edu

Y.J. Kim
Global Service Management, Sogang University, 2 Shinsu-dong, Mapo-gu, Seoul, 121–742 Korea
e-mail: yongjkim@sogang.ac.kr

G. Lee
Sogang Business School, Sogang University, 2 Shinsu-dong, Mapo-gu, Seoul, 121742 Korea
e-mail: ghlee@sogang.ac.kr

H. Abdi et al. (eds.), *New Perspectives in Partial Least Squares and Related Methods*, Springer Proceedings in Mathematics & Statistics 56, DOI 10.1007/978-1-4614-8283-3_15,

1 Introduction

To date, researchers employing PLS path modeling regularly use the non-parametric resampling procedure of bootstrapping at the standard procedure for assessing the significance of structural paths [2]. When examining the invariance of structural paths across different samples, the permutation procedure [2] or bootstrap t-tests [1, 3] is used. But, to the best of the our knowledge, there has not been any explicit discussion of how one assesses whether there are difference in impact among a set of exogenous constructs to an endogenous construct in a PLS model. Specifically, the question posed is how does one test the relative impact among a set of path estimates to a particular construct? Conversely, how does one demonstrate the impact equality or invariance of a set of predictor constructs to a dependent construct? In a literature review of top journals in Information Systems, we show that discussion based on the point estimates is the primary approach for claiming relative impact or importance of exogenous constructs. Yet, no formal statistical tests are provided. In this paper, we introduce a bootstrapping approach that is easily employed with existing PLS software packages. We also discuss the parametric approach that is part of regression based analysis. In general, we advocate the use of the PLS bootstrapping procedure since it will likely perform better relative to Type I and Type II errors under varying conditions of non-normality.

2 MIS Literature

We conducted a review of articles that performed structural path modeling in three top rated journals that the MIS discipline often submit to (i.e., MIS Quarterly, Information Systems Research, and Management Science) for the period of 2000–2011. The results included 39 articles that involve path analysis using primarily either LISREL or PLS. We then examined what the authors stated or inferred in their discussion and/or conclusion sections. Table 1 presents a sampling of our findings. In nearly all cases, the authors focused on the point estimates with discussion concerning which structural path was larger or had more impact relative to others. In the few cases in which a statistical test was applied to support such statements, the method was a parametric based t-test. Therefore, we continue with a discussion of the procedure for conducting such t-tests and follow up with our suggested bootstrap approach.

3 Approach 1: Parametric Based t-Test

Testing the difference of two independent variables in the same sample with normality assumptions requires estimating the standard error of the differences between betas (i.e., $\beta_i - \beta_j$). The standard error is estimated as follows:

$$SE_{\beta_i - \beta_j} = \sqrt{\frac{1-R_y^2}{n-k-1}(r^{ii} + r^{jj} + 2r^{ij})}$$

Table 1: Sample of articles discussing structural path differences

Information Systems Research, 2002, Vol. 13, No. 3, Kim, Lee, Han & Lee, "Businesses as Buildings: Metrics for the Architectural Quality of Internet Businesses," using LISREL, p. 250
The path coefficient from firmness to customer satisfaction was larger than that from delight to customer satisfaction in the stock brokerage domain, whereas delight had a stronger link to satisfaction than firmness in the other three domains
MIS Quarterly, 2001, Vol. 25, Bhattacherjee, "Understanding Information Systems Continuance: An Expectation Confirmation Model," using PLS, p. 364
Perceived usefulness was a stronger predictor of acceptance intention in TAM than attitude (Davis et al. 1989; Taylor and Todd 1995), while satisfaction was the stronger predictor of continuance intention in this study than perceived usefulness
MIS Quarterly, 2008, Vol. 32, Au, Ngai & Cheng, "Extending the Understanding of End User Information Systems Satisfaction Formation: An Equitable Needs Fulfillment Model Approach," using PLS, p. 53
The results of the study indicate that perceived IS performance is the most significant determining factor of EUS, with a standardized coefficient of 0.45 (H1). This is consistent with previous research findings (Suh et al. 1994; Swan and Trawick 1980) and implies that product performance as perceived by end users is still the core determinant of satisfaction. Nevertheless, equitable work performance fulfillment and equitable relatedness fulfillment do play a significant role in directly affecting satisfaction (H3 and H4), with standardized coefficients of 0.19 and 0.17, respectively. Hence there is evidence to suggest that both constructs have a more or less equal impact in affecting users' levels of satisfaction
MIS Quarterly, 2010, Vol. 34, Johnston & Warkentin, "Fear Appeals and Information Security Behaviors: An Empirical Study," using PLS, p. 560
Interestingly, while both response efficacy and self-efficacy appear to have strong predictive ability, social influence has slightly more of an effect on behavioral intent
Information Systems Research, 2006, Vol. 17, Pavlou & Dimoka, "The Nature and Role of Feedback Text Comments in Online Marketplaces: Implications for Trust Building, Price Premiums, and Seller Differentiation," using PLS, p. 405
These findings validate H1 and H2 and confirm the economic value of benevolence and credibility. Interestingly, benevolence ($b = 0.41$) had a stronger impact on price premiums compared to credibility ($b = 0.30$), ($t = 11.30$, $p < 0.001$)
Information Systems Research, 2001, Vol. 12, Plouffe, Hulland & Vandenbosch, "Research Report: Richness Versus Parsimony in Modeling Technology Adoption Decisions-Understanding Merchant Adoption of a Smart Card-Based Payment System," using PLS, p. 214
The model does a good job of explaining variance in both perceived usefulness ($R^2 = 0.282$) and intention to adopt ($R^2 = 0.327$). This explanatory power is based more on the effect of perceived usefulness (as demonstrated by the values in the final column of Table 4) than on perceived ease-of-use, a result that is consistent with earlier work
Information Systems Research, 2007, Vol. 18, Jiang & Benbasat, "Investigating the Influence of the Functional Mechanisms of Online Product Presentations," using PLS, p. 463
In terms of their relative power, the comparison of path coefficients shows that vividness exerts a stronger influence than interactivity on perceived diagnosticity (path coefficients: 0.40 versus 0.21), compatibility (path coefficients: 0.37 versus 0.14), and shopping enjoyment (path coefficients: 0.49 versus 0.26)

with $r^{ii} = \frac{1}{1-R_i^2}$ where R_i^2 is the squared multiple correlation of the $k-1$ remaining independent variables with independent variable X_i (often called the variance inflation factor or VIF) and $r^{ij} = \frac{-\beta_{ij}}{1-R_i^2} = \frac{-\beta_{ji}}{1-R_j^2}$ where $\beta_{ij}$is the standardized partial regression coefficient of X_i on X_j with the other independent variables having been partialed. Alternatively $r^{ij} = \frac{-pr_{ij}}{\sqrt{(1-R_i^2)(1-R_j^2)}}$ where pr_{ij} is the correlation between X_i and X_j with all other independent variables having been partialed out from each. The denominator is equivalent to the square root of the product of the Tolerances reported in statistical packages such as SPSS.

Hence, we can perform a t-test of significance of the differences in betas as follows:

$$t = \frac{\beta_i - \beta_j}{SE_{\beta_i - \beta_j}}.$$

4 Approach 2: Nonparametric Bootstrapping of Path Differences

A nonparametric alternative which can easily be accomplished with existing PLS software is to run N bootstrap samples of the model in question. The difference score between the paths being considered are then calculated for each bootstrap sample. Once this is calculated, a percentile bootstrap p-value can be estimated. As an example, if 1,000 bootstrap analyses were conducted, the number of path difference estimates that yielded a zero or negative difference would be an estimate of the p-value. This represents a distribution free approach for estimating path difference of independent variable effects on the same dependent variable for a single sample. Alternatively, one can go further and estimate both an acceleration and bias correction [4].

5 MIS Example: Nonparametric Bootstrapping of Path Differences Versus Parametric t-Test

As an example research, we present a model that focuses on technology acceptance using Self-Determination Theory (SDT) [5, 6]. The proposed model includes three extrinsic motivations (external, introjected, and identified regulation) and three antecedents to the extrinsic motivations (interface quality, decision support quality, and perceived network externality). In this model, behavioral intention to use an e-marketplace application is the dependent variable. This study reveals how different types of extrinsic motivation affect the behavioral intention, which may allow researchers to take a closer look at the user's cognitive process in determining behavioral intention. Secondly, by including the antecedents to the extrinsic motivators in terms of information artifacts, this study provides a tool for investigating the determinants for the motivators in the context of a networked application adoption (Fig. 1).

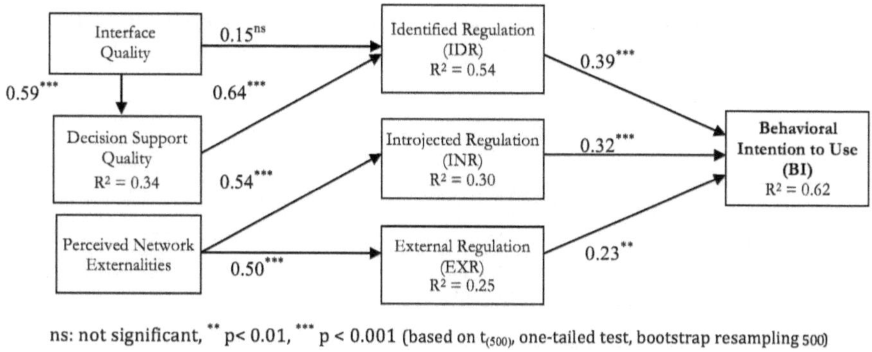

Fig. 1: PLS results of structural model assessment

The data collection was conducted using three lists of e-marketplace application users. A total of 888 subjects out of the 1,238 on the census list from a major U.S. e-marketplace company were identified as a possible sample pool; 350 subjects were used earlier for the pilot test. A total of 207 survey questionnaires were returned due to a wrong address, employees leaving the company, and a company's policy of not answering survey questionnaires. A total of 97 responses were received within a month after the questionnaire was distributed by mail. Two other lists were provided by a chemical company and a semiconductor manufacturer in Korea that run e-marketplace applications. From the lists, 300 active users (200 from the chemical company list, 100 from the semiconductor manufacturer list) were selected. The distribution of the survey questionnaire was conducted via email by an employee of each company. A total of 49 (chemical company) and 46 (semiconductor manufacturer) responses were collected . To summarize, a total of 981 final survey questionnaires were distributed. Of this number, a total of 192 surveys were returned (response rate 19.6%). Out of 192 returned survey questionnaires, 176 were usable which therefore led to a 17.9% usable response rate. In general, the three extrinsic motivations (i.e., external, introjected, and identified) are hypothesized to directly influence behavioral intention to use an e-marketplace application. Part of the study also considers the relative importance of the three extrinsic motivations, which can be assumed based on SDT [5]. SDT posits that the more self-determined or internalized a behavior is, the more persistent a nd effective the behavior is. It is conjectured that the influence of identified regulation on behavioral intention to use an application is the most significant, followed by introjected regulation, and external regulation.

H_1: Among the three extrinsic variables affecting behavioral intention to use an e-marketplace application, identified regulation has the strongest effect on behavioral intention, followed by introjected regulation, and external regulation.

For the parametric approach, we export the PLS construct scores into SPSS for a regression analysis. The unstandardized solution within rounding error matches the PLS estimates (see Table 2).

Partial Correlation Estimates are next obtained from SPSS as shown in Tables 3 and 4 using IDR and INR as the dependent variables respectively.

Table 2: Regression results using SPSS

		Unstandardized Coefficients		Standardized Coefficients			Correlations			Collinearity Statistics	
Model		B	Std. Error	Beta	t	Sig.	Zero-order	Partial	Part	Tolerance	VIF
1	(Constant)	-.002	.047		-.035	.972					
	IDR	.326	.064	.326	5.085	.000	.799	.361	.191	.343	2.917
	EXR	.244	.053	.244	4.604	.000	.716	.330	.173	.504	1.986
	INR	.401	.059	.401	6.797	.000	.799	.459	.255	.405	2.466

Dependent Variable: BI

Table 3: Partial correlation estimates obtained from SPSS with IDR as dependent variable

		Unstandardized Coefficients		Standardized Coefficients			Correlations			Collinearity Statistics	
Model		B	Std. Error	Beta	t	Sig.	Zero-order	Partial	Part	Tolerance	VIF
1	(Constant)	-.010	.060		-.173	.863					
	EXR	.176	.067	.176	2.639	.009	.617	.196	.127	.524	1.910
	INR	.639	.067	.639	9.580	.000	.760	.588	.462	.524	1.910

Dependent Variable: IDR

Table 4: Partial correlation estimates obtained from SPSS with INR as dependent variable

		Unstandardized Coefficients		Standardize Coefficients			Correlations			Collinearity Statistics	
Model		B	Std. Error	Beta	t	Sig.	Zero-order	Partial	Part	Tolerance	VIF
1	(Constant)	-.006	.055		-.105	.91					
	EXR	.357	.056	.357	6.324	.00	.690	.432	.28	.619	1.615
	IDR	.540	.056	.540	9.580	.00	.760	.588	.42	.619	1.615

Dependent Variable: INR

Using PLS path estimates, but standard error (SE) from the SPSS output, we are able to calculate the *t*-value and *p*-value for the differences between the path estimates (see Table 5).

Table 5: Hypothesis 1 test results—parametric *t*-tests

Compared construct	Difference in PLS path estimates	SE of the differences	T value	Significance
IDR vs. EXR	0.392−0.234 = 0.158	0.0713	t =2.21	$p < 0.013$
IDR vs. INR	0.392−0.320 = 0.072	0.0562	t =1.28	$p < 0.101$
INR vs. EXR	0.320−0.234 = 0.086	0.0632	t =1.36	$p < 0.087$

In contrast, Table 6 provides the results of running a bootstrap percentile analysis with 1,000 bootstrap samples.

Table 6: Hypothesis 1 test results—Bootstrap p values

Compared construct	Difference in PLS path estimates	Bootstrap percentile p value
IDR vs. EXR	0.392−0.234 = 0.158	0.080
IDR vs. INR	0.392−0.320 = 0.072	0.311
INR vs. EXR	0.320−0.234 = 0.086	0.220

In general, if one were to use a $p < 0.10$ criterion, both methods converge on the same conclusion with identified regulation (IDR) having a stronger impact than external regulation (EXR). But the results for the other two differences diverged. For the parametric test (Table 5) introjected regulation to external regulation (INR vs. EXR, $p = 0.087$) was significant at $p \leq 0.10$, while identified regulation to introjected regulation (IDR vs. INR, $p = 0.101$) might also be considered significant at the 0.10 level. In the case of bootstrapping (Table 6) the difference hypothesis (H_1) regarding identified regulation to introjected regulation (IDR vs. INR) and, introjected regulation to external regulation (INR vs. EXR) is clearly not supported. The difference between the parametric and bootstrapping results may be due to departures in normality of the items resulting in each construct score having approximately a left skewness of −3 and leptokurtosis of 21. Our recommendation under these conditions is to defer to the bootstrapping results.

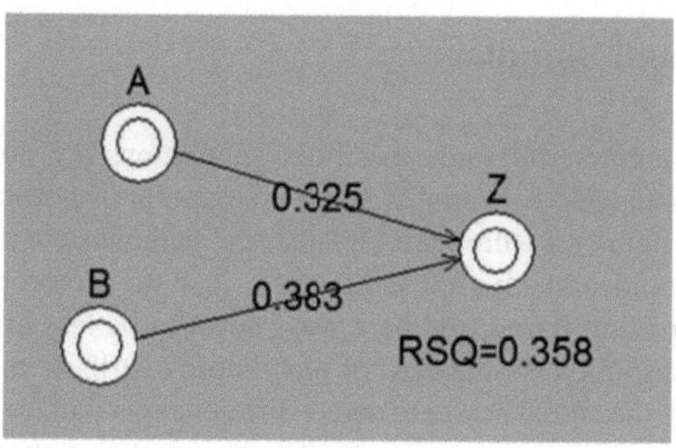

Fig. 2: PLS estimates from simulated data where population structural paths are both set at 0.50

To further support this point, we next provide a simple simulation study (see Fig. 2) where the two constructs A and B are modeled as impacting construct Z. The population paths going into Z from both A and B are specified as equivalent at 0.50. In addition, constructs A and B are correlated at 0.25 to represent the likelihood that they share common antecedent factors. All constructs are modeled with 6 item indicators where 3 indicators are set with standardized loadings of 0.60 and 3 at 0.80. The population skewness and kurtosis given to each construct was set at a less extreme level than that of our empirical sample data (see Table 7).

Table 7: Summary skewness and kurtosis statistics for simulated constructs

Constructs	Skewness	Kurtosis	Z-score skewness	P-value	Z-score kurtosis	P-value
Antecedent factor	2.33	16.206	59.113	0.000	330.884	0.000
Construct A	1.464	5.388	44.941	0.000	108.984	0.000
Construct B	1.422	5.443	44.110	0.000	111.133	0.000
Construct Z	0.976	2.382	33.786	0.000	48.642	0.000

Table 8: Partial correlation estimates obtained from SPSS with simulated data

Model		Unstandardized Coefficients		Standardized Coefficients			Correlations			Collinearity Statistics	
		B	Std. Error	Beta	t	Sig.	Zero-order	Partial	Part	Tolerance	VIF
1	(Constant)	-.010	.060		-.173	.863					
	EXR	.176	.067	.176	2.639	.009	.617	.196	.127	.524	1.910
	INR	.639	.067	.639	9.580	.000	.760	.588	.462	.524	1.910

Dependent Variable: IDR

Applying the parametric approach using the SPSS output from Table 8, we would conclude there is a significant difference in the two structural paths with a value of $t = 9.115$. In contrast, when we apply bootstrapping with 1,000 resamples, we end up with the opposite conclusion (i.e., that the two paths are not significantly different due to a p-value of 0.278).

6 Discussion and Summary

The results of the empirical study and brief simulations are not necessarily that surprising to statisticians. Essentially, if the assumption of normal distribution for the constructs or items measured are violated, the parametric t-test for differences among paths may indeed be biased and possibly lead to incorrect conclusions. The bootstrapping approach introduced here was shown to be less affected in the case of leptokurtic distribution. In the case of platykurtic distribution, the opposite will likely occur where the actual power to detect differences in paths will likely be lower using a parametric t-test as opposed to bootstrapping. We've shown earlier that researchers in the Information Systems field rarely ever conduct statistical tests when they discuss the relative impact of exogenous factors. The rare instance in which actual statistical tests are conducted, we found only parametric t-test employed. We suspect this is not necessarily unique to this discipline and therefore hope applied researchers would consider using the alternative approach presented here.

References

[1] W. W. Chin, and J. Dibbern, "A Permutation based procedure for multi-group PLS Analysis: Results of tests of differences on simulated data and a cross cultural analysis of the sourcing of information system services between germany and the USA," In V.E. Vinzi, W.W. Chin, J. Henseler and H. Wang (Eds.), *Handbook of Partial Least Squares Concepts, Methods and Applications*, pp.171–193, 2010.

[2] W. W. Chin, "How to write up and report PLS analyses," In V.E. Vinzi, W.W. Chin, J. Henseler, and H. Wang (Eds.), *Handbook of Partial Least Squares Concepts, Methods and Applications*, pp.650–690, 2010.
[3] W.W. Chin, "A permutation procedure for multi-group comparison of PLS models." Invited presentation. In M. Valares, M. Tenenhaus, P. Coelho, V. Vinzi, and A. Morineau (Eds), PLS and Related Methods, Proceedings of the PLS2003 International Symposium — "Focus on Customers," Lisbon, September 15th to 17th, pp. 33–43, 2003.
[4] B. Efron, and R. J. Tibshirani, *An introduction to the Bootstrap* New York: Chapman & Hall.
[5] E. L. Deci, and R. M. Ryan, *Intrinsic Motivation and Self-Determination in Human Behavior.* New York, NY: Plenum, 1985.
[6] R. M. Ryan, and E. L. Deci, "Self-determination theory and the facilitation of intrinsic motivation, social development, and well-being," *The American Psychologist* **55**, pp. 68–78, 2000.

Controlling for Common Method Variance in PLS Analysis: The Measured Latent Marker Variable Approach

Wynne W. Chin, Jason B. Thatcher, Ryan T. Wright, and Doug Steel

Abstract Common method variance (CMV) continues to be an important issue for social scientists. To date, methodologists have yet to agree upon a best practice for detecting and controlling for CMV. In a recent paper, the unmeasured latent marker variable approach, a frequently employed technique, was shown to be incapable of detecting or controlling CMV in PLS analyses. Unfortunately, this was the only method to date suggested for handling CMV in PLS models. To fill this gap, we introduce a measured latent marker variable (MLMV) approach and demonstrate how it is able to both detect and correct for CMV when using Partial Least Squares.

Key words: Common method variance (CMV), Unmeasured latent marker variable, Measured latent marker variable (MLMV)

W.W. Chin (✉)
Department of Decision and Information Systems, C.T. Bauer College of Business, University of Houston, Houston, TX 77204-6021, USA
e-mail: wchin@uh.edu

J.B. Thatcher
College of Business and Behavioral Science, Clemson University, Clemson, SC, USA
e-mail: jthatch@clemson.edu

R.T. Wright
School of Management, University of Massachusetts Amherst, Amherst, MA, USA
e-mail: rtwright@admin.umass.edu

D. Steel
School of Business, University of Houston-Clear Lake, Houston, TX, USA
e-mail: doug@stoglo.com

H. Abdi et al. (eds.), *New Perspectives in Partial Least Squares and Related Methods*, Springer Proceedings in Mathematics & Statistics 56, DOI 10.1007/978-1-4614-8283-3_16,

1 MLMV Approach to the Common Method

Our new MLMV approach has the potential to tease out CMVs influence on structural paths. To do so, the process requires collecting multiple unrelated measures at the same time as collecting data related to the primary research model. This contrasts with the unmeasured latent marker variable approach [1–6], which uses indicators of manifest variables from the primary research model to estimate CMVs influence. Thus, a critical aspect of the MLMV approach is to select a set of measures that reflect underlying constructs that have no nomological relationship with the particular study in question while using the same survey format and scale to reflect the common method effects. These measures in turn are modeled as capturing an underlying CMV and labeled as an MLMV. As many MLMVs are created (using the same set of measures) as there are dependent constructs in the research model where each MLMV is deployed to control for common method variance effects on each dependent construct in the research model.

To evaluate our MLMV approaches potential, we present the results of simulations that vary the form, and level, of CMV. Specifically, we demonstrate that our method accurately detects, and controls for, CMV present in structural equation models estimated with Partial Least Squares. Our research contributes to the literature on CMV by providing initial evidence of a MLMV approach that detects and controls for different levels, and forms of, method variance.

2 Guidelines to MLMV

In order to use our MLMV approach, researchers must carefully select MLMV indicators to include in their initial data collection. Ideally, researchers should consider the following guidelines when collecting data:

1. Each indicator must not be in the same domain as constructs found in the research model.
2. Each indicator must be drawn from different unit of analysis than that investigated in the research model.
3. Rather than focusing on the reliability of each indicator towards measuring their respective construct, it is more critical to ensure all unique and error variances are independent among the set of measures chosen.
4. The MLMV must include a minimum of four items. As we illustrate in our simulations, a latent marker approach is robust to using varying numbers of indicators. Although ideally, one would use 12 items to estimate an MLMV, we demonstrate that one can detect and reduce CMV by more than 70% using as few as 4 items.
5. Because the MLMV is not the primary purpose of the study, a well-designed survey should include these indicators at the end of the instrument. This would minimize the effects of respondent fatigue on the pattern of responses relevant

to the main study. Even though the measures appear at the end of the survey, their utility for evaluating CMV should not be affected.

3 Two Approaches for Applying the MLMV Items

Once a reasonable set of MLMV items are collected, two methods for minimizing CMV effect may be considered. The first, construct level correction (CLC), involves creating as many CMV control constructs as there are constructs in the model. For example, if the theoretical model consists of three constructs, you would create three CMV control constructs. Each CMV control uses the same entire set of MLMV items (mode B). CMV construct is modeled as impacting each model construct. The residuals obtained now represent the model constructs with the CMV effects removed. The second approach, item level correction (ILC), involves using the MLMV items to partial out the CMV effects at the measurement item level. Each item measure is regressed on the entire set of MLMV items. The residuals for each item now represent the construct items with the CMV effects removed. But the CMV should be replaced with an equivalent amount of random error to be equivalent to the variance of the original measures sans bias. This is necessary to obtain assessment of the reliability of the original items in capturing the underlying construct of interest. To do this, the R-square obtained from each item to MLMV regression is used. Specifically, the square root of the R-square multiplied with a number drawn from a normal distribution of mean 0 and standard deviation of 1 is added to each item residual. This represents the final ILC items used in a PLS analyses. While this second approach is more tedious, it allows for estimates of item loadings. But both approaches is meant to provide more accurate estimates of the structural paths relative to using items with CMV.

To demonstrate this, we present the results of a simulation using the same settings as Chin et al. [3] with the common method bias was set at 0.36 for each item measure and all trait loadings at 0.70. Consistent with Chin et al. results, we obtained a biased estimate of 0.741 for the structural path between two model constructs when the population parameter is 0.60 (see Fig. 1). Twelve measures that reflect the underlying method bias (i.e., MLMV) were also simulated. Concretely, this would represent questions that are unrelated to the theoretical domain being researched as well as to each other. We opted to further increase the sample size to 10,000 cases in comparison to the Chin et al. study use of 5,000 cases to guarantee further statistical stability and eliminate concerns of estimates inaccuracies due to sample size.

Now, if we include the 12 MLMV indicators as control by creating a CLC construct for the two model constructs, we see that prior inflated path of 0.741 now more closely matches the population parameter with an estimate of 0.606 (See Fig. 2). This represents the impact of construct XX on construct YY holding CLC constant. The next step is to use each CLC scores to partial out the CMV from both constructs to obtain the partial correlation between XX and YY. Table 1 shows the results where the number of MLMV items varied from 1 to 12. While a 12-item CLC effectively captured the simulated CMV, our simulation illustrates that one can use a 4 item LMV to remove 72% variance due to CMV. Given that researchers tend to have limited space on survey instruments to include additional items, these results illustrate that our MLMV approach is flexible enough to be included in surveys of varying lengths.

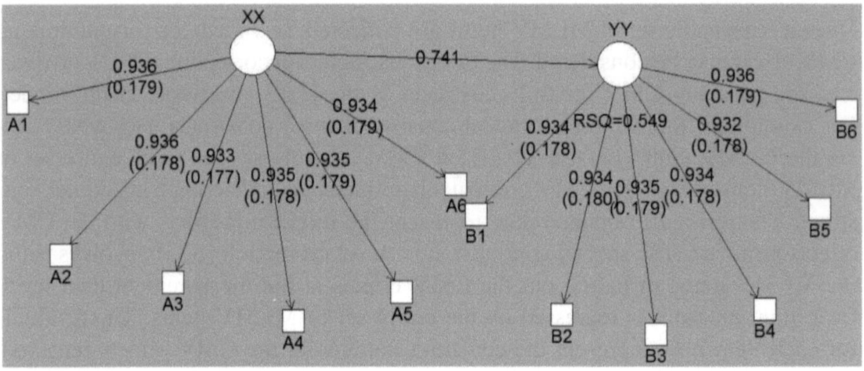

Fig. 1: PLS estimates using items CMV of 0.36

Figure 3 represents the results of using the ILC correction. In contrast to the CLC approach, the item loadings are more consistent with a PLS analyses without CMV effects. As in the Chin et al. paper [3], the population loadings are all set at 0.70. The estimated loadings varying from 0.76 to 0.789 are consistent with the tendency of PLS to overestimate the loadings by approximately 10%. Likewise, the estimated structural path of 0.552 is consistent with an approximate 10% underestimation of the population parameter of 0.60.

Figure 4 represents the results without the inclusion of noise to replace the equivalent amount of CMV removed from each item. Accordingly, the loadings are inflated to the 0.90. In turn, the path estimate of 0.602 is similar to that obtained via the CLC approach. While not presented here, it is assumed that if we replaced additional noise to match the amount of CMV removed at the construct level via the CLC approach, the resulting path estimate would be similar to Fig. 3.

Overall, both approaches seem to converge to the same results. In the case of using CLC, we obtain an accurate estimate of the path estimate at the expense of the loadings. Nonetheless, the more accurate loadings can be obtained by correlating each construct residual after partialling out the CMV with the original item measures. For the ILC, we obtain more accurate item loading estimates at the expense of the structural path. But, the path estimates can be obtained if we do not compensate for the CMV partialed out for each item.

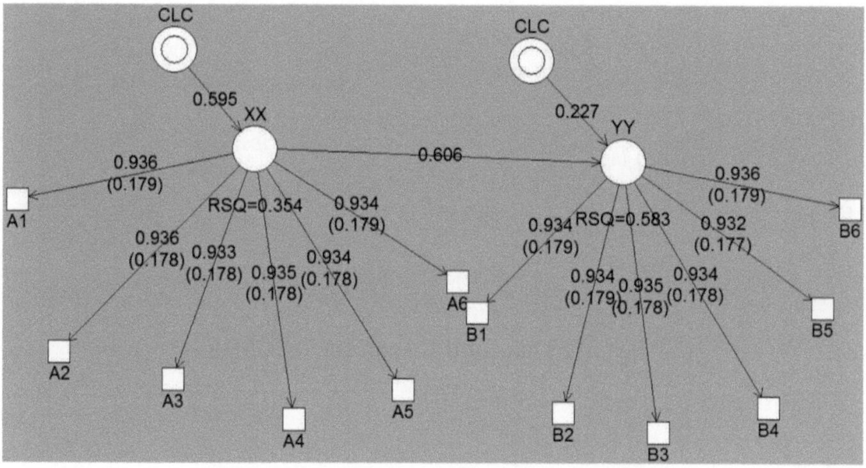

Fig. 2: First step in CLC approach for CMV

Table 1: Number of latent marker measures and percentage reduction in CMV using CLC

# of MLMV items in CLC	12	11	10	9	8	7	6	5	4	3	2	1
Structural estimate	0.600	0.604	0.608	0.613	0.616	0.619	0.624	0.630	0.639	0.652	0.670	0.696
Percent reduction (%)	100	97	94	91	89	87	83	79	72	63	50	32

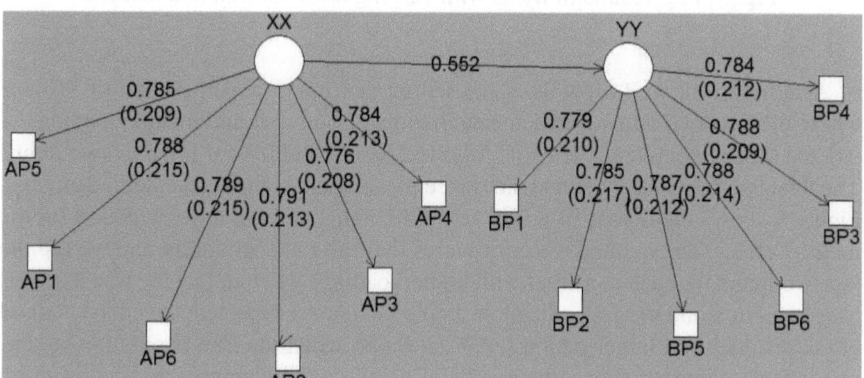

Fig. 3: Results of ILC approach for CMV with additional error added to compensate for amount of CMV removed

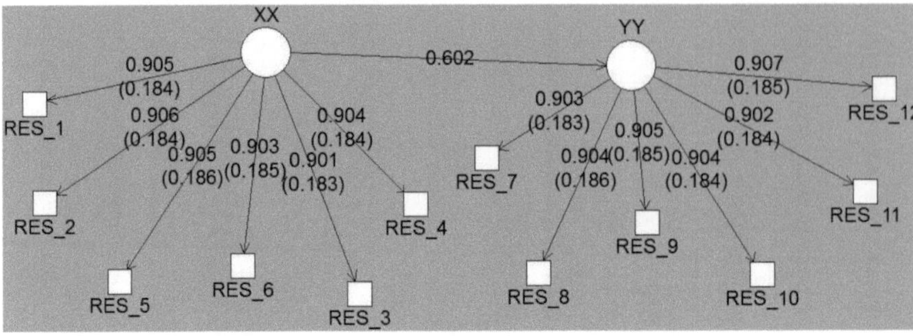

Fig. 4: Results of ILC approach for CMV

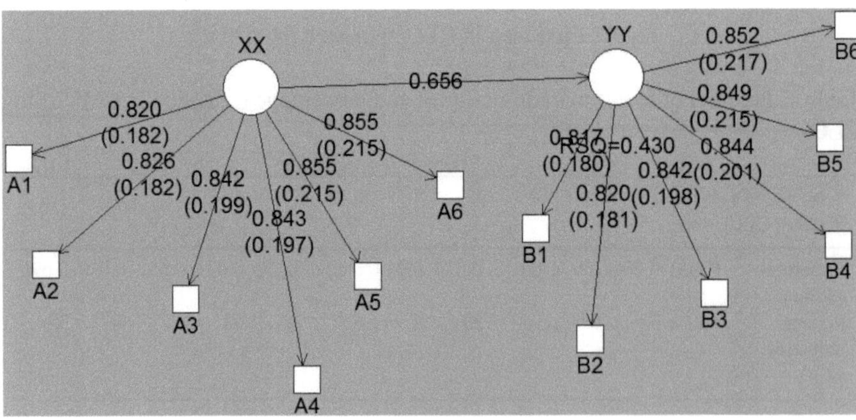

Fig. 5: PLS results of items with varying levels of trait and method

As a final test, we follow the same setting as Chin et al. [3] scenario 7. This is where the first two items for each construct had true score and method loadings of 0.8 and 0.2, respectively. This was followed by 0.7 and 0.4 for the next two items. The final two were set at 0.6 and 0.6 (i.e., equal amounts of true and method effects). Figure 5 shows the results of a PLS using the items without any correction for the method effect. The method effect obscures the trait only reliability and we see the last two items for each construct with higher loadings when in fact the true loadings based on our simulation specification should be 0.6. As expected, the added method effect resulted in a larger path estimate of 0.656 when the true population parameter is 0.60. Figure 6 shows the results of using the CLC approach. The resulting path estimate was a more realistic number 0.568. Likewise, Fig. 7 shows the ILC approach without error compensation yielded a similar 0.566 path estimate. With error compensation, Fig. 8 shows the ILC method provides more accurate estimates of the trait loading. In contrast to Fig. 5, we now see that the first two items are more

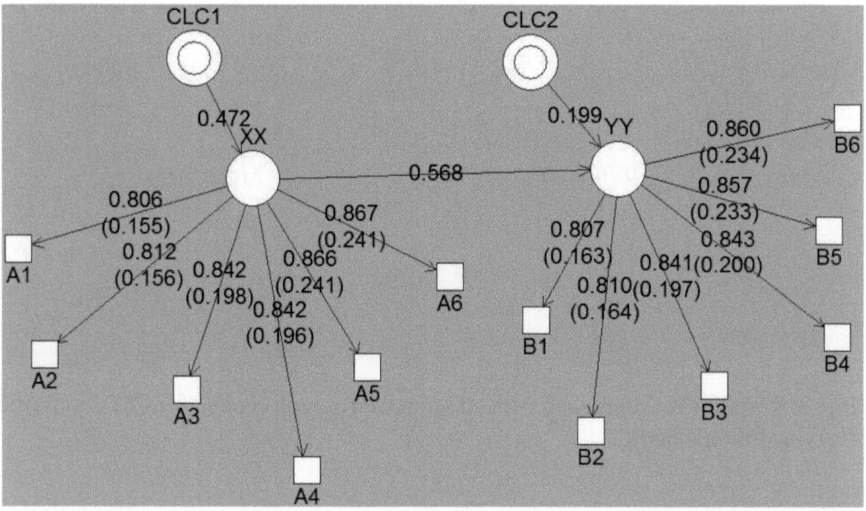

Fig. 6: Result of CLC approach for items with different trait method impact

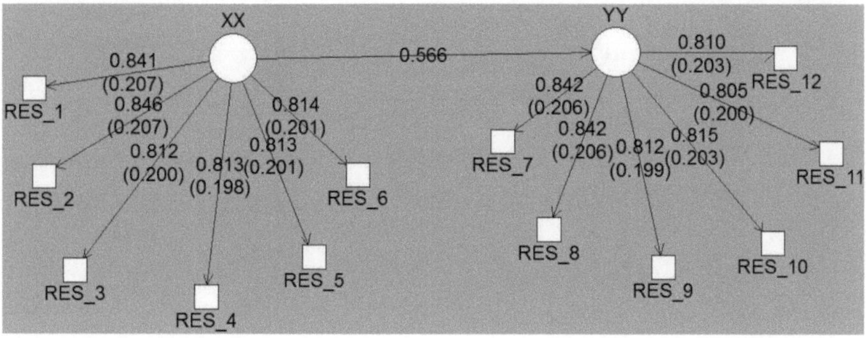

Fig. 7: Result of ILC approach with (varying trait and method): no error compensation

reliable, followed by the next two, and finishing with the last two for each construct. On average, each estimated loading is about 10% higher than the true loading consistent with the PLS algorithm. Figure 9 provides the PLS results using simulated data where the underlying model has no method effects. Instead, only the item loadings vary with 0.8, 0.7, and 0.6 set for each set of two items per construct as in the trait-method model used for Figs. 5 through 8. The 10% inflation of loadings inherent using the PLS algorithm is apparent and the path estimate is conversely underestimated.

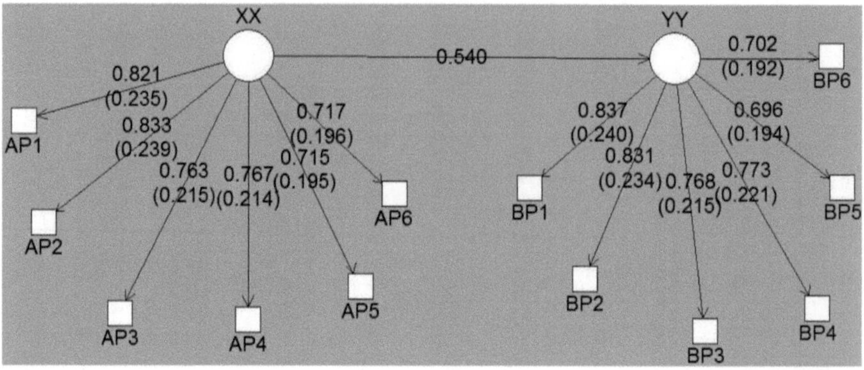

Fig. 8: Result of ILC approach with additional error replacement for CMV removed (varying trait method)

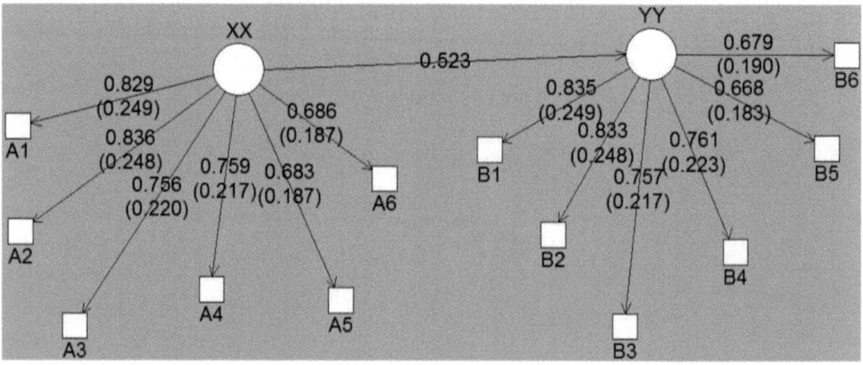

Fig. 9: PS estimate for congeneric trait only model (no CMV added)

4 Conclusion

Overall, this paper provides two new approaches for partialling out CMV in the context of PLS analyses. The CLC approach primarily corrects for structural path estimates while the ILC can correct for both structural path and item loading estimates depending on whether you replace the CMV removed with equivalent amount of random error. Nevertheless, the utility of MLMV indicators should be considered on a study by study basis. Because the appropriate MLMV should be tailored to the specific research question and sample frame, we do not suggest a universal set of items. Rather, we trust that researchers have the capacity to make reasonable judgments based on their understanding of their specific domain of inquiry following the guidelines we have set forth.

References

[1] R. P. Bagozzi, "Measurement and meaning in information systems and organizational research: Methodological and philosophical foundations," *MIS Quarterly*, **35**, 261–292, (2011)

[2] A. Burton-Jones, "Minimizing method bias through programmatic research," *MIS Quarterly*, **33**, 445–471, (2009).

[3] W. Chin, J. B. Thatcher, and R. T. Wright, "Assessing common method variance: Assessing the UMLC approach," *MIS Quarterly*, **36**, 1003–1019, (2012).

[4] P. M. Podsakoff, S. B. MacKenzie, J.-Y. Lee, and N.P. Podsakoff, "Common method biases in behavioral research: A critical review of the literature and recommended remedie," *Journal of Applied Psychology*, **88**, 879–903, (2003).

[5] R. Sharma, P. Yetton, and J. Crawford, "Estimating the effect of common method variance: The method pair technique with an illustration from TAM research," *MIS Quarterly*, **33**, 491–512, (2009).

[6] L. J. Williams, N. Hartman, and F. Cavazotte, "Method variance and marker variables: A review and comprehensive CFA marker technique," *Organizational Research Methods*, **13**, 477–514, (2010).

Part V
Comparing Groups and Uncovering Segments

Multi-group PLS Regression: Application to Epidemiology

Aida Eslami, El Mostafa Qannari, Achim Kohler, and Stéphanie Bougeard

Abstract For the investigation of the relationships between two datasets where the individuals are divided into groups a simple procedure called multi-group PLS regression is discussed. It is a straightforward extension of PLS regression to take account of the group structure. It can also be seen as an extension of multi-group principal components analysis to the case of two blocks of data. The proposed method of analysis is illustrated on the basis of a real case study pertaining to the field of veterinary epidemiology.

Key words: Partial least squares, Multi-group partial least squares, NIPALS algorithm, Multi-group principal components analysis, Epidemiology

1 Introduction

In various domains of application, data are often organized in two blocks of variables consisting of an explanatory dataset X and a dependent dataset Y. Moreover, since the measuring techniques have tremendously evolved with the times, practitioners are more than ever presented with large datasets with highly collinear

A. Eslami (✉) • S. Bougeard
Department of Epidemiology, French Agency for Food, Environmental and Occupational Health Safety (Anses), Ploufragan, France
e-mail: aida.eslami@anses.fr; stephanie.bougeard@anses.fr

E.M. Qannari
Sensometrics and Chemometrics Laboratory, ONIRIS, LUNAM University, Nantes, F-44307, France INRA
e-mail: elmostafa.qannari@oniris-nantes.fr

A. Kohler
Centre for Integrative Genetics, Norwegian University of Life Sciences, Ås, Norway
e-mail: achim.kohler@umb.no

H. Abdi et al. (eds.), *New Perspectives in Partial Least Squares and Related Methods*, Springer Proceedings in Mathematics & Statistics 56, DOI 10.1007/978-1-4614-8283-3_17,

variables. For the aim of relating dataset Y and dataset X, PLS regression [17] is nowadays very popular since it has proved to be very efficient both in terms of exploring the relationships between the two datasets and accurately predicting Y from X.

We consider the setting where, in addition to having two datasets X and Y, we know that there is a group structure among the individuals. There are numerous situations which correspond to this setting. In the field of veterinary epidemiology, the aim is often to identify risk factors (dataset X) that lead to various expressions of disease (dataset Y), all the measurements being made on animals which are grouped into farms. In environmental studies, one can be interested in explaining the abundance of species by other environmental measurements and, for this purpose, data are collected on statistical units (e.g., sites) that are organized into groups (e.g., regions).

Depending on the application field, different methods are used to analyze this kind of data. In epidemiology, the statistical procedures usually performed pertain to Generalized Linear Models (GLM). The group structure among individuals is taken into account by including repeated or random effects in Generalized Estimating Equations (GEE) [3]. These models have appealing features that justify their wide use but all the potential explanatory variables can not be included in the model because they are plagued by quasi-collinearity. It is well-known that in these circumstances, the relevance and the stability of the results obtained from GEE are impaired [2]. The multivariate analysis of covariance (MANCOVA) [5] can be viewed as an extension of GLM to the multivariate framework. However, this method, in addition to being very sensitive to the presence of quasi-collinearity among the predictive variables, is based on restrictive assumptions which are rarely fulfilled in practice. In the general framework of PLS Path Modeling, methods of analysis that take account of the presence of a group structure among individuals were proposed [1, 15]. By contrast, our strategy of analysis fits within the framework of PLS regression and is simple and straightforward.

In order to investigate the links between X and Y a first strategy of analysis is to ignore the group structure and perform PLS regression of Y upon X. However, by ignoring the group structure, it follows that the total variance recovered by the latent variables mixes up both the between and the within-group variances. A second strategy of analysis consists in applying as many PLS regressions as there are groups in the data (i.e., PLS regression applied on data from each group). Clearly, this strategy of analysis yields a large number of parameters which is likely to lead to an instability problem of the solution because of a lack of degrees of freedom to estimate all the parameters. Moreover, this strategy entails a difficulty in interpreting the outcomes and comparing the results across the groups. In order to counteract these problems, we propose to carry out a compromise PLS regression. We shall refer to it as multi-group PLS regression (MGPLS). This consists in performing PLS regression on the data from the various groups, but we impose that the vector of loadings associated to both X and Y variables are the same across the groups. As a matter of fact, MGPLS can be seen as an extension of multi-group Principal Component Analysis (MGPCA) [8]. This latter method of analysis was defined as a way to perform PCA on the data from the various groups but the vectors of loadings are assumed to be identical from one group to another. MGPLS is introduced on the basis of a maximization problem whose solution leads to an eigen-analysis problem. We also propose a NIPALS algorithm to solve this problem. Indeed, there is a great benefit in designing a NIPALS algorithm since it is much faster than the solution based on an eigen-analysis particularly when dealing with high dimensional data [16]. Moreover, it can be easily adapted to take account of missing data [14]. The method of analysis is illustrated on the basis of a real case study pertaining to veterinary epidemiology.

2 Method

2.1 Data and Notations

The datasets X and Y respectively consist in the measurement of P and Q quantitative variables on the same N individuals. Moreover, these datasets are partitioned into M groups known a priori. Let Y_m and X_m be the datasets associated with group m for $m = (1,\ldots,M)$. Each group refers to n_m individuals ($\sum_{m=1}^{M} n_m = N$). The rank of the dataset X and the maximum dimension of analysis is denoted by H. As stated in the introduction, we aim at investigating the relationships between datasets Y_m and X_m ($m = 1,\ldots,M$). For this purpose, we seek, step by step, latent variables (or components) in X_m and Y_m which are highly related. Moreover, we impose that, for each step, the loadings associated with these latent variables are identical across the groups. Let $a^{(h)}$ be the common vector of loadings associated with the dataset X and $b^{(h)}$ the one associated with the dataset Y for dimension $h = (1,\ldots,H)$. We define group components by $t_m^{(h)} = X_m a^{(h)}$ and $u_m^{(h)} = Y_m b^{(h)}$ related to group m for the hth order solution. The global components are defined by $t^{(h)} = Xa^{(h)}$ and $u^{(h)} = Yb^{(h)}$. They are directly derived from the vertical concatenation of the group components. The graphical display in Fig. 1 depicts all these elements.

2.2 First Order Solution

In a first step, we seek group components $t_m^{(1)} = X_m a^{(1)}$ and $u_m^{(1)} = Y_m b^{(1)}$ associated with the same vectors of loadings, namely $a^{(1)}$ for the X variables and $b^{(1)}$ for the Y variables. We consider the following maximization criterion:

$$\text{Max.} \sum_{m=1}^{M} n_m cov(u_m^{(1)}, t_m^{(1)}), \text{ with } \| a^{(1)} \| = \| b^{(1)} \| = 1 \tag{1}$$

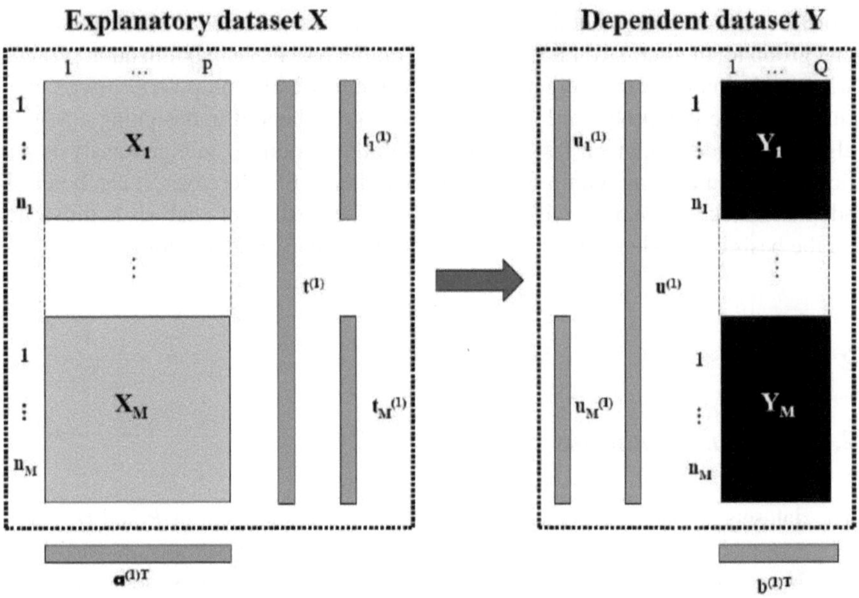

Fig. 1: Structure of two-block multi-group datasets, highlighting the relationships between the variable blocks and their associated common loadings and group components for the first dimension ($h = 1$)

Criterion (1) highlights the optimal link between the group components $t_m^{(1)}$ and $u_m^{(1)}$ for each group $m = (1,\ldots,M)$. It reflects that the group components in X_m and in Y_m have, on average, as large a covariance as possible.

The criterion to be maximized can be written as $a^{(1)T}[\sum_{m=1}^{M} X_m^T Y_m]b^{(1)}$. It follows that, for a fixed vector $a^{(1)}$, the maximum is achieved by setting:

$$b^{(1)} = \frac{\sum_{m=1}^{M} Y_m^T X_m a^{(1)}}{\| \sum_{m=1}^{M} Y_m^T X_m a^{(1)} \|} \tag{2}$$

Replacing the common loadings $b^{(1)}$ by its expression (2), we are led to maximizing the quantity:

$$a^{(1)T} \left(\frac{\sum_{m=1}^{M} X_m^T Y_m \sum_{m=1}^{M} Y_m^T X_m}{\| \sum_{m=1}^{M} Y_m^T X_m a^{(1)} \|} \right) a^{(1)} = \sqrt{a^{(1)T} \left(\sum_{m=1}^{M} X_m^T Y_m \sum_{m=1}^{M} Y_m^T X_m \right) a^{(1)}} \tag{3}$$

Under the constraints stated above, the optimal solution is achieved by setting $a^{(1)}$ to the eigenvector of $(\sum_{m=1}^{M} X_m^T Y_m)(\sum_{m=1}^{M} Y_m^T X_m)$ associated with the largest eigenvalue.

As a summing up, MGPLS consists in the following algorithm:

1. Set $a^{(1)}$ to the eigenvector of $(\sum_{m=1}^{M} X_m^T Y_m)(\sum_{m=1}^{M} Y_m^T X_m)$ associated with the largest eigenvalue;
2. Set $b^{(1)} = \frac{\sum_{m=1}^{M} Y_m^T X_m a^{(1)}}{\|\sum_{m=1}^{M} Y_m^T X_m a^{(1)}\|}$;
3. Compute group components: $t_m^{(1)} = X_m a^{(1)}$ and $u_m^{(1)} = Y_m b^{(1)}$;
4. Compute global components: $t^{(1)} = X a^{(1)}$ and $u^{(1)} = Y b^{(1)}$.

An alternative algorithm to solve the maximization problem (1) is worth considering since it does not involve an eigen-analysis solution and can easily be developed into a NIPALS algorithm. It stems from remarking that, as stated above, for a fixed vector $a^{(1)}$, the optimal solution for $b^{(1)}$ given by $b^{(1)} = \frac{\sum_{m=1}^{M} Y_m^T X_m a^{(1)}}{\|\sum_{m=1}^{M} Y_m^T X_m a^{(1)}\|}$. Likewise, for a fixed vector $b^{(1)}$, the optimal solution for $a^{(1)}$ is given by $a^{(1)} = \frac{\sum_{m=1}^{M} X_m^T Y_m b^{(1)}}{\|\sum_{m=1}^{M} X_m^T Y_m b^{(1)}\|}$. Therefore, an iterative algorithm to solve the maximization problem (1) consists in alternatively updating $a^{(1)}$ and $b^{(1)}$ until convergence. The convergence is granted by the fact that at each iteration, criterion (1) increases. Moreover, since this criterion is upper-bounded, it follows that the sequence of values corresponding to this criterion at the various steps of the iterative algorithm is convergent. This iterative algorithm can be extended into a NIPALS-like algorithm:

1. Choose an initial value for $a^{(1)}$ with $\|a^{(1)}\| = 1$
2. Compute X group components: $t_m^{(1)} = X_m a^{(1)}$
3. Compute Y group vector of loadings: $b_m^{(1)} = Y_m^T t_m^{(1)}$
4. Compute Y common vector of loadings: $b^{(1)} = \frac{\sum_{m=1}^{M} b_m^{(1)}}{\|\sum_{m=1}^{M} b_m^{(1)}\|}$
5. Compute Y group components: $u_m^{(1)} = Y_m b^{(1)}$
6. Compute X group vector of loadings: $a_m^{(1)} = X_m^T u_m^{(1)}$
7. Compute X common vector of loadings: $a^{(1)} = \frac{\sum_{m=1}^{M} a_m^{(1)}}{\|\sum_{m=1}^{M} a_m^{(1)}\|}$
8. Iterate the process starting from step 2 until convergence.

A property of MGPLS which is worth mentioning is the following. We have already stated that criterion (1) leads to maximizing with respect to $a^{(1)}$ the quantity $\sqrt{a^{(1)T} \sum_{m=1}^{M} X_m^T Y_m \sum_{m=1}^{M} Y_m^T X_m a^{(1)}}$. By recalling that $t_m^{(1)} = X_m a^{(1)}$, this latter expression can be written as:

$$\sqrt{\sum_{m=1}^{M} t_m^{(1)T} Y_m \sum_{m=1}^{M} Y_m^T t_m^{(1)}} = \sqrt{\sum_{m=1}^{M} n_m^2 \sum_{q=1}^{Q} cov^2(t_m^{(1)T}, y_{mq})} \tag{4}$$

where y_{mq} is the qth variable in group m. This means that MGPLS aims at recovering as much variation as possible in the datasets $(Y_1, \ldots, Y_M)$ by means of their associated group components $(t_1^{(1)}, \ldots, t_M^{(1)})$ which are constrained to have the same vector of loadings $a^{(1)}$.

Other properties of MGPLS can also be highlighted:

- In the particular case where there is only one group of individuals (i.e., $M = 1$), MGPLS amounts to PLS regression.
- If $X = Y$, then MGPLS is equivalent to MGPCA.
- We have already stated that $a^{(1)}$ is an eigenvector of $(\sum_{m=1}^{M} X_m^T Y_m)(\sum_{m=1}^{M} Y_m^T X_m)$ associated with the largest eigenvalue. Since $\sum_{m=1}^{M} X_m^T Y_m = X^T Y$, we can conclude that $a^{(1)}$ and $b^{(1)}$ can also be obtained by performing PLS regression of Y on X, where as indicated above Y and X are centered for each group. Notwithstanding, the interest of the original presentation and its associated NIPALS algorithm is to exhibit group components which may shed more light on the relationships between X and Y.

2.3 Higher Order Solution

Once the first order common vectors of loadings and their associated group and global components are determined, subsequent parameters can be computed following the same strategy of analysis after deflation. We chose to deflate the datasets X and Y with respect to the global component $t^{(1)} = Xa^{(1)}$. More precisely, the datasets X and Y are replaced by $X^{(1)} = P^{(1)}X$ and $Y^{(1)} = P^{(1)}Y$ where $P^{(1)} = (I - t^{(1)}(t^{(1)T}t^{(1)})^{-1}t^{(1)T})$, I being the identity matrix. Thereafter, the same procedure described in the previous section is run anew, thus leading to the second order common vectors $a^{(2)}$ and $b^{(2)}$ and their associated group components: $t_m^{(2)} = X_m^{(1)}a^{(2)}$ and $u_m^{(2)} = Y_m^{(1)}b^{(2)}$. Similarly to PLS regression, the group and global components can be expressed in terms of the original variables instead of the deflated variables. The same deflation procedure can be reiterated to compute subsequent vectors of loadings and group and global components. A stopping criterion for choosing the appropriate number of components to be included in the model will be discussed in a subsequent section.

2.4 Prediction and Selection of the Appropriate Number of Components

The usual practice in PLS regression to predict dependent variables from a dataset which contains quantitative and categorical variables is to perform a dummy coding (or 0/1 coding) of the categorical variables and, thereafter, run a PLS regression using these dummy variables together with the quantitative variables as predictors. This strategy of analysis has several shortcomings among which we single out the fact that the number of components to be introduced in the model can be large and the interpretation of the loadings vectors may not be easy [6].

From MGPLS, a prediction model can be set up as follows. Let us denote by $Y^{(0)}$ and $X^{(0)}$ the original data that is, the data before we have proceeded to centering by group. We have:

$$X^{(0)} = \bar{X} + X \text{ and } Y^{(0)} = \bar{Y} + Y \qquad (5)$$

where, as stated above, X and Y are the datasets centered by group and $\bar{X}$ (resp. $\bar{Y}$) is the dataset of the same dimensions as $X^{(0)}$ and X (resp. $Y^{(0)}$ and Y) whose entry corresponding to a given row (i.e., sample) and column (i.e., variable) is the average value of the variable under consideration for the group to which the sample (in row) belongs to.

From the outcomes of a MGPLS regression, let us consider that A (say) components have been retained and let us denote by $T_A = (t^{(1)}, \ldots, t^{(A)})$ the matrix formed of the A first global components. Dataset Y can be predicted from X using these components:

$$Y = T_A B_T^{(A)} + E_A \qquad (6)$$

where $B_T^{(A)}$ is the matrix of regression coefficients and E_A is the matrix of residuals. Similarly to what is usually done in PLS regression, this model can be re-expressed in terms of the variables in X, thus leading to:

$$Y = XB^{(A)} + E_A \qquad (7)$$

Finally, the original data $Y^{(0)}$ can be predicted from the model by:

$$Y^{(0)} = \bar{Y} + XB^{(A)} + E_A \qquad (8)$$

More precisely, for a given variable $y^{(0)}$ in $Y^{(0)}$ and a given sample i which belongs to group m, we have:

$$y_i^{(0)} = \bar{y}_m + \sum_{p=1}^{P} b_p^{(A)} x_{ip} + e_i \qquad (9)$$

where $b_p^{(A)}$ are the regression coefficients associated with variable $y^{(0)}$ and $\bar{y}_m$ is the average value of $y^{(0)}$ for group m. Obviously, this model pertains to the Analysis of Covariance (ANCOVA) class of models [7]. In this model, we analyze the effect of the quantitative variables while controlling the effect of the grouping (i.e., categorical variable). We do not assume an interaction between X and the categorical variable because, from one group to another, the regression models differ only by a constant (difference of the mean values).

For the choice of the appropriate number of components to be introduced in the model, we use a validation technique such as cross-validation [11]. The method K-fold cross-validation is applied to each group to ensure that we have sufficient samples from each group. More precisely, each group is equally or nearly equally divided into K subsets. In each iteration of the cross-validation process one subset is included in the validation set while the $K-1$ remaining subsets are put together as the training set. The training datasets are used to select the parameters of the model, namely the regression coefficient matrix B, and the root mean square error of calibration ($RMSE_C$) which reflects the fitting ability of the model. The validation set is used to compute the root mean square error of validation ($RMSE_V$) which reflects the prediction ability of the model under consideration. Both errors are computed according to:

$$RMSE^{(h)} = ||Y^{(0)} - \hat{Y}^{(0)(h)}||/\sqrt{Q} \quad \text{for} \quad h = (1, \ldots, H) \tag{10}$$

where $\hat{Y}^{(0)(h)}$ is the matrix of predicted values from a model with h components. Thereafter, the cross-validation procedure is repeated several times and the two types of errors (i.e. $RMSE_C$ and $RMSE_V$) are averaged over these repetitions. The two average errors are functions of the number $h = (1, \ldots, H)$ of latent variables to be introduced in the model. Among all these models corresponding to the various values of h, an appropriate model, with A components, which has a correct fitting ability and a good prediction ability is retained.

2.5 Alternative Methods

Although the main aim of MANCOVA is to test for differences between group means while controlling the effect of the co-variables (i.e., X variables), it is routinely used to assess the impact of the X variables on Y variables while controlling the effect of the grouping variables. However, the main disadvantages of this method of analysis over our approach are the necessity of pre-supposing some assumptions (e.g., normality, homogeneity of variances and covariances ...) and the sensitivity to multicollinearity among the X variables.

Takane [12, 13] proposed multivariate data analyses to take account of external variables on samples and variables. By comparison, our approach is more restrictive since we take account of the only information on the samples which can be expressed in terms of groups. Nonetheless, as Takane's methods of analysis pertain to canonical and redundancy analysis, they are likely to lead to unstable methods in presence of quasi-collinearity among variables. Moreover, we believe that MG-PLS can be extended to a wider scope in order to take account of more information on the samples. This extension will certainly draw inspiration from the research work by Jorgensen et al. (2004) [6] and Næs et al. (2010) [10]. The former authors discussed regression models to analyse both design (D) and predictive variables (X). In particular, they introduced a strategy of analysis called Least Squares-PLS (LS-PLS) that consists in an iterative combination of least squares and PLS regression. Although it seems that LS-PLS yields appealing outcomes, it has, nonetheless, some convergence and optimality problems [6]. More importantly, MGPLS has very strong family ties with Sequential and Orthogonalized Partial Least Squares (SO-PLS) [10]. SO-PLS method is based on a blockwise estimation of the regression parameters, where each regression is followed by an orthogonalization with respect to the blocks already fitted. More precisely, in the particular case of two blocks of data formed of a design matrix D and an explanatory block X, SO-PLS regression starts by fitting Y to the design matrix by means of PLS regression. Let us denote by E the matrix of residuals. In a second stage, X is orthogonalized with respect to the PLS components from the previous stage leading to matrix X^{orth}. Finally, the matrix of residuals E is regressed upon X^{orth} using PLS regression. SO-PLS covers a large scope of applications since it is proposed as a multi-block path modeling approach with an exploratory purpose. In the particular case where we have a design matrix $W_1 = D$ derived from the coding of a grouping variable the prediction of a dataset $Y^{(0)}$ by means of SO-PLS bears high similarities to Eq. (8). This finding bestows a mutual credit to both MGPLS and SO-PLS and shows that there is a gap to fill between the general framework offered by SO-PLS and MGPLS which tackles a very specific situation.

3 Illustration

The data which are used to illustrate MGPLS are extracted from a large case study in veterinary epidemiology. The population ($N = 105$) consists of a cohort of slaughtered broiler chicken flocks from three slaughterhouses in France [9]. The aim of this study is to assess and interpret the effect of ($P = 12$) explanatory variables on ($Q = 4$) dependent variables, while taking into account the diversity among the ($M = 3$) slaughterhouses. All these variables are described in Table 1. The categorical variables were replaced by their associated dummy (or 0/1) variables. Since the X variables are measured on different scales, they are standardized. Epidemiologists are interested in identifying risk factors which lead to the mortality of the chicken, while taking into account the diversity among the slaughterhouses.

Figure 2 shows the cumulative percentages of total variances explained by X group components in the three slaughterhouses for X (Fig. 2a) and Y (Fig. 2b) variables. The increase of the total variance in the X variables as a function of the number of components introduced in the model has the same pattern from one group to another. This is not the case for the Y variables since the first two components in group 2 explain only a small amount of total variance in this group but a jump in the total variance explained is observed from component 3 on. Group 1 shows an inverse tendency since the first two components explain a higher percentage of total variance than in group 2 but the increase of the curve depicting the cumulative percentage of total variance slows down starting from dimension 2.

Figure 3 represents the variables on the basis of the common vectors of loadings associated with X and Y variables. Similarity to what is routinely done in PLS regression, it is easy to highlight relationships between X and Y variables.

In order to assess the prediction ability of MGPLS and compare its performance to usual methods of analysis, we performed a multi-group 10-fold cross-validation. For each run, we performed MGPLS and, in parallel, we performed PLS on the original datasets by ignoring the group structure among the individuals. We also used a dummy coding of the memberships of the chicken to the three slaughterhouses. Thereafter, we performed a PLS regression of the Y variables on the matrix

Table 1: Abbreviations

Abbreviation	Explanatory variables
Soak	Cleaning step in decontamination of chicken house: yes (vs. no)
Sort	Sorting practice: yes (vs. no)
Vitamin	Vitamins and minerals during the starting period: yes (vs. no)
Homochick	Homogeneity of chicks at placement: yes (vs. no)
Nbchick	Number of chicks at placement
Homochicken	Homogeneity of chickens at the end of rearing: yes (vs. no)
Strain	Genetic strain: X (vs. other)
Locpb	Locomotor disorder observed: yes (vs. no)
Antibio	Antibiotic during the staring period
Tlairage	Average temperature of waiting time on lairage
Stress	Stress occurrence during rearing: yes (vs. no)
RainWind	Meteorological conditions during lairage: rain and/or wind (vs. neither rain nor wind)
Abbreviation	**Dependent variables**
Mort7	First-week mortality rate
Mort	Mortality rate during the rest of the rearing
Doa	Mortality rate during the transport to slaughterhouse
Condemn	Condemnation rate at slaughterhouse

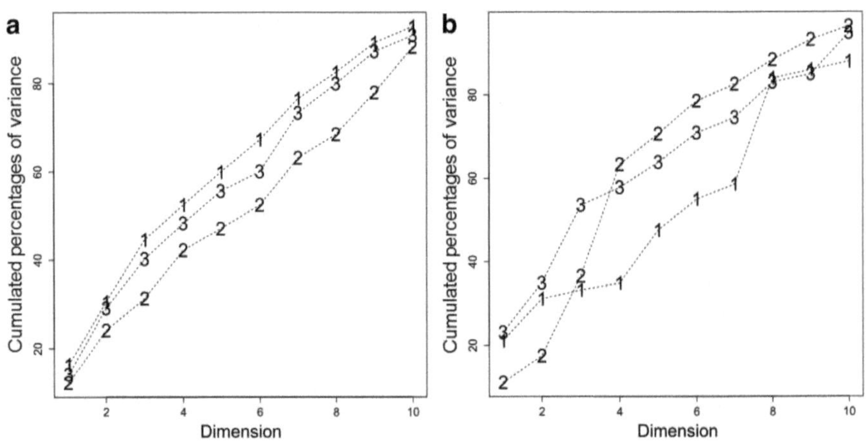

Fig. 2: (**a**) (*Left*) cumulated percentages of variance for X_m and (**b**) (*right*) cumulated percentages of variance for Y_m. The labels (*1*,*2*,*3*) refer to the three slaughterhouses

X augmented by the three dummy variables associated with the three slaughterhouses. Figure 4a (resp. Fig. 4b) shows the values of $RMSE_C$ (resp. $RMSE_V$) as a function of the number of global components introduced in the model for the three strategy of prediction considered herein. It can be seen that MGPLS has a better fitting ability since $RMSE_C$ is smaller than the other two methods for all the com-

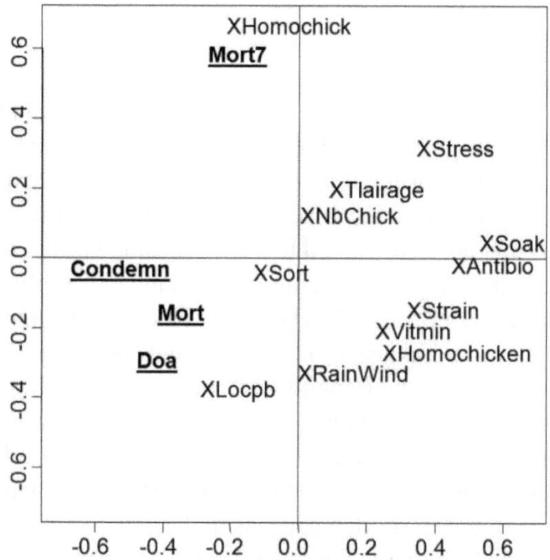

Fig. 3: Graphical display of the first two common loadings $(a^{(1)}, a^{(2)})$ and $(b^{(1)}, b^{(2)})$ of MGPLS. The dependent variables are *underlined*

ponents. From the curve depicting $RMSE_V$, it seems that a MGPLS model with two components should be retained. However, the two other methods considered herein have the same performance whether we include one or two components. As a matter of fact, we have computed $RMSE_V$ for each of the four dependent variables separately (data not shown herein) and it turned out that MGPLS outperformed the other two methods except for the last variable. Further investigations are needed to explain this finding.

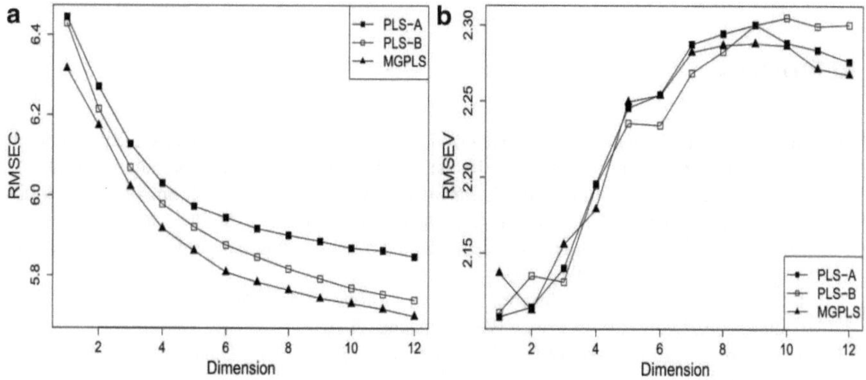

Fig. 4: (**a**) (*Left*) shows root mean square error of calibration ($RMSE_C$) and (**b**) (*right*) shows root mean square error of validation ($RMSE_V$) for MGPLS, PLS-A (ignoring the group structure) and PLS-B (X variables and three dummy variables)

4 Conclusion

MGPLS is a method of investigation of the relationships between two datasets where there is a priori known group structure. It is a straightforward extension of PLS regression to this specific setting. It can also be regarded as an extension of MGPCA [8] since it shares with this method the same underlying principle that is, seeking group components that are constrained to have the same vectors of loadings.

MGPLS is based on a simple optimization problem whose solution is either given by an eigen-analysis or a NIPALS-like algorithm. In comparison to competing methods such as MANCOVA or Takane's methods of analysis [12, 13], MGPLS makes it possible to handle the case of ill conditioned problems (i.e., highly multicollinear datasets with, possibly, less individuals than variables). Moreover, the interpretation of the outcomes is straightforward since we stick to the general framework of PLS regression which is fecund of visualization tools and indices which can be helpful for the practitioners to unveil hidden patterns in the data. Another appealing feature of MGPLS is that it can be extended in several ways. For instance, we could consider the case of more than one grouping variable, or the case of presence of interaction between the grouping variables and the predictive variables. As stated above, all these extensions will draw inspiration from SO-PLS [10] which offers an appealing and general scope to investigate the relationships between blocks of variables taking account of their interweaving relationships. Indeed, we believe that by filling the gap between our specific method of analysis (MGPLS) and the more general approach (SO-PLS) we help shedding more light on the possibilities offered by this latter approach. Another possibility of extension of MGPLS is to consider the general framework of multilevel (or hierarchical) regression [4] where the aim is to predict dependent variables from predictive variables obtained at various levels (e.g., variables measured on animals that are nested into farms and variables measured on farms). Ongoing research work on these topics is under progress.

References

[1] W. W. Chin, L. Barbara, and P. R. Newsted, "A partial least squares latent variable modeling approach for measuring interaction effects: Results from a monte carlo simulation study and voice mail emotion/adoption study," *Information Systems Research* **14**, pp. 189–217, 2003.

[2] I. Dohoo, C. Ducrot, C. Fourichon, A. Donald, and D. Hurnik, "An overview of techniques for dealing with large numbers of independent variables in epidemiologic studies," *Prev. Vet. Med.* **29**, pp. 221–239, 1997.

[3] I. Dohoo, W. Martin, and H. Stryhn, *Veterinary Epidemiologic Research*, Atlantic Veterinary College Inc., University of Prince Edward Island, 2010.

[4] J. J. Hox, *Multilevel Analysis: Techniques and Applications*, Quantitative Methodology Series, p. 304, 2002.

[5] B. Huitema, *The Analysis of Covariance and Alternatives: Statistical Methods for Experiments, Quasi-Experiments, and Single-Case Studies*, John Wiley & Sons, p. 688, 2011.

[6] K. Jorgensen, V. Segtnan, K. Thyholt, and T. Næs, "A comparison of methods for analysing regression models with both spectral and designed variables," *J. Chemometr.* **18**, pp. 451–464, 2004.

[7] G. Keppel, *Design and analysis: A researcher's handbook*, 3rd edn., I. Englewood Cliffs: Prentice-Hall, 1991.

[8] W. Krzanowski, "Principal component analysis in the presence of group structure," *J. Appl. Stat.* **33**, pp. 164–168, 1984.

[9] C. Lupo, S. L. Bouquin, L. Balaine, V. Michel, J. Peraste, I. Petetin, P. Colin, C. Chauvin, "Feasibility of screening broiler chicken flocks for risk markers as an aid for meat inspection," *Epidemiology and Infection* **137**, pp. 1086–1098, 2009.

[10] T. Næs, O. Tomica, B. H. Mevikb, and H. Martens, "Path modelling by sequential PLS regression," *J. Chemometr.* **25**, pp. 28–40, 2011.

[11] M. Stone, "Cross-validatory choice and assessment of statistical predictions," *J. Roy. Statist. Soc. Ser. B* **36**, pp. 111–147, 1974.

[12] Y. Takane, and H. Hwang, "Generalized constrained canonical correlation analysis," *Multivariate Behavioral Research* **37**, pp. 163–195, 2002.

[13] Y. Takane, and S. Jung, "Generalized constrained redundancy analysis," *Behaviormetrika* **33**, pp. 179–192, 2006.

[14] M. Tenenhaus, *La régression PLS. Théorie et pratique*, Paris: Technip. pp. 254, 1998.

[15] M. Tenenhaus, E. Mauger, and C. Guinot, "Use of uls-sem and pls-sem to measure a group effect in a regression model relating two blocks of binary variables," in *Handbook on Partial Least Squares: Concepts, Methods and Applications*, pp. 124–140, Springer, 2010.

[16] J. A. Westerhuis, T. Kourti, and J. F. MacGregor, "Analysis of multiblock and hierarchical PCA and PLS model," *J. Chemometr.* **12**, pp. 301–321, 1998.

[17] H. Wold, "Estimation of principal components and related models by iterative least squares. In P.R. Krishnaiaah (Ed.)," *Multivariate Analysis*, pp. 391–420, New York: Academic Press, 1966.

Employees' Response to Corporate Social Responsibility: An Application of a Non Linear Mixture REBUS Approach

Omer Farooq, Dwight Merunka, and Pierre Valette-Florence

Abstract We examine the effect of perceived corporate social responsibility (CSR) on employees' affective organizational commitment through the mediation of organizational trust and organizational identification. In so doing, the paper advances current understanding by positing a curvilinear relationship between CSR and organizational trust. We further suggest that employees use different processes to develop commitment to their companies' CSR initiatives. The test of the research model relies on data collected from 378 employees in South Asia. We used REBUS-PLS algorithm and identified three homogeneous employee groups that can be further differentiated in terms of work-related attitudes and behaviors.

Key words: Partial least squares path modeling, REBUS-PLS, Non-linear path modeling, Corporate social responsibility, Social exchange, Social identity

1 Introduction

Unlike previous studies that have examined the direct linear effect of perceived corporate social responsibility (CSR) on affective organizational commitment (AOC), this article examines the mediated link through organizational trust and organizational identification. In so doing, it advances current understanding by positing a

---

O. Farooq (✉)
EUROMED, Marseilles, France
e-mail: muhammadomer.farooq@euromed-management.com

D. Merunka
IAE and Cergam, University Aix-Marseille, Aix-Marseille, France
EUROMED Marseilles School of Management, Marseille, France
e-mail: dwight.merunka@iae-aix.com

P. Valette-Florence
IAE, Grenoble, France
e-mail: pvalette@upmf-grenoble.fr

H. Abdi et al. (eds.), *New Perspectives in Partial Least Squares and Related Methods*, Springer Proceedings in Mathematics & Statistics 56, DOI 10.1007/978-1-4614-8283-3_18,

curvilinear relationship between CSR and organizational trust. Social exchange and social identity theory provide the foundation for predictions that the primary outcomes of CSR initiatives are organizational trust and organizational identification, which in turn affect AOC.

2 Theoretical Framework and Research Model

This study examines the influence of CSR on AOC through the mediation of organizational identification and organizational trust. That is, we suggest two mechanisms by which CSR influences AOC: social identity and social exchange. With these two mechanisms, we propose that organizational identification and organizational trust are the direct outcomes of the firm's CSR initiatives and in turn positively affect AOC.

2.1 CSR and Organizational Identification

Social identity theory suggests that the firm's CSR actions have a direct effect on employees' organizational identification. That is, individuals strive to achieve or maintain a positive social identity [1], which they can derive from membership in different groups [4]. Hogg et al. [19] illustrate that among these groups, membership in business organizations may be the most important component, and they refer to it as organizational identification, defined as the "perception of oneness with or belongingness to an organization, where the individual defines him or herself in terms of the organization(s) in which he or she is a member [24]."

Tyler [41] further suggests that employees use the status or social standing of their organization to evaluate their self-worth. Therefore, employees prefer to identify with organizations whose images seem prestigious or whose identity enhances their self-worth and meets their need for self-enhancement [25, 36, 37]. Organizational identification thus derives from the image and perceived prestige of the organization (e.g. [43]). We argue that firm investments to support societal, environmental and consumer welfare are positively valued, such that they lead people to evaluate the organization positively. In turn, CSR actions should have a strong impact on the firm's external image. Extant research also shows that a firm's philanthropic and community development actions enhance its corporate image and external prestige (e.g. [6]). Because CSR actions enhance this image, employees feel proud to associate with the responsible company, which enhances their self-worth and self-esteem.

In addition, an employee's respect for the organization could influence his or her identification with that organization [42], because it enhances perceived status within the organization. We therefore suggest that internal CSR actions also contribute to employees' organizational identification. Because all CSR actions thus appear contribute to organizational identification, we propose:

Hypothesis 1: Employee perceptions of the firm's CSR initiatives positively influence their organizational identification.

2.2 CSR and Organizational Trust

Social exchange theory (SET) explains the relationship between CSR and organizational trust, using a dominant theoretical paradigm for understanding relationships. One of the basic tenets of SET is the rule of reciprocity [5]: if one person supplies a benefit, the receiving party should respond in kind [15]. In a social exchange, one party voluntarily provides a benefit to another, invoking an obligation to reciprocate by providing some benefit in return [44].

In direct (or restricted) exchanges, two actors grant benefits in a relation of direct reciprocity, whereas in indirect (or generalized) exchanges, each actor provides and ultimately receives benefits, but not to and from the same actor. We suggest that CSR has the capacity to induce both direct and indirect social exchanges between employees and the firm, because both forms entail some type of reciprocity [28]. For example, when the firm provides benefits to its employees beyond its legal and financial obligations, it obliges them to pay it back, directly and in kind. Furthermore, the actions the firm takes for the welfare of the society, environment, and consumers may invoke indirect social exchanges with employees. As part of the community, country, or global habitat, employees should consider societal and environmental responsibilities important; we draw this argument from Handelman et al. [16], who suggest that "a company's actions appeal to the multidimensionality of the people as not only an economic being but also as a member of a family, community and country." Handelman et al. [16] further recognize that people are conscious of not only their personal well-being but also of other stakeholder groups, of which they are actual or potential members. From this perspective, employees could indirectly reciprocate actions that a firm takes for the welfare of community, society, the environment, and consumers.

Furthermore, according to SET, trust between the parties is a primary outcome of social exchange relationships [3]. Both Blau [5] and Holmes [20] identify trust as an important outcome of favorable exchanges, and Ekeh's [12] elaboration of Levi-Strauss's thesis proposes that trust is the most important consequence of both direct and indirect reciprocity. Therefore, we propose:

Hypothesis 2: Employee perceptions of the firm's CSR initiatives positively influence their organizational trust.

We also consider a potential curvilinear relationship between CSR and organizational trust. That is, CSR may breed stakeholder cynicism (e.g. [23]) if they are skeptical of the firm's intentions for engaging in CSR [33] or suspect greenwashing. Therefore, the relationship between CSR and trust may be subject to a saturation effect, such that very high levels of CSR initiatives appear excessive and guided not by benevolence or altruism but by internal goals, such as corporate or brand image.

The credibility of the CSR initiatives then is at stake. Matheson et al. [26] suggest that increasing environmental initiatives by companies are making people more cynical; the broad saturation effect illustrates how the strength of an inducement can diminish with greater intensity of the inducement, past a saturation point. We apply this principle to CSR and employees and predict that the effect of CSR on employees' trust diminishes at higher levels of perceived CSR.

Stakeholders also express some expectations about the CSR of the firm (e.g. [46]). Stakeholders actively monitor companies' behaviors to evaluate how well they meet expectations [47], namely, if they fulfill, underfulfill, or overfulfill expectations. Literature on employees' psychological contracts shows that when an employer meets employees' expectations, psychological contract fulfillment exists, whereas if it does not, psychological contract breach occurs (e.g. [22]). We also note the strong positive relationship between psychological contract fulfillment and trust, along with a negative relationship between psychological contract breach and trust (e.g. [10]). The overfulfillment of expectations is another form of psychological contract breach [21, 29] that can minimize organizational trust. Therefore, at high levels of CSR, employees may perceive a psychological contract breach and lose trust in their organization. We suggest that overfulfillment of CSR expectations suggests the potential for greenwashing, with its associated loss of credibility and negative impact on employees' trust in the company. We therefore propose:

Hypothesis 3: The relationship between CSR and organizational trust is quadratic, such that when CSR increases beyond a saturation point, its positive influence on employees' organizational trust decreases.

2.3 Impact of CSR on AOC: Organizational Identification Mediation

In the context of social identity, AOC is a critical outcome, closely related to identification. Affective organizational commitment is "an employee's emotional attachment to, identification with, and involvement in the organization [2]," so though both organizational identification and AOC indicate psychological links between employees and the organization, the former is generally considered an antecedent of the latter. Pratt [32] specifically suggests that organizational identification is a cognitive perceptual construct that causes attitudes such as AOC. We posit in turn that employees who identify with their company are committed, because their identification maintains their external prestige and internal respect. The firm's positive external prestige, which enhances employee self-esteem and fulfills social identity needs, also keeps the employees committed to that company. Employee commitment increases with the level of CSR initiatives, because employees feel increasingly proud to identify with a firm and its positive external prestige. Therefore, CSR should affect AOC through the mediation of organizational identification, a claim that receives further support from studies that indicate a positive relationship between organizational identification and AOC [35].

Hypothesis 4a: There is a positive relationship between employees' organizational identification and affective organizational commitment.

Hypothesis 4b: Organizational identification mediates between CSR and employees' affective organizational commitment.

2.4 Impact of CSR on AOC: Organizational Trust Mediation

Organizational trust is another essential condition for AOC; Blau [5] even notes that "the establishment of exchange relations involves making investments that constitute commitment to the other party. Since social exchange requires trusting others to reciprocate, the initial problem is to prove oneself trustworthy." Firms' CSR initiates a social exchange between the firm and its employees, with both organizational trust and organizational commitment as potential outcomes. That is, organizational trust results from CSR and in turn influences commitment. Abundant research reveals that organizational trust is a strong predictor of organization commitment (e.g. [9]), so we posit:

Hypothesis 5a: There is a positive relationship between employees' organizational trust and affective organizational commitment.

Hypothesis 5b: Organizational trust mediates between CSR and employees' affective organizational commitment.

3 Methodology

We focused on local and multinational companies in the grocery, food, personal care, and household products categories in Pakistan. We selected 11 companies that publish sustainability and CSR-related information on their websites; publically available information also indicates that these companies have been involved in topical CSR issues. Employees therefore should have CSR-related perceptions about their employers.

We contacted sources in the targeted companies and sought permission and support for our data collection. The data were collected face-to-face using a questionnaire. We obtained 392 responses, though the final sample contained 378 (86 % male, 14 % female) respondents across different age groups (40 % between 18–28 years of age, 45 % 29–40 years, and 15 % older than 40 years). Regarding education, 29 % of these respondents had less than a bachelor's degree, 46 % had earned a bachelor's degree, and 25 % held master's degrees. Furthermore, 72 % of the respondents were non-management employees, and 28 % respondents were functional managers.

3.1 Measurements

To measure CSR, we rely on perceptual measures reported by the employees. Specifically, we adapt instrument developed by Turker [39, 40], which contained 16 items: four items to measure environmental CSR, three to measure societal CSR, six items for internal CSR, and three measures of product CSR. All items used seven-point Likert scales (1 = "extremely disagree" to 7 = "extremely agree"). We also pre-tested the instrument with 19 MBA students and modified the wording of a few items.

For organizational identification, we rely on a five-item revised version of Mael et al. [24] scale, which has shown good reliability in previous research. Organizational trust was measured by a three-item scale [31], and for AOC, we used Meyer et al. [27] abridged five-item scale.

3.2 Data Analyses

We selected a PLS structural equation modeling (SEM) approach, because of its minimal demands for sample size and ability to handle model complexity or violations of multivariate normality [13, 38, 45]. Our study features rather small sample sizes, particularly at the segment level, and the large, complex model involves several indicators and latent variables. For example CSR is defined as a second order formative construct with first order reflective latent indicators. Because prediction represents a major purpose of this analysis, we used a two-stage approach to deal with nonlinear relationships between CSR and trust. Although another approach to handle quadratic latent variables effect by means of squared indicators exists in literature (e.g. [30]) and seems preferable in other circumstances, we followed recent recommendations made in the case of formative latent variables [8, 17, 18]. That is, we first estimated the model with only linear terms and computed the factor scores for the latent variables. Then, we created a single indicator for the nonlinear term by transforming the linear term factor score, and we reestimated the model, including both the linear term and its indicators and the nonlinear term with its single indicator.

Hence, we used REBUS-PLS to identify homogeneous groups of employees [14]. This approach offers interesting features compared with existing response-based clustering techniques in a PLS-PM framework, such as finite-mixture PLS (FIMIX-PLS; [34]). As a distribution-free approach, REBUS-PLS is consistent with PLS basic principles, and it aims to detect sources of heterogeneity in both structural and outer models for all exogenous and endogenous latent variables. In REBUS-PLS, the distance of any unit from the model is defined by the model performance, in terms of residuals related to the structural and measurement models for all available latent variables. The measure of this distance is a sum of the squared residuals, usually referred to as a closeness measure. When homogeneous groups have been identified, multigroup comparisons support an assessment of the level of measurement invariance across segments.

After obtaining the different REBUS segments, we performed multigroup comparisons and permutation tests [7] for the full model (see Fig. 1). Regarding the measurement of the first-order latent variables, permutation tests indicate that less than 10% of the variables slightly vary across the three REBUS segments. More important, there is no difference between segments in the path coefficients related to the definition of the formative second-order CSR construct. Results indicate that the three REBUS segments behave very differently in their use of the trust or identification path toward the influence on commitment (see Table 1). We observe that all the *R*-square values obtained at the group level are higher than those computed on the pooled data. For Groups 1, 2, and 3, the trust R-squares are 0.35, 0.46, and 0.60 (0.34 at the aggregate level), those for identification are 0.44, 0.48, and 0.53 (0.37 aggregate), and the values for commitment are 0.34, 0.41, and 0.53 (0.29 aggregate), respectively. Therefore, the model constructs are better explained at each group level than at the aggregate level.

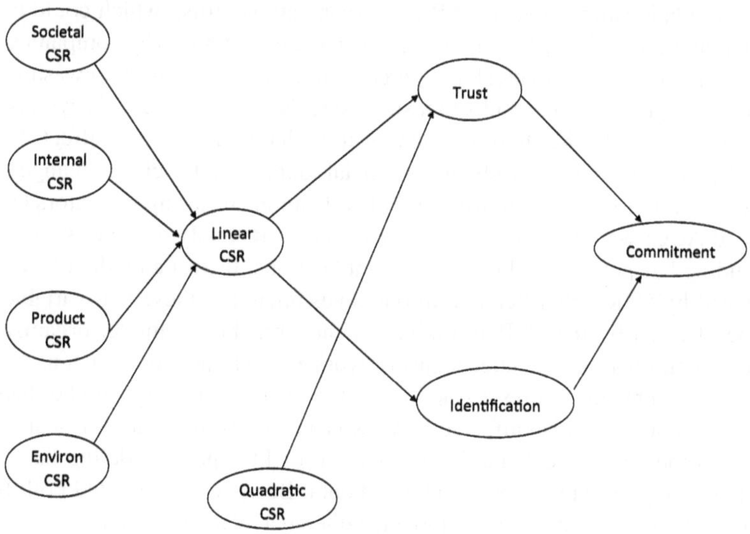

Fig. 1: Full estimated model

All in all, and according to the significant path coefficients, all the hypotheses have been validated and more particularly with regards to the mediating role of both trust and commitment. In addition, the REBUS PLS analysis shows that discovering sources of heterogeneity was worth investigating.

4 Main Contributions and Discussion

Employees use different processes to develop commitment to their companies' CSR initiatives; specifically, three homogeneous employee groups can be differentiated in terms of work-related attitudes and behaviors. The test of the research model relies on data collected from 378 employees of local and multinational companies in South Asia, with partial least squares path modeling to test both the linear and nonlinear postulated relationships. Organizational trust and organizational identification mediate between CSR and AOC, and the relationship between CSR and trust is curvilinear with a saturation point. In the employee groups, some employees derive their AOC through trust, whereas others derive it from identification. Employees who derive their AOC through organizational identification exhibit higher intrinsic motivation, knowledge-sharing behavior, readiness for change and perceptions of their participation in decision making.

At a high level of CSR, its influence on trust decreases. We uncover an inverse quadratic relationship between CSR and organizational trust, which could be a result of a saturation effect [26]. Greater environmental initiatives by companies might cause people to become cynical and perceive related actions as greenwashing. Our findings strengthen recent results that suggest CSR sometimes breeds cynicism and suspicion (e.g. [23]). We therefore demonstrate that CSR does not affect AOC directly, because mediating mechanisms (social identity and social exchange) better explain this relationship. Both trust and identification are strong mediators between CSR and AOC; we reveal the process by which CSR finally influences AOC. This new finding has potentially far-reaching implications. If trust and identification are influenced by CSR, other behavioral outcomes related to these two variables also may be affected by the CSR initiatives of the firm. For example, organizational trust and identification are antecedents of work-related outcomes, such as turnover intentions, absenteeism, job satisfaction, and motivation. Testing whether these behavioral outcomes are also affected by CSR is an interesting area for research.

From a methodological standpoint, our use of PLS path modeling has proved particularly appropriate for our model and data. Because we included both formative and reflexive constructs in the model, gathered a relatively small sample size (at the segment level), and confronted non-multivariate normality, our use of PLS path modeling was fully justified. In addition, PLS proves very useful when one wants to further utilize the latent factor scores in subsequent analyses. Whereas traditional covariance structure analysis (CSA) can hardly treat nonlinear relationships at the structural level, the use of latent scores enables modeling any kind of nonlinear links, following Ringle et al. [34]. Consequently, we were able to test the postulated non-linear relationship between CSR perceptions and organizational trust and to establish the existence of a saturation effect. The PLS approach is designed as a distribution free estimation technique [45]. Hence, contrary to the FIMX approach that deals with multivariate distribution at the latent level, the REBUS methodology does not require any specific distribution of either the measurement variables or the latent variables and then, paraphrasing Wold [45], fits into the PLS framework like hands in gloves. With PLS-REBUS, we detected sources of heterogene-

Table 1: Structural model results for pooled data and REBUS Segments: path coefficients in **bold** are statistically different across the three REBUS segments (according to the permutation tests); those in *italics* are not statistically different

Independent LV	Pooled data			REBUS segment 1 (N=122; 32 %)			REBUS segment 2 (N=160; 42 %)			REBUS segment 3 (N=96; 26 %)		
	Trust ($R^2 = 34\,\%$)			Trust ($R^2 = 35\,\%$)			Trust ($R^2 = 46\,\%$)			Trust ($R^2 = 60\,\%$)		
	Value	t	$Pr > \|t\|$	Value	t	$Pr > \|t\|$	Value	t	$Pr > \|t\|$	Value	t	$Pr > \|t\|$
Linear CSR	0.94	3.58	0.00	**0.94**	2.27	0.03	**1.97**	4.86	0.00	−1.72	−3.35	0.00
Quadratic CSR	−0.50	−1.88	0.06	−0.52	−1.27	0.21	−1.41	−3.53	0.00	**2.48**	4.77	0.00
	Identification ($R^2 = 37\,\%$)			Identification ($R^2 = 44\,\%$)			Identification ($R^2 = 48\,\%$)			Identification ($R^2 = 53\,\%$)		
	Value	t	$Pr > \|t\|$	Value	t	$Pr > \|t\|$	Value	t	$Pr > \|t\|$	Value	t	$Pr > \|t\|$
Linear CSR	0.56	13.11	0.00	*0.57*	8.04	0.00	*0.61*	10.12	0.00	**0.69**	8.39	0.00
	Commitment ($R^2 = 29\,\%$)			Commitment ($R^2 = 34\,\%$)			Commitment ($R^2 = 41\,\%$)			Commitment ($R^2 = 53\,\%$)		
	Value	t	$Pr > \|t\|$	Value	t	$Pr > \|t\|$	Value	t	$Pr > \|t\|$	Value	t	$Pr > \|t\|$
Trust	0.09	1.85	0.05	**0.12**	1.54	0.13	**0.53**	7.74	0.00	**0.75**	8.48	0.00
Identification	0.50	10.84	0.00	**0.50**	6.16	0.00	**0.19**	2.77	0.01	−0.06	−0.69	0.49

ity in all exogenous and endogenous latent variables and all model relationships. Beyond the identification of three groups and the contrasted processes through which CSR might affect AOC, the PLS-REBUS approach pursued in this research enabled us to reconsider the postulated links between two constructs. For example, we hypothesized a saturation effect between CSR and trust, which we established at the aggregate level, but the closer examination at the group level indicated that this saturation effect applied only to the identification-based and mixed groups. When employees rely on organizational trust to develop AOC, the relationship between CSR and trust becomes monotone and increasing (i.e., more CSR increases the trust level without any saturation effect). Finally, the three REBUS segments exhibit the highest R^2 and structural path coefficients between latent variables when compared to the PLS solution computed at the aggregate level.

Finally, these results have significant practical implications for firms' CSR strategies. In particular, CSR strongly influences employees' identification, trust, and AOC, which emphasizes the instrumental value of CSR and the payoff from linked corporate investments. Because employees' attitudes and behaviors constitute intangible resources that are valuable, rare, difficult to imitate, and lacking in perfect substitutes, CSR leads to intangible resources for the firms. Identification and trust significantly affect work- and job-related variables such as commitment, motivation, and turnover intentions, which are important for competitive advantages [11]. Consequently, CSR assists in creating a competitive advantage by developing a workforce that effectively carries out the firm's business strategy, leading to improved business performance. Firms with strong CSR practices in turn may develop higher productivity because of their employees' motivation, knowledge sharing, reduced absenteeism, and extra-role behavior, as well as cost benefits due to low turnover. Our results thus illustrate that the benefits of corporate contributions to communities are not restricted to external reputation and external stakeholder management but also may be reflected in the positive behaviors of internal stakeholders.

References

[1] Aberson, C. L., Healy, M., and Romero, V. (2000) Ingroup Bias and Self-Esteem: A Meta-Analysis. *Personality and Social Psychology Review*, 4: 157–173.

[2] Allen, N. J., and Meyer, J. P. (1990) The measurement and antecedents of affective, continuance and normative commitment to the organization. *Journal of Occupational Psychology*, 63: 1–18.

[3] Aryee, S., Budhwar, P. S., and Chen, Z. X. (2002) Trust as a mediator of the relationship between organizational justice and work outcomes: Test of a social exchange model. *Journal of Organizational Behavior*, 23: 267–285.

[4] Ashforth, B. E., and Mael, F. (1989) Social identity theory and the organization. *Academy of Management Review*, 14: 20–39

[5] Blau, P. M. (1964) Exchange and Power in Social Life: New York: John Wiley and Sons.

[6] Brammer, S., and Millington, A. (2005) Corporate Reputation and Philanthropy: An Empirical Analysis. *Journal of Business Ethics*, 61: 29–44.

[7] Chin, W. W., and Dibbern, J. (2010) An Introduction to a Permutation Based Procedure for Multi-Group PLS Analysis: Results of Tests of Differences on Simulated Data and a Cross

Cultural Analysis of the Sourcing of Information System Services Between Germany and the USA. In V. Esposito Vinzi et al. (eds.) Handbook of Partial Least Squares, Springer Handbooks of Computational Statistics, 171–193. Springer-Verlag Berlin Heidelberg.

[8] Chin, W. W., Marcolin, B. L. and Newsted, P. N. (2003) A partial least squares latent variable modeling approach for measuring interaction effects: results from a monte carlo simulation study and an electronic-mail emotion/adoption study. *Information Systems Research*, 14: 189–217.

[9] Cook, J. and Wall, T. (1980) New work attitude measures of trust, organizational commitment, and personal need nonfulfilment. *Journal of Occupational Psychology and Marketing*, 53: 39–52.

[10] Deery, S. J., Iverson, R. D., and Walsh, J. T. (2006) Toward a better understanding of psychological contract breach: A study of customer service employees. *Journal of Applied Psychology*, 91: 166.

[11] Datta, D. K., Guthrie, J. P., and Wright, P. M. (2005) Human resource management and labor productivity: does industry matter? *Academy of Management Journal*, 48: 135–145.

[12] Ekeh, P. P. (1974) Social exchange theory: The two traditions. Heinemann Educational.

[13] Esposito Vinzi, V. Trinchera, L., and Amato, S. (2010) PLS Path Modeling: From Foundations to Recent Developments and Open Issues for Model Assessment and Improvement. In V. Esposito Vinzi et al. (eds.) Handbook of Partial Least Squares, Springer Handbooks of Computational Statistics, 47–82. Springer-Verlag Berlin Heidelberg.

[14] Esposito Vinzi, V., Trinchera, L., Squillacciotti, S., and Tenenhaus, M. (2008) REBUS PLS: A response based procedure for detecting unit segments in PLS path modelling. *Applied Stochastic Models in Business and Industry*, 24: 439–458.

[15] Gergen, K. J. (1969) The psychology of behavior exchange: Addison Wesley Publishing Company.

[16] Handelman, J. M., and Arnold, S. J. (1999) The role of marketing actions with a social dimension: Appeals to the institutional environment. *Journal of Marketing*, 63: 33–48.

[17] Henseler, J. and Chin, W. W. (2010) A comparison of approaches for the analysis of interaction effects between latent variables using partial least squares path modeling. *Structural Equation Modeling*, 17: 82–109.

[18] Henseler, J. and Fassott, G. (2010). Testing moderating effects in PLS path models: an illustration of available procedures. In V. Esposito Vinzi et al. (eds.) Handbook of Partial Least Squares, Springer Handbooks of Computational Statistics, 713–735. Springer-Verlag Berlin Heidelberg.

[19] Hogg, M. A., and Terry, D. J. (2000) Social Identity and Self-Categorization Processes in Organizational Contexts. *The Academy of Management Review*, 25(1): 121–140.

[20] Holmes, J. G. (1981). The Exchange Process in Close Relationships; Microbehavior and Macromotives. In M. J. Lerner, and S. C. Lerner (Eds.), The Justice Motive in Social Behavior: 261–284. New York: Plenum.

[21] Lambert, L. S., Edwards, J. R. , and Cable, D.M. (2003). Breach and fulfillment of the psychological contract: A comparison of traditional and expanded views. *Personnel Psychology*, 56: 895–934.

[22] Lo, S., and Aryee, S. (2003) Psychological contract breach in a Chinese context: An integrative approach. *Journal of Management Studies*, 40: 1005–1020.

[23] Luo, X., and Bhattacharya, C. (2006). Corporate social responsibility, customer satisfaction, and market value. *Journal of Marketing*, 70: 1–18.

[24] Mael, F., and Ashforth, B. E. (1995). Loyal from day one: Biodata, organizational identification, and turnover among newcomers. *Personnel Psychology*, 48: 309–333.

[25] Maignan, I., Ferrell, O. C., and Hult, G. T. M. (1999). Corporate Citizenship: Cultural Antecedents and Business Benefits. *Journal of the Academy of Marketing Science*, 27: 455–469.

[26] Matheson, J. A., and Balichina, A. (2009). The Greenwashing Effect: Americans Are Becoming Eco-Cynical, Ecommerce Times.

[27] Meyer, J. P., Allen, N. J., and Smith, C. A. (1993). Commitment to Organizations and Occupations: Extension and Test of a Three-Component Conceptualization. *Journal of Applied Psychology*, 78: 538–551.

[28] Molm, L. D., J. L. Collett, Schaefer D.R. (2007). Building Solidarity through Generalized Exchange: A Theory of Reciprocity. *American Journal of Sociology*,113: 205–242. Molm, L. D., Takahashi, N., and Peterson, G. (2000). Risk and trust in social exchange: An experimental test of a classical proposition. *American Journal of Sociology*, 105: 1396–1427.

[29] Montes, S. D., and Irving, P. G. (2008). Disentangling the effects of promised and delivered inducements: Relational and transactional contract elements and the mediating role of trust. *Journal of Applied Psychology*, 93: 1367.

[30] Ping, R.A. (1996). Latent Variable Interaction and Quadratic Effect Estimation: A Two-Step Technique Using Structural Equation Analysis. *Psychological Bulletin*, 119: 166–175.

[31] Pivato, S., Misani, N., and Tencati, A. (2008). The impact of corporate social responsibility on consumer trust: the case of organic food. *Business Ethics: A European Review*, 17: 3–12.

[32] Pratt, M. G. (1998). To be or not to be: Central questions in organizational identification. In D. A. Whetton, and P. C. Godfrey (Eds.), Identity in Organizations, 171–208.

[33] Progressive Grocer.: Environmental Sustainability: Seeing Green. Special Report.

[34] Ringle, C. M., Wende, S., and Will, A. (2010). Finite mixture partial least squares analysis: Methodology and numerical examples. In V. Esposito Vinzi et al. (eds.) Handbook of Partial Least Squares, Springer Handbooks of Computational Statistics, 195–218. Springer-Verlag Berlin Heidelberg.

[35] Sass, J. S., and Canary, D. J. (1991). Organizational Commitment and Identification: An Examination of Conceptual and Operational Convergence. *Western Journal of Speech Communication*, 55: 275–293.

[36] Swanson, D. L. (1995). Addressing a theoretical problem by reorienting the corporate social performance model. *Academy of Management Review*, 20: 43–64.

[37] Tajfel, H. and Turner, J. C. (1985). The Social Identity Theory of Group Behavior. In H. Tajfel (Ed.), Psychology of Intergroup Relations. Cambridge: Cambridge University Press.

[38] Tenenhaus, M., Esposito Vinzi, V. E., Chatelin, Y. M., and Lauro, C. (2005). PLS path modeling. *Computational Statistics and Data Analysis*, 48: 159–205.

[39] Turker, D. (2008). How corporate social responsibility influences organizational commitment. *Journal of Business Ethics*, 89: 189–204.

[40] Turker, D. (2009). Measuring corporate social responsibility: a scale development study. *Journal of Business ethics*, 85: 411–427.

[41] Tyler, T. R. (1999). Why people cooperate with organizations: An identity-based perspective. *Research in organizational behavior*, 21: 201–246.

[42] Tyler, T. R., and Blader, S. L. (2002). Autonomous vs. comparative status: Must we be better than others to feel good about ourselves? *Organizational Behavior and Human Decision Processes*, 89: 813–838.

[43] Tyler, T. R., and Blader, S. L. (2003). The Group Engagement Model: Procedural Justice, Social Identity, and Cooperative Behavior. *Personality and Social Psychology Review*, 7: 349–361.

[44] Whitener, E. M., Brodt, S. E., Korsgaard, M. A., and Werner, J. M. (1998). Managers as initiators of trust: An exchange relationship framework for understanding managerial trustworthy behavior. *Academy of Management Review*, 23: 513–530.

[45] Wold, H. (1975). Path models with latent variables: The NIPALS approach. In H. M. Blalock, et al. (eds.), Quantitative sociology: International perspectives on mathematical and statistical modeling: 307–357.NewYork: Academic.

[46] Wood, D. J. (1991) Corporate social performance revisited. *Academy of Management Review*, 16: 691–718.

[47] Wood, D. J., and Jones, R. E. (1995). Stakeholder mismatching: A theoretical problem in empirical research on corporate social performance. *International Journal of Organizational Analysis*, 3: 229–267.

Extending the PATHMOX Approach to Detect Which Constructs Differentiate Segments

Tomas Aluja-Banet, Giuseppe Lamberti, and Gastón Sánchez

Abstract In most cases, path modeling data come from surveys or researches that contain more information (i.e., observed heterogeneity) than is used for the path models definition. For instance, in many marketing studies like those of consumer satisfaction, it is usual to collect socio-demographic variables and psycho-demographic variables such as age, gender, social-status, or consumers' habits that take no part in the path model but that can be extremely useful for segmentation purposes. In 2009, Gastón Sánchez introduced the PATHMOX methodology to incorporate the available external variables to identify different segments. The algorithm solves this problem by building a binary tree to detect those segments present in the population that cause the heterogeneity. The F-global test, based on the Fisher's F for testing the equality of two regression models, is adapted and used, as a splitting criterion, to discover whether two structural models calibrated from two different segments (i.e., two successors of a node), can be considered to be different. However PATHMOX does not identify which of the block or variables indicators are responsible for the heterogeneity. In this article we propose to extend the PATHMOX methodology to test the equality of every endogenous equation of the structural model in order to compare all path coefficients of the structural model estimated in two segments.

Key words: Heterogeneity, PLSPM, Segmentation, PATHMOX, Models comparison, Fisher's F

T. Aluja-Banet (✉) • G. Lamberti
Universitat Politecnica de Catalunya, Campus Nord UPC. C5204.
Jordi Girona 1-3, 08034 Barcelona, Spain
e-mail: tomas.aluja@upc.edu;giuseppelamb@hotmail.com

G. Sánchez
CTEG, Unité de Recherche de Sensométrie et Chimiométrie (USC INRA), ONIRIS,
Site de la Géraudière, Rue de la Géraudière BP 82225, Nantes 44322 Cedex 3, France
e-mail: gaston.stat@gmail.com

H. Abdi et al. (eds.), *New Perspectives in Partial Least Squares and Related Methods*, Springer Proceedings in Mathematics & Statistics 56, DOI 10.1007/978-1-4614-8283-3_19,

1 The PATHMOX Approach

In 2009, Gastón Sánchez proposed the PATHMOX algorithm with the purpose to develop a new segmentation approach for observed heterogeneity in Partial Least Squares Path Models (PLS-PM) ([7]). This technique adapts the principles of binary segmentation processes, to produce a segmentation tree with different path models in each of the obtained nodes.

PATHMOX does not involve any prediction purpose, but rather an identification goal (i.e., to detect different path models present in the data). For this purpose, the PATHMOX approach identifies a set of splits (based on the segmentation variables) with superior discriminating capacity in the sense of separating PLS-PM models as much as possible. Here a split criterion based on Fisher's F for testing the equality of regression models ([1–6]), has been adapted to decide whether two structural models, calibrated from two different segments (successors of a node), can be considered to be different. We will call it the *F-global test*. To identify the existence of different path models the technique performs a procedure that can be summarized in the following algorithm:

Algorithm 5 PATHMOX algorithm

Step 1. Start with the global PLS path model at the root node
Step 2. Establish a set of admissible partitions for each segmentation variable in each node of the tree
Step 3. Detect the best partition by:
- **3.1.** Compare all binary partitions in all segmentation variables
- **3.2.** Apply the F-global test, calculating for each comparison a p-value
- **3.3.** Sort the p-values in a descending order
- **3.4.** Chose as the best partition the one associated to the lowest p-value

Step 4. ***If*** (stop criteria[1] = **false**) ***then***
repeat **step 3**

1. Posible stop criteria:
a. The number of individuals in the group falls below a fixed level
b. The p-values F-global test are not significant
c. Maximum level of tree's depth attained

Before discussing the extension of the PATHMOX approach to detect the sources of heterogeneity, we illustrate in Sect. 2 how PATHMOX works. For this demonstration we consider a customer satisfaction model for three Spanish mobile carriers.

2 PATHMOX Application: Estimation of a Customer Satisfaction Model for Three Spanish Mobile Carriers

The data come from a survey collected on 87 customers and 26 questions grouped in 7 sets regarding 7 latent variables (showed in Table 1). In addition to the 26 manifest variables, 7 segmentation variables are considered: ***gender*** (*female–male*), ***age*** (*less then 25–more-equal then 25*), ***occupation*** (*employee–student*), ***education*** (*basic–high-school–university*), ***type of contract*** (*contract–prepay*), ***carrier*** (*A–B–C*), and ***switch of provider*** (*YES–NO*).

Table 1: Description of latent variables of Mobile's dataset

LV	Description
Image	Includes variables such as trustworthiness, dynamic, solidness, innovation, and caring about customer's needs
Expectation	Includes variables such as products and services provided and expectations for the overall quality
Quality	Includes variables such as reliable products and services, range of products and services, and overall perceived quality
Complaints	Includes one variable defining how well or poorly customers' complaints were handled
Value	Includes variables such as service and products, quality relative to price, and price relative to quality
Satisfaction	Includes variables such as overall rating of satisfaction, fulfillment of expectations, satisfaction relative to other phone providers
Loyalty	Includes variables such as propensity to choose the same phone provider again, intention to recommend the phone provider to friends

We begin with the calculation of the global PLS model for all customers (Fig. 1). For the main constructs of the model: *satisfaction* and *loyalty*, we have the following latent equations:

$$\textbf{satisfaction} = 0.4887 \times \textit{image} + 0.0913 \times \textit{expectation} - 0.0289 \times \textit{quality} + 0.4870 \times \textit{value}$$

$$\textbf{loyalty} = 0.2948 \times \textit{image} + 0.4889 \times \textit{satisfaction} + 0.1312 \times \textit{complaints}.$$

We can see that the main drivers of *satisfaction* are *image* and *value*, whereas, in the case of *loyalty*, are *image* and *satisfaction.*

Figure 1 shows the relation between all the latent variables, under the assumption that the model is valid for all customers. But, how can we be sure of this hypothesis? To answer this question, we apply PATHMOX to the data.

In Fig. 2, we present the obtained tree where we can observe that in fact there are three PLS-PM models. At the first split, PATHMOX defines two different models one for customers of carrier *A*, and the other, for customers of carriers *B* and *C*. We can see that the produced split is highly significant, as it gives an F-statistic of 3.5448 with has p-value smaller close to zero (i.e., smaller than 0.0001). The node of customers of carrier *A* (Node 2) is taken as final because it contains only 12

Fig. 1: Inner model of mobile data

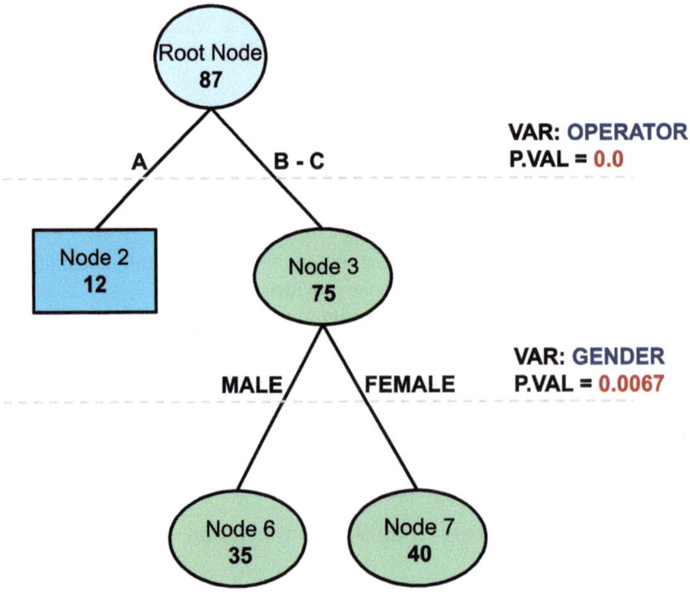

Fig. 2: PATHMOX's segmentation tree

individuals, which is below the threshold of 15 % of total population imposed by the splitting criterion.[1] The tree continues by splitting the customers of carriers *B* and *C* (Node 3). The most significant split is obtained by the variable *gender*, giving an

[1] The criterion of minimum number of individuals avoids the fragmentation of small nodes. The minimum threshold is not a prefixed parameter in PATHMOX, but must be determined by the analyst. Generally it is chosen between 10 and 15 % of the total number of individuals, depending on the total size of population, to obtain consistent groups because it does not make sense to split small nodes

F-statistic of 2.3342 with a p-value of 0.0067, obtaining a child node with the *male* customers and another with *female* customers. This ends the splitting process, as the maximum depth of two levels has been reached. Hence, at the end, we have three final segments each corresponding to distinct PLS-PM models: **Node 2** *model of carrier A customers,* **Node 6** *model of carriers B-C male customers,* **Node 7** *model of carriers B-C female customers.*

In Fig. 3, we present the PLS-PM models corresponding to the three segments. We can see that the customers of carrier A are the most influenced by the *image* of the company on either *satisfaction* or *loyalty*, whereas for *male* customers of carriers B and C, we can see that the *value* is the most important asset for *satisfaction*, in the same way it is the most important for *loyalty*. Looking at the *female* customers of carriers B and C, we can observe that *image* and *value* are more balanced on *satisfaction* and that the care service in *complaints* is important for *loyalty*.

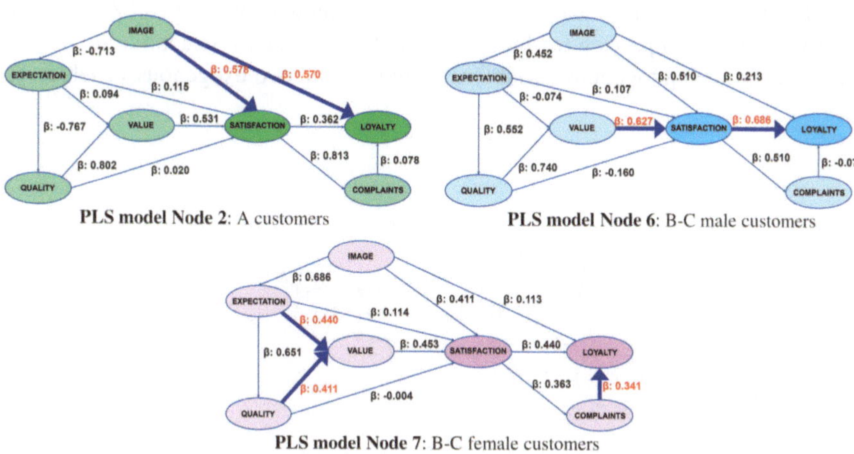

Fig. 3: PLS: PM of three PATHMOX child nodes

3 Extending the PATHMOX Approach

As we can see, the PATHMOX approach allows us to detect the existence of different path models in a data-set without identifying segmentation variables beforehand: the different segments are revealed as branches of the segmentation tree. However the F test used in PATHMOX as split criterion is a global criterion: it allows assessing whether all the path coefficients for two compared structural models are equal or not, but it does not indicate which particular endogenous equation and which path coefficients are responsible of the difference. For instance, when PAHTMOX detects a difference between two groups such as A customers and B-C customers, we do not know which ones of the six structural equations (one for each endogenous latent variable) is responsible of the detected difference. Also, if we have a significant difference in one structural equation—for instance in *satisfaction*—we do not know what path coefficient is responsible (i.e., *image*, *expectation*, *value*, or *quality*). To identify the significant distinct endogenous equation and the responsible path coefficients of the split, we introduced the F-block test and the F-coefficient test.

3.1 F-Block Test

To detect which endogenous regression (i.e., endogenous equation) is responsible for the global difference, we have extended the F-*global* test, to compare the equality of each endogenous equation of the structural model. We will call F-*block* the statistic of this comparison (or block-test). Let us consider a structural model (see Fig. 4) with two endogenous variables, η_1 and η_2 and two exogenous variables ξ_1, ξ_2:

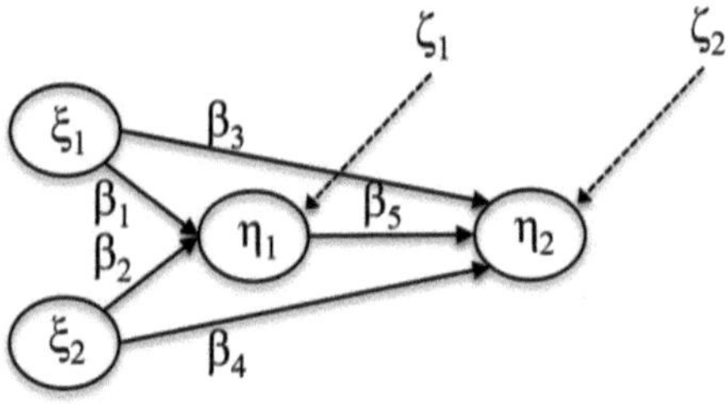

Fig. 4: Structural model of simulation study

The structural equations for both endogenous constructs are:

$$\eta_1 = \beta_1\xi_1 + \beta_2\xi_2 + \zeta_1 \quad (1)$$

$$\eta_2 = \beta_3\xi_3 + \beta_4\xi_4 + \beta_5\eta_1 + \zeta_2. \quad (2)$$

The disturbance terms ζ_1 and ζ_2 are assumed to be normally distributed with zero mean and finite variance, that is, $E(\zeta_1) = E(\zeta_2) = 0$ and $Var(\zeta_1) = Var(\zeta_2) = \sigma^2$ I. It is also assumed that $Cov(\zeta_1, \zeta_2) = 0$.

We define the following matrices:

$X_1 = [\xi_1, \xi_2]$ a column matrix with the explicative latent variables of η_1
$B_1 = [\beta_1, \beta_2]$ a vector of path coefficients for the regression of η_1
$X_2 = [\xi_1, \xi_2, \eta_1]$ a column matrix with the explicative latent variables of η_2

$B_2 = [\beta_3, \beta_4, \beta_5]$ a vector of path coefficients for the regression of η_2

Then, the structural equations are expressed as:

$$\eta_1 = X_1 B_1 + \zeta_1 \tag{3}$$
$$\eta_2 = X_2 B_2 + \zeta_2 \tag{4}$$

We assume that the parent node is divided in two child nodes or segments, one containing n_A elements and the other containing n_B observations. For each segment we compute a structural model:

$$\text{Segment } A : \eta_1^A = X_1^A B_1^A + \zeta_1^A \quad \text{and} \quad \eta_2^A = X_2^A B_2^A + \zeta_2^A \tag{5}$$
$$\text{Segment } B : \eta_1^B = X_1^B B_1^B + \zeta_1^B \quad \text{and} \quad \eta_2^B = X_2^B B_2^B + \zeta_2^B \tag{6}$$

with $\zeta_1^A \sim N(0, \sigma^2 I)$, $\zeta_2^A \sim N(0, \sigma^2 I)$, $\zeta_1^B \sim N(0, \sigma^2 I)$, and $\zeta_1^B \sim N(0, \sigma^2 I)$.

Let us assume that the F-global test gives a significant p-value. We want to investigate which equation is the endogenous equation responsible of the difference. For the sake of simplicity, we want to test whether the first endogenous equation is equal in both segments while letting the second equation free to vary. In this case, the null hypothesis, H_0, is that the endogenous equation showed in (1) is equal for segments A and B, while the alternative hypothesis, H_1, is that the all endogenous equations are different. The two hypothesis can be written as follows:

$$H_0 : \underset{[2n,1]}{\begin{bmatrix} \eta_1^A \\ \eta_2^A \\ \eta_1^B \\ \eta_2^B \end{bmatrix}} = \underset{[2n,p_1+2p_2]}{\begin{bmatrix} X_1^A & 0 & 0 \\ 0 & X_2^A & 0 \\ X_1^B & 0 & 0 \\ 0 & 0 & X_2^B \end{bmatrix}} \underset{[p_1+2p_2,1]}{\begin{bmatrix} \beta_1 \\ \beta_2^A \\ \beta_2^B \end{bmatrix}} + \underset{[2n,1]}{\begin{bmatrix} \zeta_1^A \\ \zeta_2^A \\ \zeta_1^B \\ \zeta_2^B \end{bmatrix}} \tag{7}$$

$$H_1 : \underset{[2n,1]}{\begin{bmatrix} \eta_1^A \\ \eta_2^A \\ \eta_1^B \\ \eta_2^B \end{bmatrix}} = \underset{[2n,2p_1+2p_2]}{\begin{bmatrix} X_1^A & 0 & 0 & 0 \\ 0 & X_2^A & 0 & 0 \\ 0 & 0 & X_1^B & 0 \\ 0 & 0 & 0 & X_2^B \end{bmatrix}} \underset{[2p_1+2p_2,1]}{\begin{bmatrix} \beta_1^A \\ \beta_1^B \\ \beta_2^A \\ \beta_2^B \end{bmatrix}} + \underset{[2n,1]}{\begin{bmatrix} \zeta_1^A \\ \zeta_2^A \\ \zeta_1^B \\ \zeta_2^B \end{bmatrix}} \tag{8}$$

where $n = n_A + n_B$ is the number of elements in the model containing the two nodes; p_j is the number of explicative latent variables for each j-th endogenous construct $j = 1, \ldots, J$ (in this example $J = 2$). We define the matrices X_0, and X corresponding two both hypothesis as:

$$X_0 = \underset{[2n,p_1+2p_2]}{\begin{bmatrix} X_1^A & 0 & 0 \\ 0 & X_2^A & 0 \\ X_1^B & 0 & 0 \\ 0 & 0 & X_2^B \end{bmatrix}} \quad X = \underset{[2n,2p_1+2p_2]}{\begin{bmatrix} X_1^A & 0 & 0 & 0 \\ 0 & X_2^A & 0 & 0 \\ 0 & 0 & X_1^B & 0 \\ 0 & 0 & 0 & X_2^B \end{bmatrix}} \tag{9}$$

Then we can see that $X_0 = XA$ with matrix A defined as:

$$A = \underset{[\,2p_1+2p_2, p_1+2p_2].}{\begin{bmatrix} I_{p_1} & 0 & 0 \\ 0 & I_{p_2} & 0 \\ I_{p_1} & 0 & 0 \\ 0 & 0 & I_{p_2} \end{bmatrix}} \tag{10}$$

Where I_{p_j} is the identity matrix of order p_j. We can apply **Lemma 2** from Lebart [6] to test the H_0 hypothesis by computing the following F statistic with (p_1) and $2(n-p_1-p_2)$ degrees of freedom.

$$F_{Block} = \frac{(SS_{H_0} - SS_{H_1}) \Big/ p_1}{SS_{H_0} \Big/ 2(n-p_1-p_2)} \tag{11}$$

3.2 *F*-Coefficient Test

Let us now suppose that the difference between the first structural equation in Segments 1 and 2 is significant, (i.e., Segment A differs from segment B). We want to investigate which are the responsible coefficients for this difference. Let us consider the same structural model showed in Fig. 4. For sake of simplicity we want to test the equality of coefficient β_1, of the first equation in both segments. We re-adapt the same global F test to this situation. The two hypotheses are written as follow:

$$H_0: \underset{[2n,1]}{\begin{bmatrix} \eta_1^A \\ \eta_2^A \\ \eta_1^B \\ \eta_2^B \end{bmatrix}} = \underset{[2n, 2\sum_{j=1}^{P} p_j - 1]}{\begin{bmatrix} \xi_1^A & \xi_2^A & 0 & 0 & 0 & 0 & 0 & 0 & 0 \\ 0 & 0 & \xi_3^A & \xi_4^A & \eta_1^A & 0 & 0 & 0 & 0 \\ \xi_1^B & 0 & 0 & 0 & 0 & \xi_2^B & 0 & 0 & 0 \\ 0 & 0 & 0 & 0 & 0 & 0 & \xi_3^B & \xi_4^B & \eta_1^B \end{bmatrix}} \underset{[2\sum_{j=1}^{P} p_j - 1, 1]}{\begin{bmatrix} \beta_1 \\ \beta_2^A \\ \beta_3^A \\ \vdots \\ \beta_5^B \end{bmatrix}} + \underset{[\,2n,1]}{\begin{bmatrix} \zeta_1^A \\ \zeta_2^A \\ \zeta_1^B \\ \zeta_2^B \end{bmatrix}} \tag{12}$$

$$H_1: \underset{[\,2n,1]}{\begin{bmatrix} \eta_1^A \\ \eta_2^A \\ \eta_1^B \\ \eta_2^B \end{bmatrix}} = \underset{[\,2n, 2\sum_{j=1}^{P} p_j]}{\begin{bmatrix} \xi_1^A & \xi_2^A & 0 & 0 & 0 & 0 & 0 & 0 & 0 & 0 \\ 0 & 0 & \xi_3^A & \xi_4^A & \eta_1^A & 0 & 0 & 0 & 0 & 0 \\ 0 & 0 & 0 & 0 & 0 & \xi_1^B & \xi_2^B & 0 & 0 & 0 \\ 0 & 0 & 0 & 0 & 0 & 0 & 0 & \xi_3^B & \xi_4^B & \eta_1^B \end{bmatrix}} \underset{[2\sum_{j=1}^{P} p_j, 1]}{\begin{bmatrix} \beta_1^A \\ \beta_2^A \\ \beta_3^A \\ \vdots \\ \beta_5^B \end{bmatrix}} + \underset{[\,2n,1]}{\begin{bmatrix} \zeta_1^A \\ \zeta_2^A \\ \zeta_1^B \\ \zeta_2^B \end{bmatrix}} \tag{13}$$

Denoting X_0 the design matrix of the null hypothesis and X the design matrix of the alternative hypothesis, we have $X_0 = XA$, where:

$$A = \begin{bmatrix} 1\,1\,0\,0\,0\,0\,0\,0\,0 \\ 0\,0\,1\,1\,1\,0\,0\,0\,0 \\ 1\,0\,0\,0\,0\,1\,0\,0\,0 \\ 0\,0\,0\,0\,0\,0\,1\,1\,1 \end{bmatrix}_{[2p_1+2p_2, p_1+2p_2]} \tag{14}$$

Then, as before, applying **Lemma 2** from Lebart [6], we can test the H_0 hypothesis computing the following F-coefficient statistic with 1 and $2(n\sum_{j=1}^{P} p_j)$ degrees of freedom.

$$F_{Coefficient} = \frac{(SS_{H_0} - SS_{H_1}) \big/ 1}{SS_{H_0} \big/ 2(n\sum_{j=1}^{P} p_j)} \tag{15}$$

4 Simulation

In order to evaluate the sensitivity of the split criterion used in F-global, F-block and F-coefficient test, we run a series of Monte Carlo simulations. We have evaluated the performance of the test under different experimental conditions.The factors of the experimental design are the following: distribution of data, difference between coefficients, sample of size, variances of the endogenous error terms. We have considered the same structural model of (1) with two endogenous variables, η_1 and η_2, and two exogenous variables ξ_1, ξ_2.

4.1 Experimental Factors

4.1.1 Data Distributions

To detect the behavior of the tests with different data distributions, we generate the exogenous constructs ξ_1 and ξ_2 as realizations from a beta distribution $\beta_{(a,b)}$. In order to take into account both symmetry and skewness in distributions for the latent variables, three cases of parameters a and b for the beta distribution are considered: $\beta_{(6,6)}$, $\beta_{(9,4)}$, $\beta_{(9,1)}$.

4.1.2 Differences Between Coefficients

The model has been estimated in two segments A and B, varying the levels of the difference between path coefficients, that is, they can be *EQUAL* in both segments, or the difference can be *SMALL*, *MEDIUM* and *LARGE*, meaning that we have added +0, +0.3, +0.5, and +0.8 respectively to the corresponding path coefficients of segment A. All the coefficients in the model have been modified in the same way.[2]

[2] When, for example, the *SMALL* difference scheme is considered, 0.3 is added to all the coefficients for segment A

4.1.3 Size

We consider five sample sizes as the total number of cases: $\{100, 200, 400, 500$ and $1{,}000\}$. We take balanced segments in all cases.

4.1.4 Variance of Endogenous Terms

We assume that the error term ζ follows a normal distribution with zero mean and three different levels of variance. The levels are chosen such that the variance of ζ accounts for 10, 20, 30, 50 and 90 % of the total variance of η.

In total, we have $3 \times 4 \times 5 \times 5 = 300$ scenarios, which are the number of possible combinations of sample sizes, beta distributions, noise levels, and difference between coefficients. We run ten repetitions for each experimental condition.

5 Results of the Simulation Study

We present the results of the simulations in two parts. In the first part we want to verify if the three F-tests have the same sensitivity to detect heterogeneity. The behavior is evaluated with respect to: real difference between coefficients, distribution of data, sample size and levels in error variance terms. In the second part we want to verify if the three F-tests are consistent with respect to the distribution of data, sample size and levels in error variance terms.

5.1 Analysis of the Behavior of F-Global, F-Block, and F-Coefficient

We use the simulation to illustrate the behavior of these three tests regarding the different sample sizes (100, 200, 400, 500, and 1,000), the different distributions, and the different levels in error variance terms of the endogenous construct. In Fig. 5 we present the boxplots of the p-values according to the different levels of experimental factors: differences between path coefficients, sample size, variance of disturbance terms, and data distribution. The results indicate that the three tests present a very similar behavior. We can see that:

1. There is a clear effect of the differences between the path coefficients in the two segments: the more different the path coefficients the more sensitive the tests.
2. There is a clear effect of sample size: the larger the sample size the more sensitive the tests.
3. There is no apparent effect of the data distribution on the sensitivity of the tests.
4. There is a clear effect of the level of noise: the larger the level of noise is, the less sensitive the tests are.

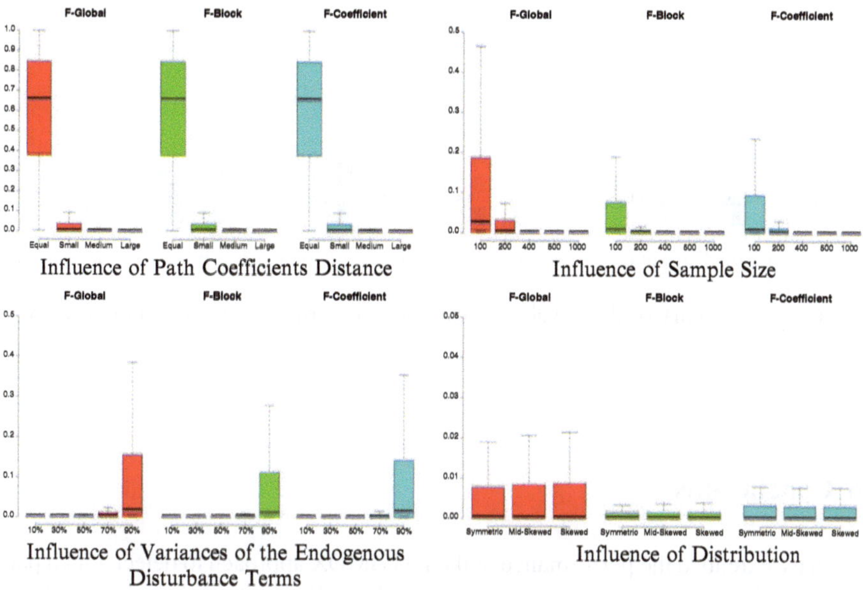

Fig. 5: Influence of different data generating conditions on the significance of the three F-tests at the aggregated level

5.2 Comparison Between F-Global, F-Block and F-Coefficient Tests

Analyzing the behavior of F-global, F-block and F-coefficient with respect to the distribution of data, levels of variances of the endogenous disturbance terms, and sample size, we want to verify whether the three F-tests are consistent.[3] For each comparison we calculate the error ratio as the proportion of discordances, (i.e., cases in which we obtain opposite p-values (for example the F-global is significant whereas the F-block is not significant or vice-versa, or the F-block is significant whereas the F-coefficient is not significant or vice-versa). In Fig. 6, we present the distribution of the error ratio considering the different factors previously mentioned. We observe that on average, we have a 5 % of error ratio discordance when comparing outcome of the F-global with F-block test, and 7 % comparing the F-block with F-coefficient test. We can clearly see two trends: the error ratio decreases when the sample size increases, the error ratio increases when the level of error variance terms increases. The form of distribution have no impact on the error ratio (i.e., the error ratio is almost constant).

[3] When the F-coefficient test finds a significant difference between two path coefficients, should it imply that the F-block test—comparing the two endogenous equations containing the significant coefficient—also gives a significant p-value. Likewise the F-global test should give a significant p-value comparing the PLS model that contains this endogenous equation.

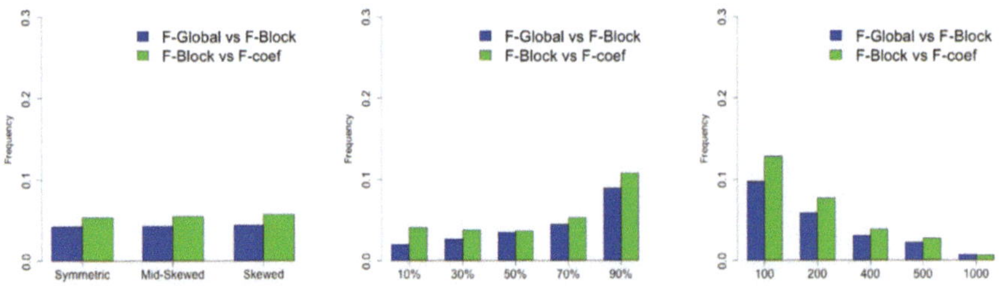

Error ratio by distribution Error ratio by variance error terms Error ratio by by sample size

Fig. 6: Summary of the p-value distribution error ratio of the simulation study

6 Conclusions

We have extended the performance of the PATHMOX approach to detect which particular path coefficients of a structural model (F-coefficient test) and which endogenous equation (F-Block test) are responsible for an observed difference between two models detected by the F-Global test of PATHMOX. We have seen how the F-test of model comparison of Lebart ([1, 6]) can be adapted for these purposes. In the performed simulation we have seen that a concordance does not always not always exists between these three aforementioned tests (F-coefficient, F-block, and F-global). We have also seen, however, that for large samples the concordance is almost perfect. In addition, we have seen that the original distribution of the data does not affect the sensitivity of the tests. However we have seen that sample size clearly affects the results of the tests and, obviously, that the difference of path coefficients in both segments clearly affects the significance of the tests.

References

[1] Chow, G.C. (1960) Test of equality between sets of coefficients in two linear regressions. *Econometrica*, 28(3): 591–605;

[2] Cleveland, W.S. (1979) Robust Locally Weighted Regression and Smoothing Scatterplots. Journal of the American Statistical Association, 74, 829–836;

[3] Chin, W.W. (2000) Frequently Asked Questions - Partial Least squares PLS-Graph. Available from http://disc-nt.cba.uh.edu/chin/plsfaq/plsfaq.htm;

[4] Chin, W.W. (2003) A Permutation Based Procedure for Multi- Group Comparison of PLS Models. In: *Proceedings of the PLS03 International Symposium*, 33–43. M. Vilares., M. Tenenhaus, P. Coelho, V. Esposito Vinzi, A. Morineau (Eds), Decisia;

[5] Gaston Sanchez (2009) PATHMOX Approach: Segmentation Trees in Partial Least Squares Path Modeling;

[6] Lebart, L., Morineau, A. and Fenelon, J.P. (1979) Traitement des donneés statistiques. Paris. Dunod;

[7] Wold, H. (1982b) Soft modeling: The Basic Design and Some Extensions. In: Systems under indirect observation: Causality, structure, prediction. Part II, 1–54. K.G. Jöreskog & H. Wold (Eds). Amsterdam: North Holland;

Part VI
Management, Operations and Information Systems

Integrating Organizational Capabilities to Increase Customer Value: A Triple Interaction Effect

Gabriel Cepeda, Silvia Martelo, Carmen Barroso, and Jaime Ortega

Abstract The aim of this paper is to contribute to the strategic management literature by identifying empirically possible combinations of three organizational capabilities and to analyze whether the possible interaction between them leads to the creation of superior customer value. We aim to determine how the interaction between three capabilities (i.e., market orientation, knowledge management and customer relationship management) is and the potential effects of this relationship for increasing customer value. In order to test this question, we model a triple interaction effect following an orthogonalization approach using partial least squares (PLS). We used data from Spanish banking industry. Surprisingly, the triple effect explains more variance of customer value than the alternative operationalizations of the three organizational capabilities linked to customer value creation.

Key words: Organizational capabilities, Customer value, Interaction effects, Orthogonalization approach

1 Introduction

Given the increasing intensity of competition in business and the strong trend towards globalization, attitudes towards the customer are very important; their role has changed from that of a mere consumer to one of consumer, co-operator, co-producer, co-creator of value and co-developer of knowledge and competencies [1]. Furthermore, the complex competitive environment in which firms operate has led to the increase in customer demand for superior value [2]. Therefore, more and more firms see customer value as a key factor when looking for new ways to achieve and maintain a competitive advantage [3, 4].

G. Cepeda (✉) • S. Martelo • C. Barroso • J. Ortega
University of Seville, Seville, Spain
e-mail: gabi@us.es; smartelo@us.es; barroso@us.es; joguti@us.es

H. Abdi et al. (eds.), *New Perspectives in Partial Least Squares and Related Methods*, Springer Proceedings in Mathematics & Statistics 56, DOI 10.1007/978-1-4614-8283-3_20,

A firm's organizational capabilities are of vital importance for increasing customer value creation. Thus, a firm should focus on improving those capabilities that view the customer as its key component, in order to maximize the value created for them. We will emphasize the next three capabilities: 'market orientation' (MO), 'knowledge management' (KM) and 'customer relationship management' (CRM). It is of interest that, although all these capabilities are developed by the companies and could therefore be considered to be internal in nature, a relationship with the customer and the capabilities associated with market orientation require strong external contact for them to be developed.

After reviewing the existing literature, it is clear that each of these three capabilities is linked to customer value. The primary aim of market-oriented firms, firms that manage their knowledge or those that manage customer relationships is to offer superior customer value. However, there is no single or intermittent influence that is important, but rather, the effect of the three capabilities has to be global and sustainable (i.e., permanent). According to [5], merely possessing valuable and rare resources and capabilities, does not guarantee the development of competitive advantage or the creation of value; firms must be able to manage them effectively. It follows therefore that value can also be created by recombining existing resources and capabilities [6]. It should be possible to reconfigure organizational capabilities so that the firm can be continually creating value, and this is where dynamic capabilities (DC) come into play.

We have not come across any other papers in the previous literature that deal with this relationship between the three proposed organizational capabilities, or any that consider its influence on customer value. We will address this gap in the literature by stating that customer value will be increased if there is interaction between the three proposed capabilities (MO, KM and CRM). The idea is to see how the three proposed capabilities jointly influence customer value. We will also state that the interaction between them can constitute a DC (viewed as a "black box"), which allows a firm to maintain its competitive advantage. Specifically, our research question is: If the customer demands superior value, how should a firm combine its existing capabilities in order to offer this superior value?

In short, the aim of this paper is to contribute to the strategic management literature by identifying empirically possible combinations of the three proposed organizational capabilities and to analyze whether the possible interaction between them leads to the creation of superior customer value. We aim to determine how the interaction between the three capabilities (MO, KM and CRM) is and the potential effects of this relationship for increasing customer value.

We therefore propose the model and hypothesis illustrated in Fig. 1.

Hypothesis 1: The interaction between MO, KM and CRM is positively related to customer value creation.

In order to test this question, we model a triple interaction effect following an orthogonalization approach [7, 8] using partial least squares (PLS). The interaction term is composed of the three aforementioned capabilities (MO, KM and CRM) impacting on customer value.

2 Methodology

2.1 Data Collection

The research hypotheses were tested within the Spanish banking industry, including retail and commercial banks and savings banks serving the general public, representing around 18 % of the national GDP.

We have chosen this industry because we consider banks to be the type of business that simultaneously demonstrates the four organizational capabilities proposed in our model (MO, KM, CRM and customer value creation). Banking is a very knowledge-intensive industry and therefore an appropriate one in which to identify, analyze and evaluate these capabilities. The increasingly intense competition within the financial service industry means that banks need to recognize the need to look for new ways of creating customer value. Alongside the competitiveness of the industry, the relative intangibility of their products/services creates the need to capture and retain customers by offering them something extra, through MO, KM and CRM. These aspects indicate that this industry is best suited to our study.

It is important to point out the significant crisis in the financial services industry, both currently and at the time we carried out the study. The crisis has forced many countries to apply severe measures to reduce the impact on their financial services industry. Numerous banks and insurance companies have been taken over or capitalized, company mergers as a rescue measure have multiplied and crashes have increased. The full extent of the crisis is still unknown, since events have occurred at an unusually high speed, leading to enormous changes within a short time, mainly subsequent to the collapse of Lehman Brothers in September 2008.

At the time of the study there was a total of 85 banks operating in Spain; of which 40 were commercial/retail banks and 45 were savings banks.

The low number of entities that comprise the banking industry in Spain can be viewed either as an advantage or a disadvantage. On the one hand, it allows us to look at the whole population rather than a particular sample of it. But, on the other

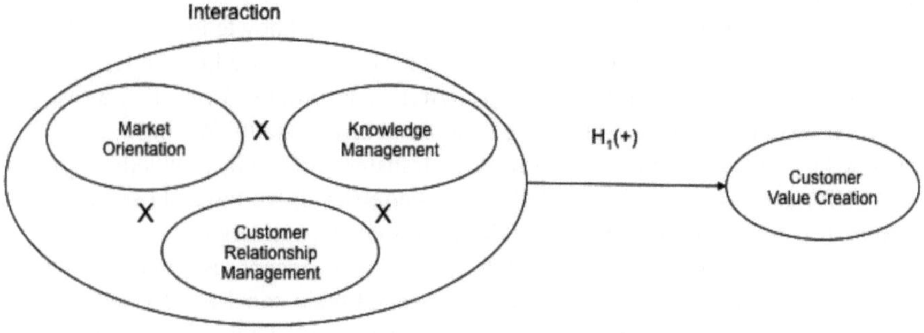

Fig. 1: The conceptual model

hand, we are forced to work with a small sample size that can lead to problems in the analysis of data, as we will see later.

The response rate to the questionnaire was high, at around 90 %, with 76 of the 85 bodies responding. It is important to note that all of the completed questionnaires were valid.

We also gathered data from customers in order to gain a more precise picture of the value generated through the three capabilities (i.e. MO, KM, and CRM). A pool of customer data (with a minimum of 30 customers) was obtained for each of the 76 banks to observe standard customer behavior regarding perceived value. We decided to integrate these two sets of data (bank and customer) to test the hypotheses in our theoretical model.

Furthermore, because the data sample (76) is very close to the real population in Spanish banking industry (85), we used the factor correction suggested by [9] to adapt the standard error generated.

2.2 Measures

We would point out that all the constructs included in the questionnaires have been measured against existing scales in the literature and we have therefore used instruments whose validity and reliability has already been proven in other research.

To measure MO we used the Narver and Slater [10] 15-item scale (the so-called MKTOR scale), consisting of three dimensions: customer orientation (CO), competitor orientation (COO) and interfunctional coordination (IC). We believe the MKTOR scale is appropriate for our study, with its emphasis on customer orientation, since the customer is the main object of our study. We also believe that it is appropriate to use the MKTOR scale because our study should have a strategic perspective and we believe that the cultural focus of this scale is better suited to our study than the behavioral focus of Jaworski and Kohli's [11] MARKOR scale. After cleaning the data, only 11 items were used.

We have created our own scale to measure KM, taking items from various scales used in previous investigations. From our literature review, we identified four key dimensions that affect KM processes: knowledge creation, knowledge transfer, knowledge application and knowledge storage/retrieval. To measure knowledge creation, we have chosen an absorptive capacity (AC) scale proposed by [12], as we believe this would add to the conceptual richness of our study. We have used Gold et al.'s [13] scales to measure knowledge transfer (KT) and knowledge application (KA) and, to measure knowledge storage/retrieval, we have chosen the scale to measure organizational memory (OM) proposed by [14]. Organizational memory refers to the processing of saved knowledge, and this concept resembles our idea of knowledge storage and retrieval. The final cleaned scale consists of 9 items for the creation dimension, 10 items for the transfer dimension, 10 items for the application dimension and 4 items for the storage/retrieval dimension.

To measure CRM, we have used Reinartz et al.'s [15] scale, which measures the initiation (IN), and maintenance and termination (MT) phase of the CRM processes. We consider this to be a very intuitive scale and easy to understand in practice. Due to the high number of items (the original scale consists of 39 items), we have selected items closest to the concepts, ideas and objectives of our study and have created a CRM scale consisting of 12 items (7, and 5 items, respectively). A group of experts, using a Delphi method, judged whether those 12 items were the most appropriate for the objectives of the study and the final, cleaned scale consists of 7 items.

In the case of the customer value creation capability, and after a review of the scales developed in previous investigations, we opted for the scale proposed by [16]. The lack of measurement proposals for the creation of customer value makes it more difficult to select the most appropriate instrument for this construct. We have used Hooley et al.'s [16] scale because we consider that it is complete and refers to the creation of value for customers, as opposed to other proposals, which analyze value creation for all the stakeholders.

We then created the double interaction terms and the triple interaction term using Little et al.'s [8] orthogonalization approach based on Lance's [17] residual centering regression approach. The approach involves a three-step procedure in which the double interaction terms are first regressed on their own components via ordinary least squares and then residuals of this regression are used instead of the respective double interaction terms in tests of the structural model. The triple interaction term is then also regressed on its three components and the double interaction components via ordinary least squares and the residuals of this regression are also used instead of the respective triple interaction term in tests of the structural model. Following the suggestion by [7] derived from Monte Carlo simulations, we chose the orthogonalization approach over alternatives such as a product indicator, because the former delivers the most accurate point estimates for interaction effects. Moreover, it has a high prediction accuracy, which is of focal interest for studies using structural equation path models mainly for predication purposes, such as customer value indexes (e.g. [18]).

2.3 Results

We simultaneously tested our model an its hypothesis using partial least squares (PLS); a structural equation modeling technique which uses a principal component-based estimation approach [19]. PLS was chosen because of the characteristics of our model and sample. The model uses formative indicators, the sample size is relatively small (76 cases), and the data are non-normal. It is not possible to run these models using other techniques of structural equation models (e.g. the covariance-based model performed by LISREL or AMOS) (see for example, [20]). For hypothesis testing, we used the bootstrapping procedure recommended by [19] with 500 resamples, using 76 cases each. We tested the hypotheses using SmartPLS

(Version 2.3, [21]). The structural model contains the three capabilities (MO, KM, and CRM), and customer value creation.

The means, standard deviations, internal consistency and reliability estimates, and the paired correlation coefficients of all constructs appear in Table 1. The scale reliability of all reflective measures is satisfactory, with composite reliability (CR) ranging from 0.82 to 0.88.

Table 1: Descriptive statistics and discriminant validity

	Mean	SD	AVE	CR	1	2	3	4
1. Market orientation	5.52	0.94	0.65	0.88	**0.81**			
2. Knowledge management	5.33	0.88	0.80	0.88	0.80	**0.89**		
3. Customer relationship management[a]	5.48	1.33	n.a	n.a	0.51	0.46	**n.a**	
4. Customer value creation	5.25	1.21	0.63	0.82	0.29	0.42	0.23	**0.79**

n.a not applicable because they are formative measures, *Mean* the average score for all of the items included in this measure, *SD* Standard deviation, *AVE* average variance extracted; the bold numbers on the diagonal are the square root of the average variance extracted, shared variances are given in the lower triangle of the matrix, *CR* composite reliability
[a]Formative scales

In all the measurements, Bagozzi and Yi's [22] composite reliability index and Fornell and Larcker's [23] average variance extracted index are higher than the evaluation criteria of 0.7 for composite reliability and 0.5 for the average variance extracted.

Discriminant validity was determined by calculating the shared variance between pairs of constructs (i.e. the lower triangle of the matrix in Table 1) and verifying that it was lower than the average variances extracted for the individual construct (i.e. the diagonals in Table 1). The shared variances between pairs of all possible scale combinations indicated that the variances extracted were higher than the associated shared variances in all cases [23]. The shared variances, means and standard deviations are shown in Table 1.

We also sought formative dimensions by examining the weights [24], which provide information on the contribution of each indicator to its respective construct. Weights do not need to exceed any particular benchmark because a census of indicators is required for a formative specification [20]. The concern with regard to formative dimensions is the potential multicolinearity with overlapping dimensions, which might produce unstable estimates [24]. Results of a colinearity test show that the variance inflation factor (VIF) scores of each second-order construct for all dimensions are far below the commonly accepted cut-off of 3 (<1.92).

Finally, common method bias might influence some of the relationships formulated in our model. To rule out the existence of such a bias, we used methods suggested by [25], who recommend procedural remedies when including formative constructs. We therefore applied these to protect respondent anonymity and reduce evaluation apprehension by assuring subjects that there were no right or wrong answers;

to improve the scale items with a pre-test to a set of experts; and to counterbalance question order.

2.4 Hypothesis Testing

Once the psychometric properties of the measures had been checked, the next step was to evaluate the hypothesized relationship developed from our consideration of the relevant literature (see Fig. 1), discussed in the text as H_1.

We provide three models to test Hypothesis 1: (1) the direct model, which includes the main effects of the three capabilities (MO, KM, and CRM) on customer value creation; (2) the second model shows the effect on customer value creation of the interaction between each of the three organizational capabilities (MO × KM; MO × CRM; KM × CRM); and (3) the theoretical model, which includes the interaction of the three organizational capabilities (MO × KM × CRM). The PLS results for the three models, are shown in Table 2.

Table 2: Summary of results from partial least squares analyses[a]

Path from	To	Direct model	Second model	Theoretical model
		Path coefficient $(t)^b$	Path coefficient $(t)^b$	Path coefficient $(t)^b$
MO	Customer value creation	$-0.17^*(-2.54)$	$-0.16^*(-2.30)$	$-0.16^*(-2.53)$
KM	Customer value creation	$0.52^{**}(9.33)$	$0.51^{**}(7.95)$	$0.51^{**}(8.94)$
CRM	Customer value creation	$0.08^*(2.05)$	$0.07(1.56)$	$0.08(1.63)$
MO × KM	Customer value creation		$0.07(1.58)$	$0.07(1.37)$
KM × CRM	Customer value creation		$0.00(0.04)$	$0.00(0.02)$
MO × CRM	Customer value creation		$-0.05(-0.80)$	$-0.06(-0.85)$
MO × KM × CRM	Customer value creation			$0.28^{**}(7.77)$
R^2, Customer value creation		0.19	0.19	0.27
F for increment in R^2			0.00	7.41^{***}

$^*p < 0.05$; $^{**}p < 0.001$; $^{***}p < 0.01$

[a]Values of t were calculated through bootstrapping with 500 resamples with 76 cases per sample

[b]t values were adapted by correction factor because of the sample was very close to the entire population of the banking industry

According to Table 2, we find a significant link between the triple interaction construct (MO × KM × CRM) and "value creation" ($\beta = 0.28$, $p < 0.001$), which supports hypothesis 1, but the path between the three double interaction terms is not significant, as shown in Table 2. This provides more arguments to support the impact of the triple interaction on customer value creation, using empirical arguments to prove that the triple interaction is the best way for these organizational capabilities to create value for customers.

With regard to variance explanation, we find that the theoretical model that contains the triple interaction term explains 27 % of value creation. Both the direct model (which only includes the direct link between the three individual organizational capabilities and value creation) and the second model (which includes the three organizational capabilities, the three double interaction terms and value creation) explain 19 % of the variance of value creation. We therefore conclude that the 8 % difference in the variance explanation can be attributed to the simultaneous interaction of the three capabilities. We also estimated the ratio F^2 suggested by [19], to provide the level of significance of the improvement. When F^2 is greater than 0.02, the improvement is significant. In our case F^2 was 0.11.

3 Conclusion and Implication for Researchers

In recent years, customers have become the focus of attention, and every firm seeks to satisfy them in one way or another. Some firms are market-oriented to create superior customer value through the culture and behaviors that this orientation promotes. Other firms prefer to manage their knowledge, while others focus on creating and maintaining long-term relationships with their customers.

Understanding what it is that customers value in an offer, creating value for them and then managing it over time, have long been recognized as essential elements of a firm's business strategy. Customer value emerged in the 1990s as an area of increasing interest for firms, both at an academic and a professional level. On the one hand, service marketing literature focuses on the demand perspective of value; customer value and its perception. On the other hand, service management literature considers that the distinctive competence is value creation and the firm's capabilities for it.

Organizational capabilities are considered to be highly valuable attributes in a firm. Firms therefore want to be perceived as entities that can demonstrate a set of outstanding capabilities [26]. Very often, firms invest heavily in resources and capabilities, but not enough in the capabilities required to select, develop and deploy them efficiently [27]. According to these authors, firms ignore the development of the DC required to make these investments successful. When a firm possesses VRIN resources but does not use any DC, it cannot maintain its superior performance [28]. Firms' competitive advantages in the current environment are not derived simply from the distinctive resources and capabilities they possess but also from the way they are used [29, 30].

We argue that the three proposed capabilities form a distinctive competence for firms and when they are combined a series of changes take place, which transform this distinctive competence into a DC for the firm. The high speed of change in the environment and the increasing strength of the competition make it all the more important for a firm's combinations of resources and capabilities to be difficult to imitate.

One of the main limitations of our study is that the investigation was carried out at a single point in time, which is a particular limitation because customer value is a dynamic construct. Our study was carried out in a single industry (the Spanish banking industry), which does not allow us to generalize the results attained to other economic industries. Furthermore, our model focuses on the three capabilities that we consider to be the most important for customer value creation. We have chosen these capabilities (MO, KM and CRM) because they are the organizational capabilities mentioned most often in the existing literature as having the greatest influence on customer value [1, 31, 32]. We would point out that it is of course possible to include other capabilities in our model. The explanatory power is therefore limited to the variables we have considered.

Finally, it is important to stress the situation that the industry was in at the time of the study. Although we believe that this situation provided an ideal opportunity for our study, it also created problems when collecting data for the empirical investigation. Because of the high degree of turbulence in the industry at the time and the fact that the industry and its problems and uncertainties were the subject of much discussion, some managers were wary of giving out data.

We consider that this investigation provides a starting point for future investigations relating to customer value creation or maintenance in the current environment, where the customer is daily more demanding and the competition is stronger. Possible future investigations might be an extension of the timescale of our study and an expansion into other economic industries, in order to be able to generalize the results; and an extension of the model to introduce, for example, other capabilities that a firm possesses that might influence customer value creation.

References

[1] Wang, Y. H., Lo, P., Chi, R., and Yang, Y., "An Integrated Framework for Customer Value and Customer-Relationship-Management Performance: A Customer-Based Perspective from China," *Managing Service Quality*, 14, 169–182, 2004.

[2] Sánchez, R., Iniesta, M. A., and Holbrook, M. B., "The Ccnceptualisation and measurement of consumer value in services," *International Journal of Market Research*, 51, 93–113, 2009.

[3] Woodruff, R.B., "Customer Value: The Next Source for Competitive Advantage," *Journal of Academic Marketing Science*, 25, 139–153, 1997.

[4] Woodruff, R.B., & Gardial, S.F., *Know Your Customer: New Approaches to Understanding Customer Value and Satisfaction*. Cambridge, Mass: Blackwell Business, 1996.

[5] Sirmon, D.G., Hitt, M.A., and Ireland, R.D., "Managing Firm Resources in Dynamic Environments to Create Value: Looking inside the Black Box," *Academic Management Review*, 32, 273–292, 2007.

[6] Morrow, J.L., Sirmon, D.G., Hitt, M.A., and Holcomb, T.R., "Creating Value in the Face of Declining Performance: Firm Strategies and Organizational Recovery," *Strategic Management Journal*, 28, 271–283, 2007.

[7] Henseler, J., & Chin, W. W., “A Comparison of Approaches for the Analysis of Interaction Effects Between Latent Variables Using Partial Least Squares Path Modeling.” *Structural Equation Modeling*, 17, 82–109, 2010.
[8] Little, T.D., Bovaird, J.A., and Widaman, K.F., “On the merits of orthogonalizing powered and product terms: Implications for modeling interactions among latent variables,” *Structural Equation Modeling*, 13, 497–519, 2006.
[9] Malhotra, N.K., and Birks, D.F., *Marketing Research: An Applied Approach*. Essex, UK: Financial Times/Prentice Hall, 2006.
[10] Narver, J.C., & Slater, S.F., “The effect of a market orientation on business profitability,” *Journal of Marketing*, 54, 20–35, 1990.
[11] Jaworski, B.J., & Kohli, A.K., “ Market orientation: Antecedents and consequences,” *Journal of Marketing*, 57, 53–70, 1993.
[12] Jansen, J.J.P., Van den Bosch, F.A.J., and Volberda, H.W., “Managing potential and realized absorptive capacity: How do organizational antecedents matter?” *Academic Management Journal*, 48 999–1015, 2005.
[13] Gold, A. H., Malhotra, A., and Segars, A. H. (2001). Knowledge management: An organizational capabilities perspective. *Journal of Management Information Systems*, 18, 185–214.
[14] Chou, T.-C., Chang, P. -L., Cheng, Y.P., and Tsai, C.-T., “A path model linking organizational knowledge attributes, information processing capabilities, and perceived usability,” *Information Management*, 44, 408–417, 2007.
[15] Reinartz, W., Krafft, M., and Hoyer, W.D., “The customer relationship management process: Its measurement and impact on performance,” *J Marketing Res*, 41, 293–305, 2004.
[16] Hooley, G. J., Greenley, G. E., Cadogan, J. W., and Fahy, J., “The performance impact of marketing resources,” *Journal of Business Research*, 58, 18–27, 2005.
[17] Lance, C. E., “Residual centering, exploratory and confirmatory moderator analysis, and decomposition of effects in path models containing interactions,” *Applied Psychological Measurement*, 12, 163–17, 1988.
[18] Fornell, C., “A national customer satisfaction barometer: The Swedish experience,” *Journal of Marketing*, 56, 6–21. 1992.
[19] Chin, W. W., “Issues and opinion on structural equation modeling,” *MIS Quarterly*, 22, 1–12, 1998.
[20] Diamantopoulos, A., & Winklhofer, H., “Index construction with formative indicators: An alternative to scale development,” *Journal of Marketing Research*, 38, 269–277, 2001.
[21] Ringle, C., Wende, S., and Will, A., *SmartPLS, version 2.0 M2*. www.smartpls.de, 2005.
[22] Bagozzi, R. P., & Yi, Y., “On the evaluation of structural equation models,” *Journal of Academic Market Science*, 16, 74–94, 1988.
[23] Fornell, C., & Larcker, D. F., “Evaluating structural equation models with unobservable variables and measurement error,” *Journal of Marketing Research*, 18, 39–50, 1981.
[24] Mathieson, K., Peacock, E., and Chin, W. W., “Extending the technology acceptance model: The influence of perceived user resources,” *The Data Base for Advances in Information Systems*, 32, 86–112, 2001,
[25] Podsakoff, P. M., MacKenzie, S. B., Lee, J. -Y., and Podsakoff, N. P., “Common method biases in behavioral research: A critical review of the literature and recommended remedies,” *Journal of Applied Psychology*, 88, 879–903, 2003.
[26] Schreyögg, G., and Kliesch-Eberl, M., “How dynamic can organizational capabilities be?” Towards a dual-process model of capability dynamization. *Strategic Management Journal*, 28, 913–933, 2007.
[27] Maklan, S., and Knox, S., “Dynamic capabilities: The missing link in CRM investments,” *European Journal of Marketing*, 43, 1392–1410, 2009.
[28] Ambrosini, V., ND Bowman, C., “What are dynamic capabilities and are they a useful construct in strategic management?” *International Journal of Management Reviews*, 11, 29–49, 2009.
[29] Luo, Y. (2000). “Dynamic capabilities in international expansion,” *Journal of World Business*, 35, 355–378, 2000.

[30] Teece, D. J., "Capturing value from knowledge assets: The new economy, markets for know-how, and intangible assets," *California Management Review*, 40, 55–79, 1998.
[31] McNaughton, R. B., Osborne, P., Morgan, R. E., and Kutwaroo, G., "Market orientation and firm value," *Journal of Marketing Management*, 17, 521–542, 2001.
[32] Vorakulpipat, C., & Rezgui, Y., "Value creation: The future of knowledge management," *Knowledge Engineering Review*, 23, 283–294, 2008.

Satisfaction with ERP Systems in Supply Chain Operations

Michael J. Murray, Wynne W. Chin, and Elizabeth Anderson-Fletcher

Abstract A key reason for implementing an enterprise resource planning (ERP) system is the ability it provides an organization to synchronize and automate the flows of material, information and cash through the supply chain. Viewed from this perspective ERP systems can be seen as an enabling technology to achieve better supply chain integration, which should result in better decision making and improved financial performance. Yet much of the debate regarding the value of ERP systems focuses on their implementation costs and the corresponding difficulty measuring the benefits generated by these projects. The significant level of total global spending on ERP systems—estimated currently at $253.7 billion—provides the motivation behind this study. We seek to understand how effective these systems are in providing the information needed by decision makers in production and operations management roles. This is a necessary step in determining what benefits can be achieved by these systems. To do this we developed surveys through a compilation of several pilot interviews with plant managers and production supervisors in various industries. For both of these management roles, functional areas under their responsibility were identified and questions were formulated to assess: (1) the usefulness of various functionalities of the ERP system within the manager's functional area and the manager's opinion of the effectiveness of the ERP system in that area, and (2) the manager's opinion of ERP performance in a functional area and his/her overall satisfaction with the ERP system. We used Partial Least Squares (PLS) methodology to analyze the responses from the survey. The results indicate that the majority of plant managers use ERP systems in manufacturing, cost control, inventory & logistics activities and in reporting, as if they were still using MRPII

M.J. Murray (✉) • E. Anderson-Fletcher
University of Houston, 4800 Calhoun Road, 260-G Melcher Hall, Houston, TX 77204-6021, USA
e-mail: michael.murray@mail.uh.edu; EFletcher@uh.edu

W.W. Chin
Department of Decision and Information Systems, C. T. Bauer College of Business, University of Houston, Houston TX 77204-6021, USA
e-mail: wchin@uh.edu

H. Abdi et al. (eds.), *New Perspectives in Partial Least Squares and Related Methods*, Springer Proceedings in Mathematics & Statistics 56, DOI 10.1007/978-1-4614-8283-3_21,

(Manufacturing Resource Planning) systems. They do not seem to be making use of the additional capabilities that ERP systems have over and above those found in MRP II systems. Production supervisors appear to be using ERP systems more evenly across their areas of responsibility. For production supervisors, as in the case of plant managers, reporting is the area where ERP performance has the highest impact on overall satisfaction of the user with the ERP system. Finally, the results indicate that there are several avenues for improvement in the way the current ERP systems support daily operations of these professionals, most notably in the area of analytics and providing better business intelligence.

Key words: Enterprise systems, Supply chain management, Data analytics, Business intelligence, Empirical research methods

1 Introduction

Until recently most research involving enterprise resource planning (ERP) systems has focused on perceived overall firm-level performance improvement or implementation issues like cost, time, and success [1, 3, 8, 10, 13]. However from a supply chain perspective a more interesting question is how can the adoption of the ERP system actually improve supply chain performance? It has been shown that organizations that had high levels of information system usage generally tended to have better manufacturing performance [15]. While functional fit and user acceptance are important in achieving near term benefits, in the long term operational benefits depend more on process integration, optimization, improved access to information and business process improvement [12]. For example, a study of Taiwanese IT firms showed that three benefits provided by ERP systems in particular (operational process integration, customer and relationship management, and manufacturing planning and control integration) enhanced supply chain performance in that industry [14].

In principle, investments in ERP and other enterprise systems provide an organization the ability to synchronize and automate the flow of material, business processes, information and cash throughout the supply chain which should result in better decision making, improved financial performance, and higher stockholder returns. While there is some research supporting this contention [7, 11], other studies have yielded mixed results [6] and it is not clear what factors affect these results. This uncertainty, coupled with the significant level of investment in ERP and manufacturing intelligence systems over the past decade (currently estimated at over $250 billion annually [5]) provides the motivation behind this study. We seek to understand how effective these systems are in providing the information needed by decision makers in supply chain operations roles to make better decisions. This is a necessary step in determining whether the investment in an ERP system can improve supply chain and firm performance. By focusing on the managers responsible for the daily production operations of the enterprise we seek to understand how well the technology solves the problems faced by these managers and offer insight into ways the technology can be improved.

Our study differs from previous studies in several important ways. First, we specifically targeted our survey to supply chain operations. Second, we focused on those aspects of the systems that comprise the manufacturing intelligence capabilities that operations managers need, such as the ability to monitor inventories, report on labor costs, etc. Finally, we introduce a facet based approach using both formative and reflective indicators in a Partial Least Squares (PLS) analysis to determine which factors influenced satisfaction within the supply chain operations group.

2 Research Methodology

To develop the research instrument we first conducted interviews with several plant managers and production supervisors at 15 large companies (defined as having annual sales greater than $200 million) in the Chemicals, Aerospace & Defense, Industrial Machinery & Construction, High Tech, Life Sciences and Consumer Products industries. The companies were selected based on size, industry, and their willingness to participate. The purpose of the interviews was to identify (a) principal areas of responsibility, (b) the major tasks in each area of responsibility, and (c) the information requirements necessary to perform these tasks. Our interviews established that the primary responsibilities for production supervisors are in the areas of manufacturing, inventory management, labor and personnel, and cost control, while plant managers had responsibilities in health, safety and environmental, and facility maintenance in addition to the four areas listed above. Under our definitions most production supervisors would have a reporting relationship to the plant manager. So while the production supervisors are responsible for complying with safety, health and environmental performance, and facilities and maintenance procedures, they do not have financial management responsibility for these areas.

Next, we developed a facet based model for explaining overall satisfaction with ERP system performance (see Fig. 1). We propose that overall satisfaction with the ERP system performance in supply chain operations is dependent upon satisfaction with how the system supports the needs of the operations manager in each functional area, or facet, of responsibility. This in turn is predicated on how effectively the system provides the information needed for decision making in those functional areas. Thus, our model consists of emergent constructs using formative measures to assess the relative impact and effectiveness of the ERP system in performing the various tasks in each area of responsibility, latent constructs with measures that reflect the satisfaction with ERP system performance for that particular area of responsibility, and a global latent construct with reflective measures for satisfaction with the overall ERP system performance.

Based on this model we developed a set of questions for the survey instrument. The surveys have an introductory section asking about company demographics followed by sections corresponding to the areas of responsibility determined for the management role. Each section consists of questions on a 7-point Likert scale (1 = strong disagree, 4 = neutral, 7 = strongly agree) about the effectiveness of ERP

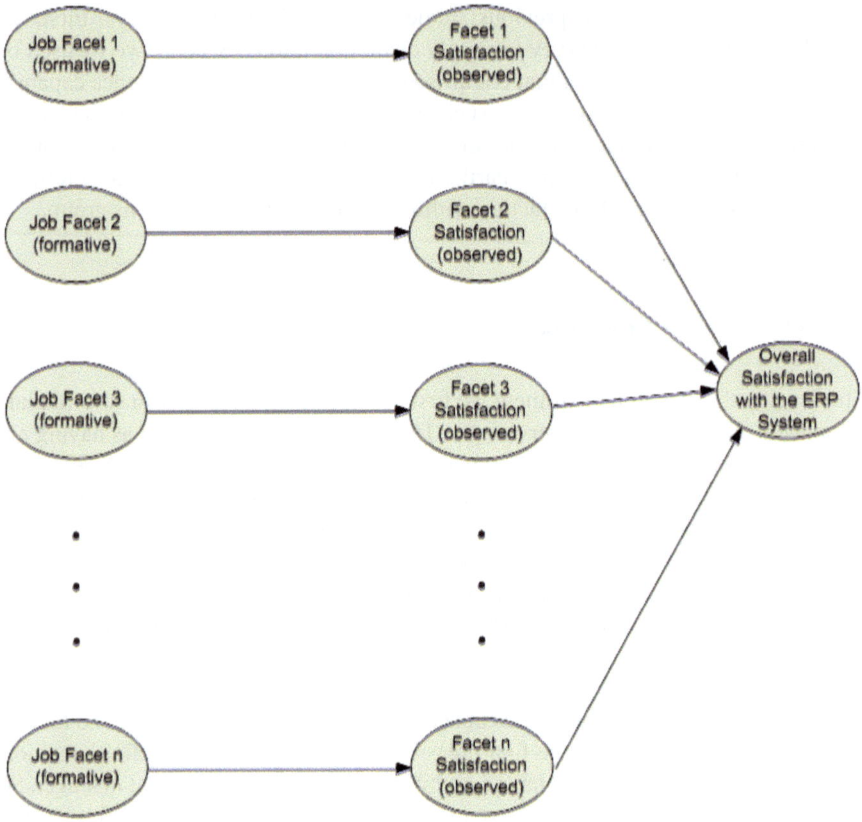

Fig. 1: The proposed satisfaction model

system in performing various tasks within a particular area (facet) of responsibility. Also included in each section was a question asking to what extent the managers used the following tools in performing their tasks in each of their areas of responsibility: ERP Systems, Spreadsheets, Advanced Planning Systems, Manufacturing Execution Systems, Phone, Paper, and Other. Respondents could check multiple options in answering this question. The final section in each survey was a "General" one including a set of questions repeating the inquiry into the satisfaction of the user with ERP system performance in user's various areas of responsibility, and two questions asking the user's overall satisfaction with the ERP system (the surveys are provided in Appendices "Plant Manager Survey Questions" and "Production Supervisor Survey Questions"). These questions were reviewed by a cross-section of the managers and supervisors we initially interviewed who provided feedback on the content and clarity.

In our research design, each question is an indicator that is linked to a particular construct or factor. For example, in the plant manager survey the question, "The current information system supports me very well in the management of my plant," is a direct indicator of the plant manager's overall satisfaction with the ERP system. This is often termed as a "reflective" indicator of the satisfaction construct, because we would expect that an increase or decrease in the actual level of satisfaction would result directly in an increase or decrease in the response to this measure as reported by the plant manager.

An indicator can also be "formative." A change is this indicator implies a certain level of impact on the overall construct, but the converse where if we see a change in the construct does not imply a change in that particular indicator. Referring again to the plant manager survey question, "Our current information system is an effective tool to track deviations from on-time delivery targets," is one of many reasons that may influence the plant manager's perceived effectiveness of the ERP system in managing the manufacturing area. But a change in the level of the manageress satisfaction may be due to other factors/reasons and we cannot necessarily infer that a change occurred for that particular reason. In other words, a drop in the manager's overall satisfaction with the ERP system in the manufacturing area need not imply any change in the manager's perception of how well the system allows her to track on-time deliveries.

A business services provider was engaged to send email invitations to participate in the survey to a large number of individuals at companies with sales greater than $200 million in the target industries. The service provider did not provide individual names or identifying information, and only guaranteed that the email addresses on their list represented individuals in manufacturing roles. Realizing that many of the recipients on that list would not hold positions as plant managers or production supervisors, the decision was made to send a second email to members of APICS (The Association for Operations Management). APICS members were chosen based on the belief that they would have proportionally more members in the targeted roles. Unfortunately, many of the addresses on the APICS list were obsolete or otherwise unusable (we were able to control for company size and industry using the company demographic information in the survey). Because it was not possible to identify how many (if any) email recipients were on both lists, we were unable to determine exactly how many unique survey invitations were sent, nor can we identify how many reached their intended target audience of plant managers and production supervisors. Data was collected for a relatively short period of 4 weeks, which resulted in a convenience sample of 156 usable responses (75 responses for Plant Managers and 81 for Production Supervisors). The number of responses for both roles was adequate to analyze the models that we have developed based on the initial interviews.

3 Results

Partial Least Squares (PLS) methodology was used to analyze the survey data. PLS makes no assumptions about multivariate normality in the data, can be used for theory confirmation as well as for suggesting where relationships may (or may not) exist and it avoids problems that often occur in covariance-based SEM analyses, such as inadmissible solutions and factor indeterminacy [2, 4]. Finally, we chose PLS for its ability to estimate formative measures directed (i.e., our F-constructs) since these measures would result in identification constraints that hampers the use a covariance-based approach such as LISREL [9]. The path coefficients obtained from a PLS analysis are standardized regression coefficients, while the loadings of items on individual constructs are factor loadings. Factor scores created using these loadings are equivalent to weighted composite indices. Thus, PLS results can be easily interpreted by considering them in the context of regression and factor analysis. As such, it allows us to assess how well the set of items used to measure one of our latent constructs, O-production, for example, actually served to predict the overall satisfaction with the ERP.

3.1 Survey Responses

Descriptive statistics for all indicators in both models is presented in Table 1. What is most striking about the survey responses from both plant managers and production supervisors is their overall neutral attitude about how well the enterprise system supports them in the various facets of their job responsibilities. With the exception of inventory management their attitudes clustered around either the "slightly disagree" or "slightly agree" side of neutral for all facets, while they tended to "slightly agree" that enterprise systems supported the inventory management facet of their responsibility. This neutrality may indicate that managers in supply chain operations rely on only a few features and functions of their enterprise systems to perform their job responsibilities.

3.2 Construct Validity and Reliability

Convergent validity of the scales was established by performing a bootstrap analysis on the raw data. The t-values for the path loadings of the outer model for all indicators were greater than 1.96 for both models (see Tables 2 and 3). Discriminant validity was demonstrated in two ways: (1) by examining crossloadings for the observed constructs (O-constructs) for both models (Table 4), and (2) by examining the ratio of the square root of the average variance extracted for each O-construct to the correlations of that construct to all the other constructs for both models (Table 5).

3.3 Plant Manager Model Results

The proportion of the variation explained (r^2) in overall satisfaction with the ERP System by the hypothesized PLS model is 80.8%. This is an excellent r^2 value

Table 1: Descriptive statistics

Plant manager model			
Construct	Number of items	Average	Std dev
Formative			
Manufacturing	9	4.93	0.53
Inventory	7	5.16	0.63
Health, safety & environ.	4	3.48	0.14
Cost controls	10	4.77	0.62
Facilities maintenance	4	3.57	0.10
Labor & personnel	6	3.97	0.35
Reporting	5	4.40	0.88
Observed			
Manufacturing	2	4.74	0.27
Inventory	2	5.11	0.25
Health, safety & environ.	2	3.44	0.18
Cost controls	2	4.47	0.47
Facilities maintenance	2	3.54	0.09
Labor & personnel	2	3.88	0.04
Reporting	4	4.91	0.53
Overall	3	4.33	0.29

Production supervisor model			
Construct	Number of items	Average	Std dev
Formative			
Manufacturing	20	4.41	0.48
Inventory Mgt	5	5.29	0.55
Labor & personnel	6	3.78	0.24
Cost control	8	4.09	0.47
Reporting	4	4.50	0.43
Observed			
Manufacturing	2	4.65	0.48
Inventory Mgt	2	5.04	0.40
Labor & personnel	2	3.79	0.22
Cost control	2	4.16	0.08
Reporting	2	4.40	0.04
Overall satisfaction	2	4.42	0.40

indicating that the model is a good predictor of overall satisfaction with the enterprise system. As would be expected, plant managers' satisfaction with the various facets of the system (i.e., manufacturing, inventory management, etc.) is strongly dependent on the how well those facets satisfy their needs. In each instance the path loadings indicate a strong, statistically significant relationship, with $p < 0.005$ in all cases (see Fig. 2).

What is more interesting, however, is how the level of satisfaction with each facet translates into overall satisfaction with the enterprise system. These results indicate that three of the facets hypothesized to affect overall satisfaction are statistically significant: satisfaction with manufacturing performance ($p < 0.05$), satisfaction with

Table 2: Indicator loadings: plant manager model

Outer model	Loading	Mean	Std error	T-stat	Outer model	Loading	Mean	Std error	T-stat
Formative indicators					Reflective indicators				
F-Manufacturing					O-Manufacturing				
MF01	0.8068	0.7739	0.0765	10.5461	MS01	0.8923	0.8915	0.0391	22.8486
MF02	0.5627	0.5283	0.1332	4.2243	MS02	0.9171	0.9194	0.0208	44.0380
MF03	0.7913	0.7627	0.0715	11.0616	O-InvMgt				
MF04	0.6742	0.6428	0.1041	6.4734	IS01	0.9382	0.9381	0.0137	68.5565
MF05	0.8232	0.8009	0.0583	14.1247	IS02	0.9341	0.9353	0.0165	56.5475
MF06	0.6885	0.6584	0.1036	6.6448	O-Safety				
MF07	0.6546	0.6207	0.1069	6.1222	SS01	0.9486	0.9480	0.0123	77.1310
MF08	0.7872	0.7221	0.0788	9.9864	SS02	0.9398	0.9392	0.0201	46.7677
MF09	0.8683	0.8469	0.0669	12.9879	O-CostControl				
F-InvMgt					CS01	0.8542	0.8493	0.0363	23.5477
IF01	0.8021	0.7771	0.0658	12.1828	CS02	0.8759	0.8656	0.0330	26.5083
IF02	0.6855	0.6382	0.1096	6.2547	O-Facilities				
IF03	0.8436	0.8291	0.0603	13.9965	FS01	0.9498	0.9515	0.0159	59.5975
IF04	0.8237	0.8067	0.0569	14.4686	FS02	0.9442	0.9457	0.0198	47.6697
IF05	0.8489	0.8290	0.0469	18.0856	O-Labor				
IF06	0.8530	0.8312	0.0354	24.1035	LS01	0.9319	0.9323	0.0137	68.1279
IF07	0.7752	0.7553	0.0736	10.5261	LS02	0.9220	0.9194	0.0235	39.1542
F-Safety					O-Reporting				
SF01	0.8657	0.8615	0.0483	17.9370	RS01	0.8920	0.8896	0.0240	37.2188
SF02	0.9290	0.9263	0.0281	33.0029	RS02	0.8810	0.8770	0.0272	32.3654
SF03	0.9298	0.9220	0.0274	33.8775	Overall_Satisfaction				
SF04	0.9936	0.9890	0.0079	126.5068	OS01	0.9300	0.9308	0.0136	68.2170
F-CostControl					OS02	0.9061	0.9076	0.0296	30.6142
CF01	0.6390	0.5982	0.1227	5.2089	OS03	0.9559	0.9543	0.0117	81.5962
CF02	0.8355	0.8132	0.0484	17.2585					
CF03	0.7186	0.7129	0.0774	9.2896					
CF04	0.8270	0.8025	0.0609	13.5740					
CF05	0.6820	0.6605	0.0848	8.0382					
CF06	0.6778	0.6290	0.0722	9.3902					
CF07	0.6045	0.5759	0.0963	6.2750					
CF08	0.6379	0.6093	0.0905	7.0494					
CF09	0.4574	0.4185	0.1275	3.5885					
CF10	0.6495	0.6159	0.0813	7.9897					
F-Facilities									
FF01	0.9694	0.9663	0.0136	71.5074					
FF02	0.9645	0.9619	0.0214	45.0801					
FF03	0.8355	0.8164	0.0634	13.1758					
FF04	0.8981	0.8916	0.0445	20.1732					
F-Labor									
LF01	0.8721	0.8558	0.0413	21.0945					
LF02	0.8869	0.8700	0.0449	19.7516					
LF03	0.8484	0.8368	0.0571	14.8451					
LF04	0.7519	0.7533	0.0605	12.4186					
LF05	0.8582	0.8555	0.0473	18.1508					
LF06	0.9365	0.9272	0.0282	33.2112					
F-Reporting									
RF01	0.7127	0.6943	0.1073	6.6430					
RF02	0.7781	0.7525	0.0894	8.7051					
RF03	0.8422	0.8328	0.0773	10.8980					
RF04	0.7709	0.7520	0.0879	8.7656					
RF05	0.8818	0.8486	0.0763	11.5626					

Table 3: Indicator loadings: production supervisor model

Outer model	Loadings	Mean	Std error	T-stat	Outer model	Loadings	Mean	Std error	T-stat
Formative indicators					Reflective indicators				
F-Manufacturing					O-Manufacturing				
MF01	0.7080	0.6919	0.0799	8.8567	MS01	0.8909	0.8890	0.0327	27.2629
MF02	0.7660	0.7478	0.0741	10.3444	MS02	0.9221	0.9237	0.0121	76.1992
MF03	0.7070	0.6958	0.0717	9.8601	O-InvMgt				
MF04	0.7272	0.7128	0.0566	12.8449	IS01	0.9152	0.9110	0.0265	34.4831
MF05	0.7174	0.7002	0.0705	10.1702	IS02	0.9239	0.9242	0.0162	56.9274
MF06	0.8051	0.7918	0.0556	14.4712	O-Labor				
MF07	0.7757	0.7617	0.0540	14.3545	LS01	0.9100	0.9119	0.0198	45.9022
MF08	0.7142	0.7002	0.0571	12.5091	LS02	0.9012	0.9031	0.0248	36.3844
MF09	0.6716	0.6611	0.0649	10.3553	O-CostControl				
MF10	0.7078	0.6899	0.0675	10.4895	CS01	0.9616	0.9623	0.0115	83.5728
MF11	0.6778	0.6757	0.0634	10.6872	CS02	0.9661	0.9664	0.0095	101.7249
MF12	0.5795	0.5713	0.0858	6.7567	O-Reporting				
MF13	0.7529	0.7358	0.0657	11.4631	RS01	0.9188	0.9195	0.0176	52.2958
MF14	0.7220	0.7051	0.0680	10.6200	RS02	0.9058	0.9077	0.0210	43.1530
MF15	0.7255	0.7124	0.0559	12.9835	Overall_Satisfaction				
MF16	0.8029	0.7846	0.0495	16.2060	OS01	0.9531	0.9530	0.0126	75.4704
MF17	0.8106	0.7934	0.0445	18.1986	OS02	0.9579	0.9584	0.0091	105.3975
MF18	0.7489	0.7332	0.0573	13.0694					
MF19	0.7940	0.7794	0.0683	11.6331					
MF20	0.8590	0.8404	0.0418	20.5659					
F-InvMgt									
IF01	0.6916	0.6785	0.0837	8.2655					
IF02	0.7627	0.7396	0.0904	8.4380					
IF03	0.8118	0.8016	0.0622	13.0450					
IF04	0.8529	0.8394	0.0597	14.2794					
IF05	0.9169	0.9049	0.0325	28.2104					
F-Labor									
LF01	0.8837	0.8820	0.0399	22.1717					
LF02	0.8170	0.8188	0.0747	10.9441					
LF03	0.8947	0.8932	0.0374	23.9037					
LF04	0.8438	0.8423	0.0480	17.5952					
LF05	0.9319	0.9268	0.0257	36.2129					
LF06	0.9244	0.9206	0.0270	34.2973					
F-CostControl									
CF01	0.7737	0.7673	0.0593	13.0537					
CF02	0.7621	0.7563	0.0691	11.0362					
CF03	0.8926	0.8860	0.0460	19.4119					
CF04	0.8356	0.8300	0.0527	15.8534					
CF05	0.9003	0.8933	0.0344	26.1621					
CF06	0.8197	0.8126	0.0607	13.5089					
CF07	0.8185	0.8133	0.0530	15.4459					
CF08	0.9276	0.9213	0.0354	26.1868					
F-Reporting									
RF01	0.8905	0.8838	0.0422	21.1006					
RF02	0.8985	0.8912	0.0446	20.1418					
RF03	0.9125	0.9040	0.0379	24.0489					
RF04	0.8603	0.8435	0.0533	16.1381					

inventory management ($p < 0.05$) and reporting satisfaction ($p < 0.005$). Of the three, satisfaction with the reporting features of the enterprise system has the greatest effect on satisfaction.

Table 4: Outer model loadings and cross loadings

Plant manager model															
	F-Manuf.	F-InvMgt	F-Safety	F-CostCont.	F-Facilities	F-Labor	F-Report	O-Manuf.	O-InvMgt	O-Safety	O-CostCont.	O-Facilities	O-Labor	O-Report	Overall_Satisf.
MS01	0.7973	0.5805	0.4604	0.5226	0.6055	0.4502	0.5464	**0.8923**	0.4829	0.3975	0.5225	0.5795	0.4815	0.5394	0.5853
MS02	0.7443	0.6655	0.3290	0.7085	0.4690	0.4886	0.6776	**0.9171**	0.6826	0.3738	0.7516	0.4956	0.5679	0.8003	0.8333
IS01	0.6851	0.8771	0.5210	0.6494	0.5097	0.4418	0.6451	0.5870	**0.9382**	0.4707	0.5629	0.5149	0.4626	0.5819	0.6539
IS02	0.6731	0.8034	0.3412	0.6357	0.3700	0.3872	0.6422	0.6315	**0.9341**	0.3606	0.6353	0.3741	0.4202	0.6371	0.6923
SS01	0.5107	0.4216	0.9594	0.3517	0.7390	0.6162	0.5824	0.3804	0.4200	**0.9486**	0.2666	0.7088	0.5965	0.4281	0.3797
SS02	0.4794	0.3890	0.8133	0.5103	0.7487	0.6602	0.6402	0.4243	0.4202	**0.9398**	0.4264	0.7860	0.6793	0.5433	0.5034
CS01	0.5216	0.3812	0.2548	0.7211	0.3719	0.4674	0.5088	0.6257	0.3720	0.3155	**0.8542**	0.4227	0.4683	0.5871	0.6219
CS02	0.6263	0.6250	0.2752	0.7746	0.3597	0.2949	0.6312	0.6064	0.7215	0.3140	**0.8759**	0.4057	0.3169	0.6071	0.6732
FS01	0.5499	0.4553	0.7336	0.5002	0.9475	0.6894	0.7485	0.5693	0.4481	0.7433	0.4177	**0.9498**	0.6891	0.5391	0.4754
FS02	0.5578	0.4486	0.7203	0.5367	0.8471	0.6058	0.6408	0.5492	0.4535	0.7530	0.4898	**0.9442**	0.5923	0.5374	0.5452
LS01	0.4323	0.3907	0.6209	0.3887	0.6835	0.9138	0.7035	0.4615	0.3998	0.6489	0.3527	0.6222	**0.9319**	0.5024	0.4075
LS02	0.4706	0.4580	0.5316	0.5642	0.6231	0.7605	0.6758	0.6239	0.4776	0.5988	0.4868	0.6351	**0.9220**	0.5968	0.5499
RS01	0.5848	0.5412	0.4612	0.6689	0.5441	0.6188	0.8051	0.6135	0.5612	0.5077	0.6039	0.5361	0.6112	**0.8920**	0.6606
RS02	0.7277	0.6217	0.3485	0.6305	0.4455	0.3783	0.5448	0.7165	0.5931	0.3976	0.6207	0.4703	0.4329	**0.8810**	0.8351
OS01	0.7396	0.6691	0.3528	0.7096	0.4563	0.4596	0.6325	0.7839	0.6903	0.4092	0.7287	0.4847	0.5379	0.8282	**0.9300**
OS02	0.8097	0.7164	0.5003	0.7020	0.5778	0.5020	0.6576	0.7681	0.6800	0.5313	0.7220	0.5886	0.5313	0.8177	**0.9559**
OS03	0.6777	0.6611	0.3196	0.5922	0.3806	0.3400	0.4542	0.6498	0.6338	0.3487	0.6358	0.4200	0.3499	0.6934	**0.9061**

Production supervisor model											
	F-Manuf.	F-InvMgt	F-Labor	F-CostCont.	F-Report	O-Manuf.	O-InvMgt	O-Labor	O-CostCont.	O-Report	Overall_Satisf.
MS01	0.8269	0.6784	0.5889	0.5798	0.7065	**0.8909**	0.6685	0.6118	0.6098	0.6278	0.6305
MS02	0.8512	0.7516	0.6751	0.7775	0.7549	**0.9221**	0.7705	0.7845	0.7723	0.8499	0.8721
IS01	0.7045	0.8564	0.5090	0.6478	0.6503	0.6895	**0.9152**	0.5727	0.6276	0.5809	0.6078
IS02	0.8047	0.7524	0.5893	0.7580	0.7713	0.7744	**0.9239**	0.7080	0.8049	0.7895	0.8093
LS01	0.6852	0.6087	0.8950	0.7485	0.7230	0.6701	0.5930	**0.9100**	0.7334	0.6426	0.6699
LS02	0.7226	0.6487	0.7350	0.8166	0.7047	0.7384	0.6736	**0.9012**	0.7993	0.7481	0.7777
CS01	0.7255	0.7293	0.6524	0.9044	0.7956	0.7376	0.7210	0.7652	**0.9616**	0.7353	0.6960
CS02	0.7433	0.7441	0.7230	0.9160	0.8181	0.7436	0.7834	0.8616	**0.9661**	0.7971	0.7936
RS01	0.7440	0.6792	0.6315	0.7254	0.8906	0.7433	0.7415	0.7179	0.8017	**0.9188**	0.8093
RS02	0.7475	0.6368	0.6049	0.6611	0.7262	0.7606	0.6193	0.6797	0.6453	**0.9058**	0.8547
OS01	0.7811	0.6950	0.6725	0.6896	0.7409	0.7648	0.6984	0.7276	0.6900	0.8539	**0.9531**
OS02	0.8416	0.6926	0.7199	0.7744	0.7950	0.8355	0.7775	0.7954	0.7870	0.8861	**0.9579**

Table 5: Inter-construct correlations and reliability measures. Squared correlations of latent variables: plant manager model

Composite reliability	Avg. var. extracted		F-Manuf.	F-InvMgt	F-Safety	F-CostCont.	F-Facilities	F-Labor	F-Report	O-Manuf.	O-InvMgt	O-Safety	O-CostCont.	O-Facilities	O-Labor	O-Report	O_Satisf.
0.9172	0.5558	F-Manuf.	1														
0.9284	0.6503	F-InvMgt	0.6209	1													
0.9627	0.8661	F-Safety	0.3110	0.2247	1												
0.8941	0.4635	F-CostCont.	0.4803	0.4437	0.1612	1											
0.9556	0.8437	F-Facilities	0.3461	0.2190	0.6245	0.2532	1										
0.9447	0.7410	F-Labor	0.2545	0.1924	0.4605	0.2302	0.5243	1									
0.8980	0.6389	F-Report	0.4048	0.4069	0.3920	0.4998	0.5548	0.5917	1								
0.9003	0.8187	O-Manuf.	0.7215	0.4774	0.1861	0.4708	0.3465	0.2701	0.4629	1							
0.9342	0.8765	O-InvMgt	0.5263	0.8066	0.2134	0.4712	0.2218	0.1964	0.4727	0.4230	1						
0.9427	0.8916	O-Safety	0.2754	0.1848	0.8865	0.2053	0.6201	0.4555	0.4175	0.1807	0.1979	1					
0.8561	0.7484	O-CostCont.	0.4428	0.3441	0.0941	0.7488	0.1784	0.1907	0.4373	0.5061	0.4087	0.1322	1				
0.9456	0.8968	O-Facilities	0.3419	0.2278	0.5895	0.2991	0.9003	0.4691	0.5402	0.3490	0.2265	0.6238	0.2287	1			
0.9243	0.8593	O-Labor	0.2364	0.2084	0.3883	0.2609	0.4980	0.8203	0.5540	0.3393	0.2226	0.4539	0.2028	0.4594	1		
0.8801	0.7859	O-Report	0.5451	0.4286	0.2098	0.5377	0.3128	0.3196	0.5846	0.5606	0.4232	0.2621	0.4766	0.3231	0.3494	1	
0.9512	0.8666	O_Satisf.	0.6397	0.5375	0.1788	0.5192	0.2605	0.2211	0.3972	0.6271	0.5164	0.2163	0.5615	0.2893	0.2640	0.7076	1

Squared correlations of latent variables: production supervisor model

Composite reliability	Avg. var. extracted		F-Manuf.	F-InvMgt	F-Labor	F-CostCont.	F-Report	O-Manuf.	O-InvMgt	O-Labor	O-CostCont.	O-Report	O_Satisf.
0.9603	0.5492	F-Manuf.	1										
0.9049	0.6574	F-InvMgt	0.6165	1									
0.9552	0.7806	F-Labor	0.4869	0.3634	1								
0.9514	0.7109	F-CostCont.	0.5694	0.5906	0.5814	1							
0.9388	0.7933	F-Report	0.6582	0.5498	0.5547	0.6549	1						
0.9023	0.8220	O-Manuf.	0.8564	0.6255	0.4901	0.5720	0.6514	1					
0.9163	0.8456	O-InvMgt	0.6755	0.7625	0.3579	0.5865	0.6001	0.6354	1				
0.9012	0.8201	O-Labor	0.6033	0.4812	0.8133	0.7452	0.6215	0.6033	0.4876	1			
0.9632	0.9290	O-CostCont.	0.5809	0.5844	0.5106	0.8920	0.7012	0.5905	0.6105	0.7147	1		
0.9085	0.8324	O-Report	0.6677	0.5211	0.4596	0.5789	0.7904	0.6785	0.5595	0.5875	0.6333	1	
0.9545	0.9130	O_Satisf.	0.7224	0.5271	0.5318	0.5886	0.6470	0.7028	0.5980	0.6365	0.5994	0.8297	1

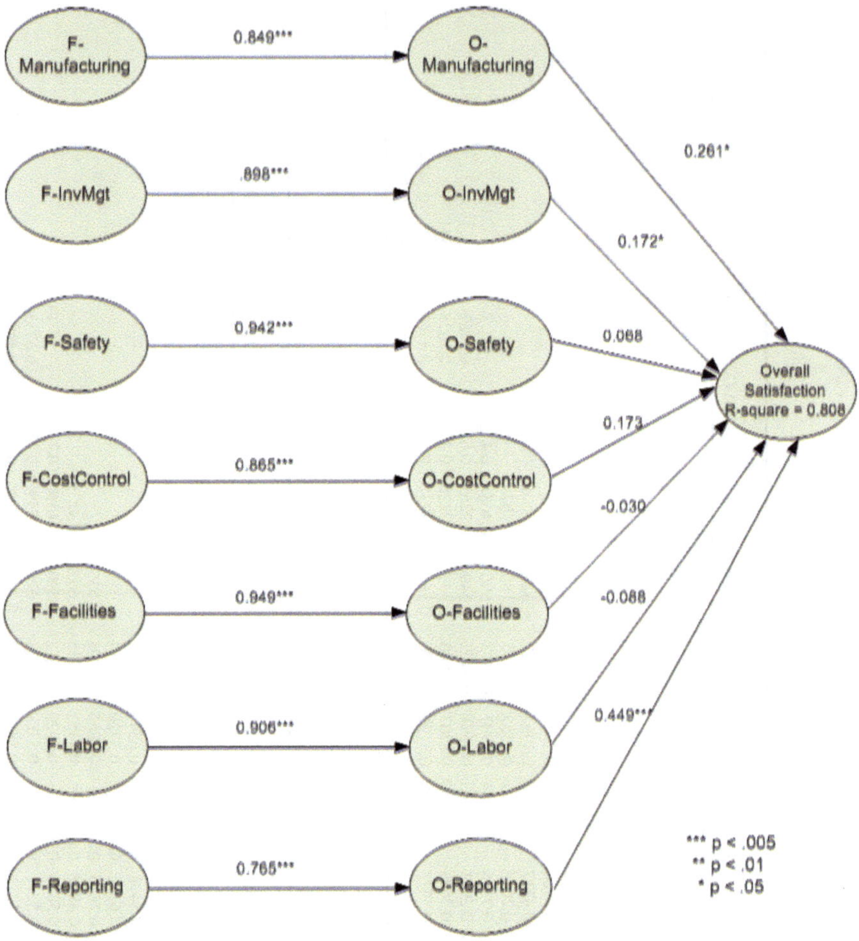

Fig. 2: Plant manager model results

3.4 Production Supervisor Model Results

The proportion of the variation explained in overall satisfaction with the ERP System by the hypothesized model is 87.1%. Similar to the plant manager model this is an excellent r^2 value and indicates that the model is a good predictor of overall satisfaction with the enterprise system. Like the plant managers, production supervisors satisfaction with the various facets of the enterprise system also depends on the how well those facets satisfy their needs. In each instance the path loadings indicate a strong statistically significant relationship, with $p < 0.005$ in all cases (Fig. 3).

However the results indicate that satisfaction with only two of the facets hypothesized to affect overall satisfaction are statistically significant: satisfaction with labor

($p < 0.01$) and satisfaction with reporting ($p < 0.005$). As with the plant manager model, satisfaction with the reporting features has the strongest impact on satisfaction with the enterprise system.

4 Discussion

Our findings show how plant managers and production supervisors view satisfaction with enterprise systems through the lens of their individual facets of responsibility. Yet while these systems encompass a variety of facets, neither plant managers nor production supervisors are making use of the additional capabilities that ERP systems have over and above those found in legacy systems such as manufacturing resource planning (MRP II) systems. Furthermore overall satisfaction with the system is linked to a few key indicators, most importantly satisfaction with the report generating/reporting features of the system.

Because reporting plays such a prominent role in assessing satisfaction with enterprise systems, we feel it warrants a closer look. Table 6 summarizes the answers to the question, "To what extent do you use the following tools in performing your tasks in [this areas of responsibility]?" Examining these results shows that the most common tool for reporting and analysis for plant managers in virtually all areas of functional responsibility are spreadsheets, followed by reports generated by the enterprise system. In some instances the difference in usage is quite large; for example, plant managers indicated that spreadsheets are used predominantly for health, safety and environmental reporting and analysis (71.6%) versus enterprise systems (10.8%). For production supervisors usage is more evenly divided between spreadsheets and enterprise system reports, although there are some notable differences such as in inventory reporting and analysis which relies principally on enterprise systems (73.8%) as opposed to spreadsheets (52.4%).

The high utilization of spreadsheet analysis versus enterprise system reports in most functional areas seems to indicate that the enterprise system does not fulfill the reporting and analysis needs in those facets. This could be the result of several factors: difficulty in extracting data from the enterprise system, lack of flexibility in enterprise system reports, timeliness in generating reports, and greater familiarity with spreadsheet tools in general. This highlights the need to provide better operating analytics (i.e., business intelligence) to managers and supervisors in supply chain operations.

5 Conclusion

In conclusion, this research presents a methodology that provides both an overall and detailed understanding of users' opinions towards ERP systems. It allows for an initial estimate of the overall performance of the ERP as well as general evaluations

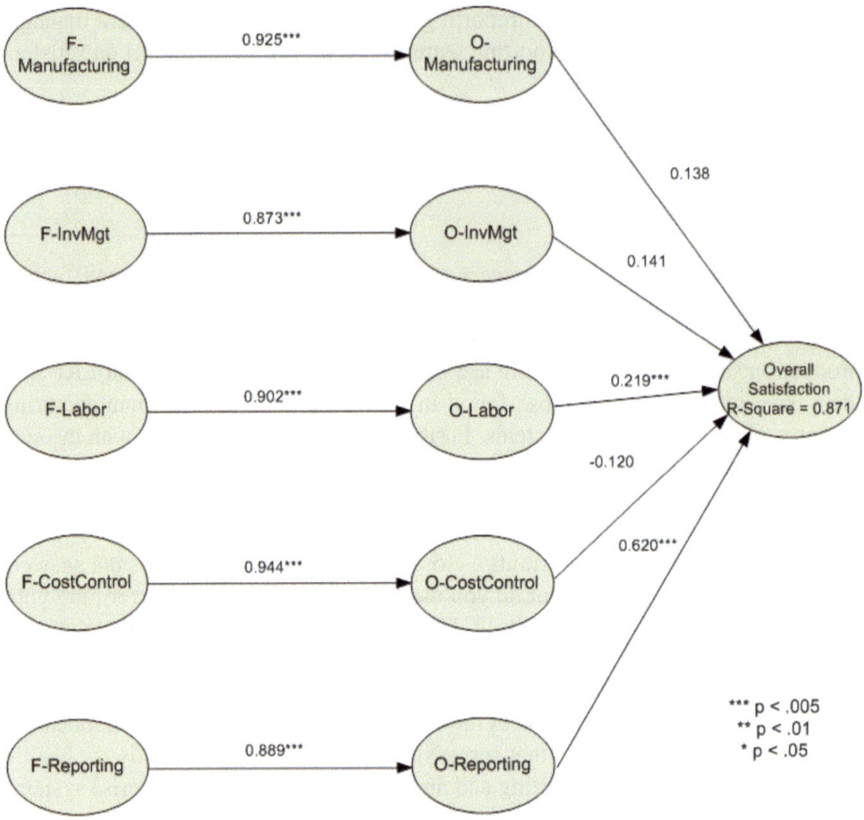

Fig. 3: Production supervisor model results

for key subsystems that are seen as affecting the current level of performance. This is followed up by examining those specific ERP features that are most influential in affecting each of the subsystems. Our approach to developing models and survey structures resulted in very satisfactory results in terms of internal consistency and validity of our results. We believe this approach can be extended to international environments and different industries where research findings will find immediate applicability.

An important contribution of this research comes from the design and application of the survey instrument in a supply chain operations context. To our knowledge this research is the first attempt to gage the satisfaction with ERP systems by managers in specific operational roles. Other research streams have only tried to measure overall user satisfaction, primarily among financial and corporate executive management. By focusing on the managers responsible for the daily production operations of the enterprise, we seek to understand how well the technology solves the problems faced by these managers and offer insight into ways the technology can be improved.

Table 6: Utilization of reporting and analysis tools by function

	Reporting and analysis tool						
	SpreadSheet	ERP	Paper	Phone	Manf. ex.	Other	Adv. plan.
Plant managers							
Cost controls	**76.9**	61.5	38.5	23.1	20.5	11.7	6.4
Facilities & maintenance	**64.9**	23	33.8	20.3	14.9	24.4	2.7
Inventory	70.4	**71.1**	25.9	23.5	24.7	8.4	6.2
Labor & personnel	**69.9**	29.1	45.2	20.5	9.6	24.9	1.4
Manufacturing	**77.1**	71.1	39.8	36.1	16.9	12	8.4
Reporting	**78.1**	64.4	42.5	19.2	24.7	14	6.8
Safety health & environ.	**71.6**	10.8	25.7	50	8.1	18	1.4
Averages	**72.7**	47.3	35.9	27.5	17.1	16.2	4.8
Production supervisors							
Cost controls	48.7	**52.6**	25.6	11.5	19.2	18.1	2.6
Inventory	52.4	**73.8**	23.8	25	26.2	14.4	11.9
Labor & personnel	**50.6**	40.5	32.9	16.5	10.1	30.6	2.5
Manufacturing	**70.1**	69	13.8	28.7	33.6	12.2	35.6
Reporting	**63.6**	57.1	23.4	13	26	22.1	5.2
Averages	57.1	**58.6**	23.9	18.9	23.0	19.5	11.6

There is no clear answer yet to the question of whether ERP systems effectively meet the needs of supply chain operations. While users are satisfied with some of the functionalities, they are not as happy with others. To understand shortcomings in a way that will be helpful to both developers and users of such software, detailed, functional role-based analysis is a viable approach. Our results provide meaningful information to software vendors as well as to practitioners in assessing their experience within their particular environment. Further research in different professional roles, such as quality or project managers, would extend the understanding of usefulness of ERP systems can be conducted to increase insights in ERP effectiveness.

Acknowledgements The authors wish to thank Sukran N. Kadipasaoglu who contributed to the survey design and collected the data for this research.

References

[1] K. Amoako-Gyampah and A.F. Salam, "An extension of the technology acceptance model in an ERP implementation environment," *Information & Management,* **41**, pp.731–745, 2004.

[2] W. W. Chin, "How to write up and report PLS analyses," In V.E. Vinzi, W.W. Chin, J. Henseler, and H. Wang (Eds.), *Handbook of Partial Least Squares Concepts, Methods and Applications*, pp.650–690, 2010.

[3] S.W. Chou and Y.C. Chang, "The implementation factors that influence the ERP (enterprise resource planning)," *Decision Support Systems,* **46**, pp.149–157, 2008.

[4] C. Fornell and F.L. Brookstein, "Two structural equation models: LISREL and PLS applied to consumer exit-voice theory," *Journal of Marketing Research,* **19**, pp.440–452, 1982.

[5] G. Gartner, "Market Trends: Enterprise Software Markets Are Shifting From Bricks and Mortar to 'BRIC and Mortals' Worldwide," 2011.

[6] K.B. Hendricks, V.R. Singhal, and J.K. Stratman, "The impact of enterprise systems on corporate performance: A study of ERP, SCM, and CRM system implementations" *Journal of Operations Management,* **25**, pp.65–82, 2007.

[7] L.M. Hitt, D.J. WU, and X. Zhou, "Investment in Enterprise Resource Planning: Business Impact and Productivity Measures" *Journal of Management and Information Systems,* **19**, pp.71–98, 2002.

[8] F.R. Jacobs and E. Bendoly, "Enterprise resource planning: Developments and directions for operations management research," *European Journal of Operational Research,* **146** pp. 233–240, 2003.

[9] R.C. MacCallum and M.W. Browne, "The use of causal indicators in covariance structure models: Some practical issues," *Psychological Bulletin,* **114** pp.533–541, 1993.

[10] F.F.H. Nah, X. Tan, and S.H. Teh, "An empirical investigation on end-users' acceptance of enterprise systems," *Information Resources Management Journal,* **17** pp.32–53, 2004.

[11] C. Ranganathan and C.V. Brown, "ERP investments and the market value of firms: Toward and understanding of influential ERP project variables," *Information Systems Research,* **17** pp.145–161, 2006.

[12] P.B. Seddon, C. Calvert, and S. and Yang, "A multi-project model of key factors affecting organizational benefits from enterprise systems," *MIS Quarterly,* **34** pp.305–328, 2010.

[13] T.M. Somers, and K.G. Nelson, "The impact of strategy and integration mechanisms on enterprise system value: Empirical evidence from manufacturing firms," *European Journal of Operational Research,* **146** pp.315–338, 2003.

[14] Y.-F. Su and C. Yang, "A structural equation model for analyzing the impact of ERP on SCM," *Expert Systems with Applications,* **37** pp.456–469, 2010.

[15] Q. Tu, "Measuring organizational level IS usage and its impact on manufacturing performance," In. *Proceedings of the Eighth Americas Conference on Information Systems*, Dallas, TX pp. 2188–2194, 2002.

Plant Manager Survey Questions

Manufacturing	*"Our current information system is an effective tool to:"*
MF-01	Prepare production plans to meet demand forecast/customer orders
MF-02	Track the $ volumes shipped from my plant
MF-03	Track demand versus volumes shipped
MF-04	Track deviations from on-time delivery targets
MF-05	Project capacity issues (too much/too little) in the near future
MF-06	Track product yields
MF-07	Prepare/contribute to Sales & Operations Planning
MF-08	Track planned production output versus actual production output
MF-09	Track manufacturing lead times
MS-01	Monitor production activities in general
Inventory	*"Our current information system is an effective tool to:"*
IF-01	Monitor the $ value of inventories
IF-02	Monitor inventory turns
IF-03	Monitor inventory quantities on hand
IF-04	Project inventory issues (too much/too little) in the near future
IF-05	Monitor variances in raw material inventory targets
IF-06	Monitor variances in work-in-process inventory targets
IF-07	Monitor variances in finished goods inventory targets
IS-01	Monitor inventory metrics in general
Health, safety & env.	*"Our current information system is an effective tool to:"*
SF-01	Track environment-related metrics in my plant
SF-02	Track the Recordable Incident Rate in my plant
SF-03	Monitor plant operations to ensure conformance to environmental emission standards
SF-04	Monitor the status of corrective actions related to Safety Health & Environmental issues
SS-01	Monitor Safety, Health & Environmental metrics in general
Cost control	*"Our current information system is an effective tool to:"*
CF-01	Monitor variances from the budget
CF-02	Track cost metrics for operations in real time
CF-03	Track product costs
CF-04	Monitor labor cost metrics
CF-05	Monitor purchased material costs
CF-06	Monitor personnel costs
CF-07	Monitor maintenance costs
CF-08	Monitor utility costs in my plant
CF-09	Monitor variations from the capital budget
CF-10	Plan future budgets
CS-01	Track costs that are of interest to me in real-time
Facilities	*"Our current information system is an effective tool to:"*
FF-01	Track equipment failure rates
FF-02	Monitor the maintenance schedule on major equipment
FF-03	Track asset utilization
FF-04	Track downtimes on major equipment
FS-01	Monitor facilities- and maintenance-related activities in general
Labor & personnel	*"Our current information system is an effective tool to:"*
LF-01	Monitor training requirements in my plant
LF-02	Monitor number of hires and fires in my plant
LF-03	Provide me with the information to carry out performance evaluations for my personnel
LF-04	Monitor employee attendance/absence
LF-05	Track corrective action notices
LF-06	Track personnel resources and availability in my plant
LS-01	Track personnel-related information in general
Reporting	*"Our current information system is an effective tool to:"*
RF-01	Report on Safety, Health & Environment-related metrics
RF-02	Report on facilities & maintenance-related metrics
RF-03	Report on production-related metrics
RF-04	Report on cost metrics
RF-05	Report on labor and personnel metrics
RS-01	Report on all the key performance metrics for my plant

General	
RS-02	I can easily generate reports using the current information system
OS-01	The current information system is an effective tool that supports me in performing my tasks and fulfilling my responsibilities
MS-02	I can easily monitor manufacturing activities in my plant using the current information system
LS-02	I can easily monitor personnel activities in my plant using the current information system
SS-02	I can easily monitor Safety, Health & Environmental activities using the current information system
IS-02	I can easily monitor inventory-related metrics in my plant using the current information system
CS-02	I can easily monitor cost metrics using the current information system
FS-02	I can easily monitor facilities and maintenance activities using the current information system
OS-02	The current information system supports me very well in the management of my plant
OS-03	I am very happy with the planning tools I have in my current information system

Production Supervisor Survey Questions

Manufacturing	*"Our current information system is an effective tool to:"*
MF-01	Provide all the information I need (such as open orders, inventory positions, available capacity, etc.)
MF-02	Make changes to the current production schedule
MF-03	Show the impact of schedule changes on critical operational metrics (such as materials, due dates, capacity, etc.)
MF-04	Monitor open orders and their progression on the shop floor
MF-05	Monitor the actual production output versus the planned production output in a given week
MF-06	Monitor production variations from the planned product mix
MF-07	Monitor variations in material usage from the standards in the routings
MF-08	Monitor variations in labor hours from the standards in the routings
MF-09	Monitor variations in machine hours from the standards in the routings
MF-10	Monitor actual lead times versus planned lead times
MF-11	Monitor the capacity of critical resources and incorporating capacity constraints into the production schedule
MF-12	Monitor variations in planned versus actual process steps on a customer order
MF-13	Track whether completed orders are shipped on time
MF-14	Provide component visibility and traceability when needed
MF-15	Track machine utilizations in my area of responsibility
MF-16	Track labor absorption hours based on work orders in my area of responsibility
MF-17	Plan downtimes and incorporate this information into the production schedule
MF-18	View equipment maintenance/calibration schedules
MF-19	Track product yields in my area of responsibility
MF-20	Track scrap rates in my area of responsibility
MS-01	Track production information in general
Inventory	*"Our current information system is an effective tool to:"*
IF-01	Monitor material shortages that may affect scheduled production
IF-02	Monitor finished good inventory levels
IF-03	Monitor work-in-progress (WIP) inventory levels
IF-04	Monitor purchased material levels
IF-05	Monitor deviations from inventory targets
IS-01	Track material- and inventory-related information in general
Labor & personnel	*"Our current information system is an effective tool to:"*
LF-01	Provide the information necessary to carry out performance evaluations for the personnel under my supervision
LF-02	Track the availability of personnel resources in my area of responsibility
LF-03	Monitor shift turnovers in my area of responsibility
LF-04	Track training requirements in my area of responsibility
LF-05	Monitor employee attendance/absence in my area of responsibility
LF-06	Track corrective action notices in my area of responsibility
LS-01	Track my personnel metrics in general

Cost control	*"Our current information system is an effective tool to:"*
CF-01	Track training costs in my area of responsibility
CF-02	Track utility consumption in my area of responsibility
CF-03	Track overhead costs in my area of responsibility
CF-04	Track unit production costs in my area of responsibility
CF-05	Track deviations from operational cost metrics in my area of responsibility
CF-06	Track personnel and labor costs in my area of responsibility
CF-07	Track variations from budget in my area of responsibility
CF-08	Track costs that are of interest to me in real-time
CS-01	Monitor cost metrics in my area of responsibility in general
Reporting	*"Our current information system is an effective tool to:"*
RF-01	Report on cost metrics in my area of responsibility
RF-02	Report on labor metrics in my area of responsibility
RF-03	Report on production metrics in my area of responsibility
RF-04	Report on inventory metrics in my area of responsibility
RS-01	Report on all the performance metrics that fall under my area of responsibility
General	
RS-02	The current information system allows me to easily and effectively generate reports
OS-01	The current information system is an effective tool that supports me in performing my tasks and fulfilling my responsibilities
MS-02	I can easily monitor production metrics in my area using the current information system
LS-02	I can easily monitor personnel activities in my area of responsibility using the current information system
IS-02	I can easily monitor inventory metrics using the current information system
CS-02	I can easily monitor costs in my area of responsibility using the current information system
OS-02	The current information system supports me very well in the supervision of production in my area of responsibility

An Investigation of the Impact Publicly Available Accounting Data, Other Publicly Available Information and Management Guidance on Analysts' Forecasts

Michael R. Newman, George O. Gamble, Wynne W. Chin, and Michael J. Murray

Abstract There have been a number of studies that indicate that analysts recommendations are superior to other forecasts, such as those by time-series models, and add economic benefit, adjusted for transaction costs, to clients who first receive and then use analysts' forecasts. There is also academic literature documenting the use of accounting information in valuing firms by analysts and others, the use of financial information from other sources than the firm itself by analysts, the impact of management guidance on decisions made by analysts, and the concept of herd behavior among analysts. The majority of studies about analysts have used sell-side analyst data to reach their findings. However, there has been little research involving buy-side analysts, analysts who are employed by institutional investors to provide stock purchase recommendations to their employers for internal investment decision making purposes. The research there has studied investments by institutional investors, many of which employ buy-side analysts. The purpose of this study is to add to the literature by investigating what information buy-side analysts use in arriving at their stock investment recommendations. This study also investigates whether or not buy-side analysts are predominantly influenced by the information they receive from publicly available accounting data, other available public information, other analysts or management guidance. The data for this investigation is being obtained from a survey of buy-side analysts. A list of 130 analysts was prepared and asked to take the survey. The use of a survey to gather the data was consistent with the use in prior studies. The PLS approach to structural equation analysis was used to assess the measurement model because it can be used for theory confirmation and

M.R. Newman (✉) • G.O. Gamble • M.J. Murray
University of Houston, 4800 Calhoun Road, 260-G Melcher Hall,
Houston, TX 77204-6021, USA
e-mail: MichaelN@uh.edu; ggamble@uh.edu; michael.murray@mail.uh.edu

W.W. Chin
Department of Decision and Information Systems, C. T. Bauer College of Business,
University of Houston, Houston, TX 77204-6021, USA
e-mail: wchin@uh.edu

H. Abdi et al. (eds.), *New Perspectives in Partial Least Squares and Related Methods*, Springer Proceedings in Mathematics & Statistics 56, DOI 10.1007/978-1-4614-8283-3_22,

suggest possible relationships, and because it is more suitable for prediction since it assumes that all measured variance can be explained in a study. The SEM-based method has been described as a coupling of two traditions: an econometric perspective focusing on prediction and a psychometric emphasis that models concepts as latent (unobserved)variables that are indirectly inferred from multiple observed measures (alternately termed as indicators or manifest variables). This method allows for the performance of path analytic modeling and has been referred to as a second generation of multivariate analysis. The results of this study further the academic literature concerning analysts by investigating what information buy-side analysts use to arrive at their overall stock investment recommendations and by the use of the PLS approach.

Key words: Buy-side analysts, Sell-side analysts, Analysts stock investment recommendations, Structural equation modeling, Partial least squares

1 Introduction

Our global financial market has created a need for information about firms that will allow us to minimize the risk associated with investing in them. Financial analysts are used to fill that need [1, 8, 16, 18, 26, 27, 33]. Financial analysts are employed by brokerage houses which sell stocks and mutual funds, and institutional buyers (such as the aforementioned mutual funds, retirement plans, insurance policies and pension plans) who hire their own "proprietary" financial analysts. Although both buy-side analysts and sell-side analysts issue financial (earnings) forecasts and buy/sell recommendations on publicly traded firms, their clients are different [14, 23].

Sell-side analysts are employed by firms that "sell" securities and investment advice, e.g. brokerage firms. Sell-side analysts issue publicly available earnings forecasts and stock purchase recommendations to their firm's customers. These forecasts and stock purchase recommendations can be found in any number of sources that report financial information such as the Wall Street Journal, Yahoo Financial and the I/B/E/S data base.

On the other hand, buy-side analysts are employed by companies that "buy" securities for internal investment decision making purposes, namely to earn profits from investing in security markets. Buy-side analysts are employed to provide forecasts, offer investment advice and make stock purchase recommendations exclusively for their employers. Their private forecasts and stock investment recommendations seldom become publicly known and many are very secretive about the "private information" they use to arrive at their company's investment decisions.

As a result of this disproportionate availability of information with regard to sell-side versus buy-side analysts, more academic research has been conducted surrounding the behavior of sell-side analysts (especially given the availability of data through data bases such as I/B/E/S), as opposed to the behavior of buy-side analysts. Thus, there appears to be a real void in the academic literature regarding the behavior of buy-side financial analysts.

The purpose of this study is to determine the relative importance of different information sources on buy-side analysts' stock purchase recommendations. Representational faithfulness will also be evaluated in terms of whether or not it has an impact on the relative importance assigned to a particular information source. The specific research questions addressed in this study were:

- What sources of information do buy-side analysts use to make their stock recommendation?
- Do buy-side analysts find the sources of information they use to make their stock recommendation to be accurate and unbiased?
- Is a buy-side analyst's attitude towards a company affected by the sources of information used?
- Does the degree to which a buy-side analyst trusts management of a firm affect their attitude or stock purchase recommendation?
- Does a buy side analyst's attitude toward a firm influence their stock purchase recommendation for that firm?

The study is significant in the sense that it adds to the accounting literature by investigating to what extent buy-side analysts' overall stock investment recommendations are influenced by various sources of information. The sources that we evaluate in this study are the information buy-side analysts receive from publicly available accounting data, other available public information, other analysts, and management guidance, as well as trust of management.

2 Research Methodology

We first examined the academic literature to develop the research models and survey instrument. A survey [4, 10, 30, 31] was then developed and sent to known buy-side analysts. We received 135 usable responses which were then evaluated using the Partial Least Squares methodology.

The selection of the subjects for this study (i.e., buy-side analysts currently employed by institutional investors) was guided by the academic literature. The literature finds that institutional investors (e.g., employers of buy-side analysts) prefer to own stock in large firms with high visibility [2, 5, 11, 12, 19, 22, 35].

A list of 703 buy-side analysts was compiled from a number of sources that met this criteria, primarily from analyst meetings sponsored by public companies and from referrals (from Investor Relations Departments and other buy-side analysts). The list was then culled to eliminate bad email addresses and each buy-side analyst was then contacted to verify if they were still buy-side analysts and asked if they would be willing to participate in a survey. We had response from 227 analysts. One hundred forty analysts participated in the survey. Of these, 117 identified themselves as "buy side" and gave complete sets of data.

The study was anonymous and the participants received no compensation. The identity of the participants is not known but the survey tool gave an identifying number to eliminate duplications/multiple answers by the respondents. Some respondents did enclose information on how to send them copies of the results, but these participants are not disclosed on any reports or to any parties by the researcher. Others sent emails indicated they had responded.

The sample size needed for this experiment, calculated to be 60, was determined based on the amount of power needed given the relationships that were described earlier in this section. The actual sample of 135 (18 "sell side" and 117 "buy-side") surpasses this minimum requirement [7, 9, 15]. Three main models (and a total of seven models) were then developed to test the data received from the survey.

2.1 Development of the Models

As key areas for research were identified by the literature review, definitions were developed, each definition was decomposed into component parts (the key facets of the question) in order to understand the underlying meaning, and several questions were developed in order to select those that had the most congruence/meaning. The models used in this study were created using SEM and this facet based approach to develop the questions that were asked in the survey.

The literature confirms that analysts make financial forecasts, provide investment advice and make stock recommendations [29] and that sell-side analysts is usually provided to major investors at no cost [14]. Based on this evidence, we began with the "Analyst Stock Recommendation." We defined this variable as the "extent or degree to which an analyst would make a recommendation involving the purchase or sale of a company's common stock" and then developed the questions for this variable as well as its reflective and formative variables.

The literature next led to another potential direct relationship based on the question of whether or not the buy-side analyst formulates his or her stock recommendation based on these variables or if the relationship is one between the buy-side analyst and his/her attitude towards the firm he/she is analyzing [14]. We defined "attitude" as the "affective evaluation an analyst has towards this firm's financial performance" and developed the questions for this variable as well as for its reflective and formative variables.

The literature identifies several sources of information that sell-side analysts use to arrive at their stock recommendation including publicly available accounting data, other publicly available information, management guidance, and the opinions of other analysts [10, 14, 24, 32, 36, 37, 40].

We next defined each of these sources and developed the questions for each main variable as well as its reflective and formative variables (including source questions) to help address the variables of information analysts use to make their decisions. "Trust of management" was defined based on evidence in the literature that the quality of a company's financial reports affect the way analysts view it and questions were developed for this variable [3, 5, 21, 25, 39, 41].

The question of representational faithfulness was also an area for contribution according to the literature and included in our research instrument [13, 14, 34]. We defined "representational faithfulness" and developed the questions for this variable as well as the reflective and formative variables (including source questions) for three of the sources of information identified in the literature (publicly available accounting data, other publicly available information and management guidance).

As discussed earlier, there is ample academic evidence to suggest that analysts use publicly available accounting data, other publicly available information, and, management guidance, and are influenced by other analysts' opinions and their trust of management. However, there is no consensus on the "usefulness" (e.g., degree to which an analyst uses each of them in their decision making process) of each source, even though such a relationship is purported [36]. So, we next developed questions for the "Usefulness" variable for each of the three sources of information (publicly available accounting data, other publicly available information, and, management guidance). Finally, we defined "evaluation activity" based on the potential for earnings management and developed the questions for this variable [29].

Additional survey questions were developed based on concerns raised by the academic literature which suggests that the size of the firm a sell-side analyst works for, the number of companies they follow, their experience level and the availability of adequate resources can affect an analyst's stock recommendation. Three Models were developed: the Stock Purchase Recommendation Model, the Attitude Model and the Full Model.

The Stock Purchase Recommendation Model was developed based upon the findings in the academic literature that suggest that financial analysts use publicly available accounting data, other publicly available information, management guidance, the opinions of other financial analysts and their trust of management as a basis for the recommendation (buy, hold or sell) they give on a stock purchase decision (see Fig. 1).

It was then modified (The Stock Purchase Recommendation Model with Interaction Effects) to test if Representational Faithfulness and Usefulness (Degree of Use) of these sources is important in this process (see Fig. 2).

The Attitude Model was developed based upon the findings in the academic literature that suggests that financial analysts' attitudes may influence their stock purchase recommendation. This model tested the effect that publicly available accounting data, other publicly available information, management guidance, the opinions of other financial analysts and their trust of management has on their attitude towards the organization about which they are making a stock purchase recommendation (see Fig. 3).

It too was then modified (The Attitude Model with Interaction Effects) to test if representational faithfulness and usefulness (Degree of Use) of these sources is important in this process (see Fig. 4).

By examining both models, we tested to see if the information a buy-side analyst receives from these sources (publicly available accounting data, other publicly available information, management guidance, the opinions of other financial analysts

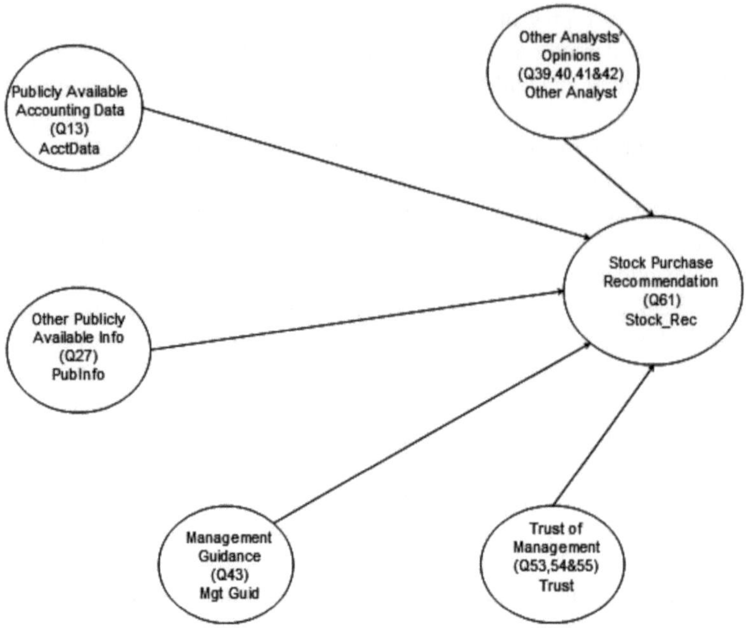

Fig. 1: Stock purchase recommendation model

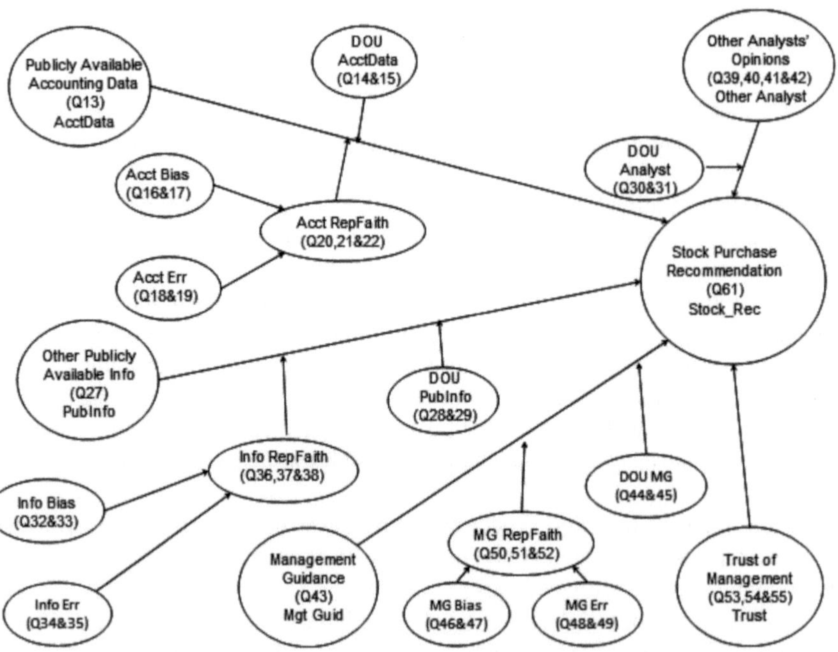

Fig. 2: Stock purchase recommendation model with interaction effects

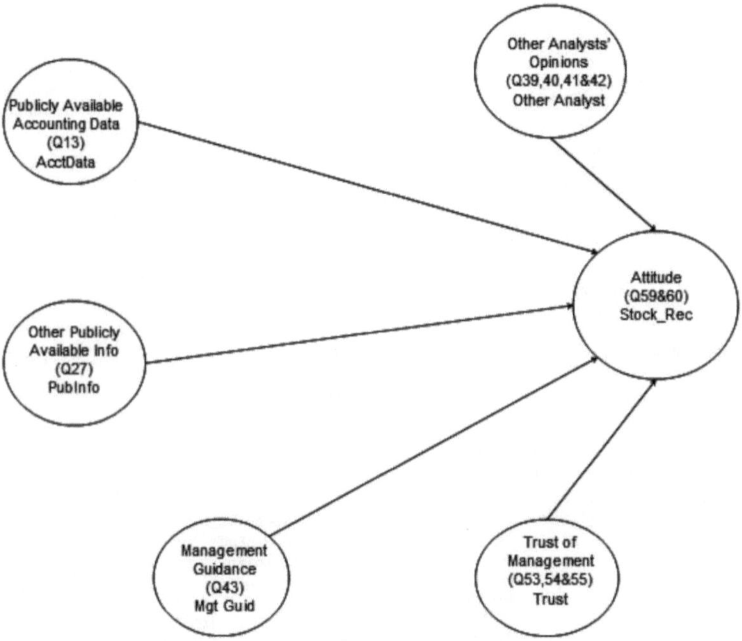

Fig. 3: Attitude model

and the analyst's trust of management) have similar or different influence on his/her stock purchase recommendation or his/her attitude towards the company.

Our third model, the Full (Combined) Model, tested to see if attitude did indeed have an effect on an analyst's stock purchase recommendation. It was tested in three ways. First, we added "attitude" as a construct to the Stock Purchase Recommendation Model (see Fig. 5).

Next we tested it by adding "attitude" as a construct to the Stock Purchase Recommendation Model with Interaction Effects (see Fig. 6).

Finally, we took the Stock Purchase Recommendation Model and added the Attitude Model as a construct to see the extent of the effect that the constructs the academic literature identified have on the stock purchase recommendation decision (see Fig. 7).

2.2 Survey Design

As was previously discussed, the structural model was developed after an extensive review of the existing academic literature. Key research areas were identified, constructs were established, and key questions were developed based on the key facets of each construct.

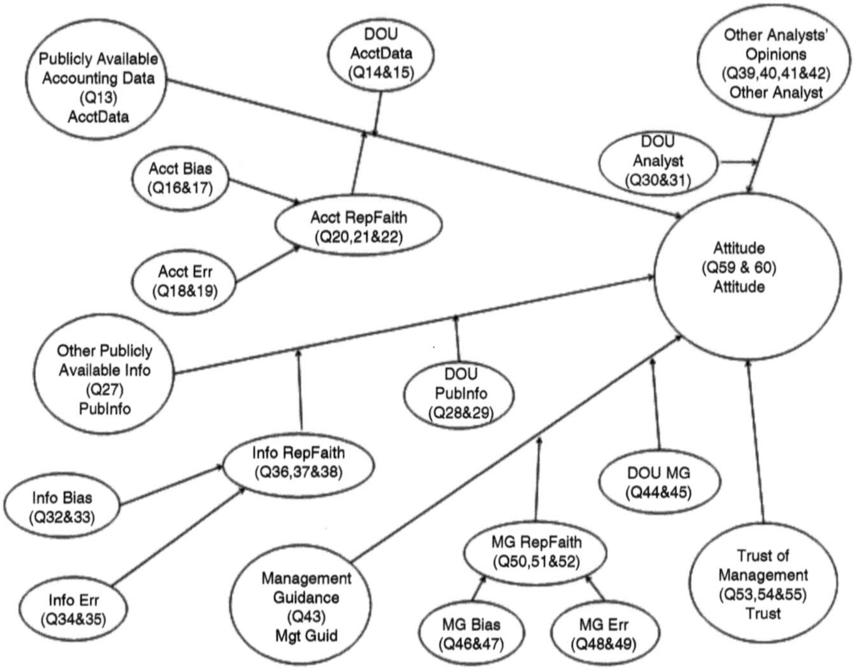

Fig. 4: Attitude model with interaction effects

The study included the appropriate use of negation in some cases, and asked questions several ways to better measure each construct. The items were measured using mostly seven-point Likert-type scales (with anchors such as "strongly disagree" to "strongly agree" and measurements ranging from "−3" to "+3" where "0" was neutral) and some eleven-point Likert-type scales (with anchors such as "pessimistic" to "optimistic" with a range of "0" to "+10"). A pilot study involving subject matter experts (including buy-side analysts) was conducted to clarify questions and to measure consistency and validate these items. Some changes in wording were made and some questions were eliminated as a result of the pilot study in order to refine the survey instrument.

The methodological approach used to test the relationships involved a survey that was accessed on the Web. A number of papers have been written about the use of Web based surveys. Recommendations made in this academic literature were incorporated into the survey design, especially as it pertained to increasing survey response and the development of the survey tool. Kaplowitz et al., state that "For special populations that regularly use the Internet, the Web has been found to be a useful means of conducting research [28]."

Selection of the relationships that were tested was appropriate since each was based on evidence of such relationships in the academic literature among financial

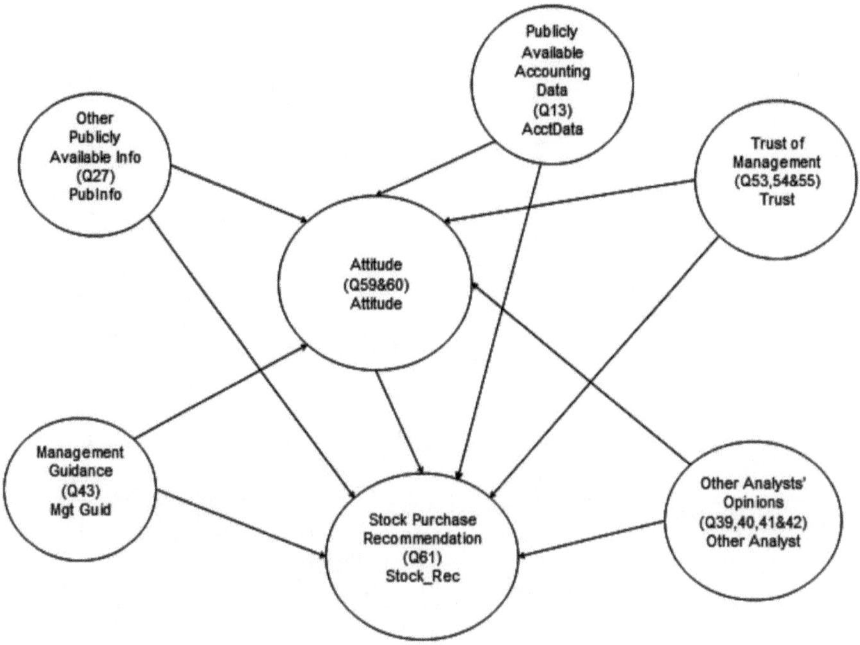

Fig. 5: Full (combined) model

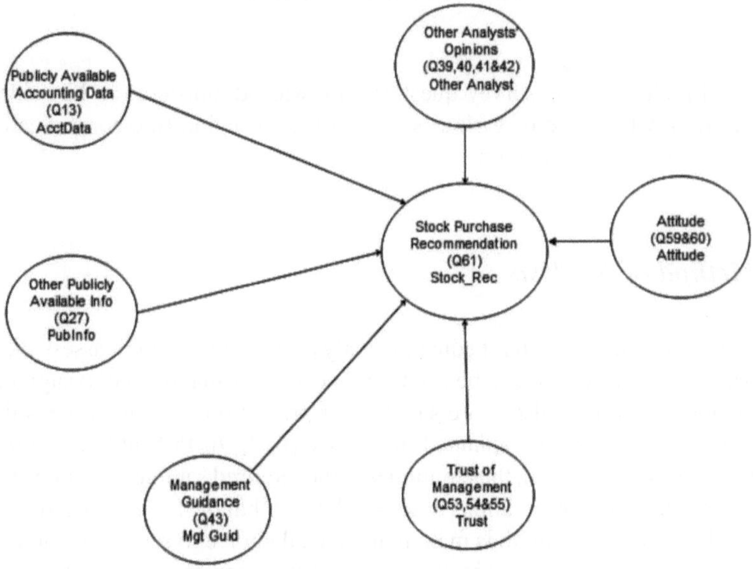

Fig. 6: Combined model

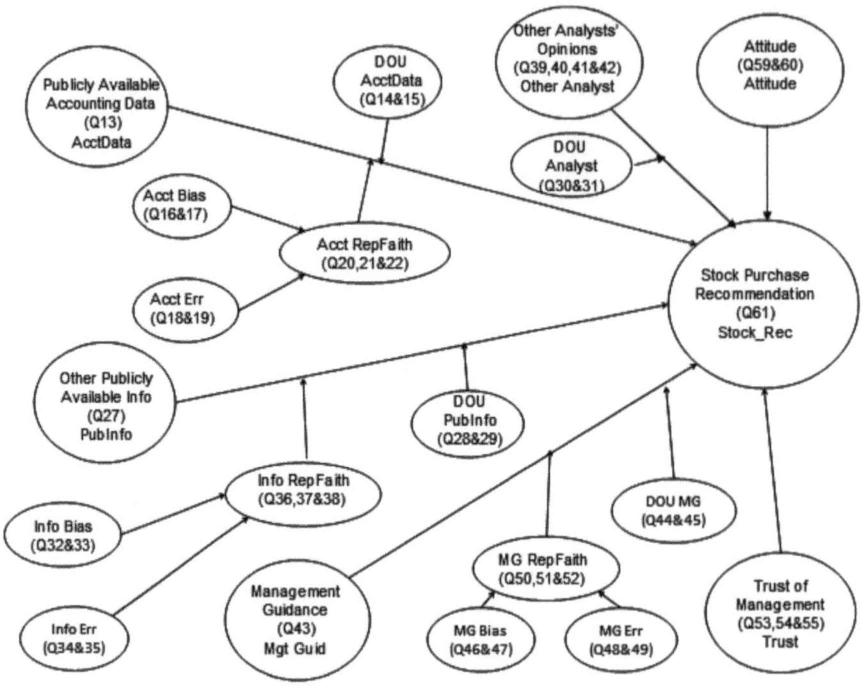

Fig. 7: Combined model with interaction effects

analysts. In addition, the survey questionnaire was administered to buy-side analysts, a group whose use of various sources of information (the constructs in this study) is not public information.

2.3 Method of Analysis

The PLS approach to structural equation analysis was employed to assess the measurement model because it can be used for theory confirmation and to suggest possible relationships. It is also more suitable for prediction since it assumes that all measured variance can be explained in a study [6, 7, 9]. PLS utilizes a principal component-based approach, thus minimizing the required sample size. PLS only requires a sample of 60 observations based on the fact that the largest construct in this study had 6 items (10×6), thus making it an ideal choice to assess the model. The number of items is determined by the greater of the number of measures forming a construct or the number of constructs it requires to predict the dependent construct. In contrast, if we were to use a covariance based SEM model, a sample of 280 (28 predictors times 10) would be required.

The analysis of data was completed by first assessing the measurement model and then assessing the structural model. In this case, significant tests were assessed using boot strap analysis with 1,000 re-samples. No assumptions of normality were made in accordance with the partial least squares (PLS) approach used in this study.

Individual item loadings, internal consistency and discriminant validity were assessed for the measurement model using PLS. The structural model and our hypotheses were tested by examining the path coefficients and their respective statistical significance. The predictive power of the model was based on the explained variance in the dependent constructs.

3 Results

A total of 703 analyst names and email addresses were collected over the last 8 years from professional meetings and associations. Emails were sent to these analysts asking if they would be willing to participate in a survey. Of the 227 responses, 135 gave complete sets of data, 18 of whom classified themselves as "sell-side" analysts and 117 as analysts analyzing stocks for their own portfolios ("buy-side analysts").

3.1 Model Evaluation

The 117 completed questionnaires were examined for outliers. A frequency Test was run using SPSS that examined the raw data file for skewness, kurtosis and outliers. No significant outliers were identified although there were occasions where one or two analysts selected an extreme value in answer to a specific survey question. Since none of the observations appeared to be extreme, all responses were used for analysis.

3.2 Reliability Assessment

Individual item loadings and internal consistency of reliabilities were examined as a test of reliability. Consistent with the academic literature, the reliability of the constructs exceeded 0.7 in all but one of the constructs (use of other publicly available information, which had a 0.678 reliability) [38].

A second test of reliability was then performed using the Composite Reliability Index. This test is considered to be much more accurate than Cronbach's Alpha. According to Chin, "use of this formula, which does not assume equal loadings or error terms among the measures, typically provides more accurate estimates of the composite reliability [7]." All of the composite reliabilities exceeded the minimum acceptable value of 0.70 including other publicly available information with a 0.862 Composite Reliability [20].

3.3 Convergent and Discriminant Validity Assessment

Convergent validity is shown when the t-values of the Outer Model Loadings are above 1.96. The *t*-values of the loadings are, in essence, equivalent to t-values in least-squares regressions. Each measurement item is explained by the linear regression of its latent construct and its measurement error [17]. The *t*-values in all the convergent validity assessments are much greater than 1.96 indicating high convergent validity.

Discriminant validity was assessed in two ways: one test is based on how well the items load on their own construct as compared to other constructs in the model, and, a second based on the fact that the average variance shared between the constructs and their measures is greater than the variance shared between the constructs themselves. The first test of discriminant validity requires that loadings should be high and cross-loadings should be low compared to the loadings. The individual cross-loadings of the items all exceeded 0.82 and the factor cross-loadings of all the items shared by the constructs were low, showing clear discriminant validity [7]. The second test of discriminant validity is to show that the average variance extracted (AVE) for each construct is greater than the square of the correlations of that construct to all the others in the model [7, 15, 17], which was true. Thus, both approaches demonstrate discriminant validity; results of the other models tested gave similar results.

3.4 Structural Model Evaluation

Each structural model was tested by examining the path coefficients (and their respective statistical significance) and the r^2 to determine the predictive power of the model, based on the explained variance in the dependent constructs. Each hypothesis was tested using PLS Graph 3.0. Path coefficients, along with their degree of significance based on T-values, were calculated using a boot-strapping procedure. The latent scores derived from PLS Graph were then loaded into SPSS to further test the significance of the path coefficients. SPSS was then used to calculate the interaction terms. The PLS coefficients are different from the SPSS coefficient combination of the interactions, rounding and the fact that the results of the boot-strapping is based on an average of 1,000 re-samples.

The significance of the path (betas) and the adjusted r^2 was used to measure significance in each model. Each was then evaluated by first looking at the main effect model without interactions and then by looking at it with interactions included. Table 1 contains a summary of all the findings for the seven models.

Table 1: Summary of results for models 1 through 7

Summary of results (N = 117)							
	Model #1	**Model #2**	**Model #3**	**Model #4**	**Model #5**	**Model #6**	**Model #7**
Intercept	9.7E-006 (1.000)	−0.037 (0.666)	0.000 (0.998)	0.045 (0.496)	0.000 (0.998)	−0.061 (0.410)	−2.03E-005 (1.000)
Main effects							
Publ. Avail. Acct. Data(PAAD)	**0.393 (0.000)**	**0.497 (0.000)**	**0.450 (0.000)**	**0.437 (0.000)**	0.083 (0.380)	**0.210 (0.057)**	0.083 (0.383)
Other Publ. Avail. Info. (OPAI)	0.114 (0.273)	−0.021 (0.851)	0.112 (0.164)	0.021 (0.812)	0.040 (0.653)	−0.041 (665)	0.040 (0.657)
Mgmt. Guid. (MG)	−0.014 (0.873)	0.018 (0.869)	−0.079 (0.260)	−0.082 (0.355)	0.039 (0.613)	0.048 (0.616)	0.038 (0.625)
Other Anal. Opinions (OAO)	**0.258 (0.002)**	**0.197 (0.040)**	**0.195 (0.002)**	**0.194 (0.011)**	**0.124 (0.089)**	0.087 (0.297)	**0.124 (0.091)**
Trust of Mgmt. (TM)	0.084 (0.312)	**0.172 (0.097)**	**0.383 (0.000)**	**0.334 (0.000)**	**−0.175 (0.032)**	−0.082 (0.401)	**−0.179 (0.030)**
Interaction effects							
DOU PAAD		**0.175 (0.035)**		**0.192 (0.005)**		0.086 (0.228)	
DOU OPAI		0.122 (0.108)		0.003 (0.962)		**0.133 (0.041)**	
DOU MG		0.009 (0.928)		**0.151 (0.072)**		−0.125 (0.178)	
DOU OAO		−0.010 (0.900)		−0.083 (0.208)		0.021 (0.770)	
RF PAAD		0.096 (0.278)		0.060 (0.390)		0.057 (0.453)	
RF OPAI		−0.011 (0.896)		−0.045 (0.477)		0.020 (0.772)	
RF MG		−0.058 (0.542)		**−0.186 (0.015)**		0.067 (0.424)	
Attitude					**0.685 (0.000)**	**0.679 (0.000)**	**0.684 (0.000)**
Moderating effects							
Publ. Avail. Acct. Data(PAAD)							**0.450 (0.000)**
Other Publ. Avail. Info. (OPAI)							0.112 (0.164)
Mgmt. Guid. (MG)							−0.079 (0.260)
Other Anal. Opinions (OAO)							**0.195 (0.002)**
Trust of Mgmt. (TM)							**0.383 (0.000)**
R2 and adjusted R2	0.398/0.371	0.527/0.434	0.642/0.626	0.700/0.641	0.569/0.546	0.659/0.587	0.565/0.541

Note: Shows Standardized coefficient (significance level) for ; *DOU* degree of usage, *RF* representational faithfulness

3.4.1 Model 1: Stock Purchase Recommendation

The r^2 of the structural model used to measure the main effects of the Stock Purchase Recommendation Model is 0.398 with an adjusted r^2 of 0.371. As can be seen in Fig. 8, two significant effects were found: use of publicly available accounting data has a path of 0.393 with a significance of 0.000, and, opinions of other analysts has a path of 0.258 with a significance of 0.002.

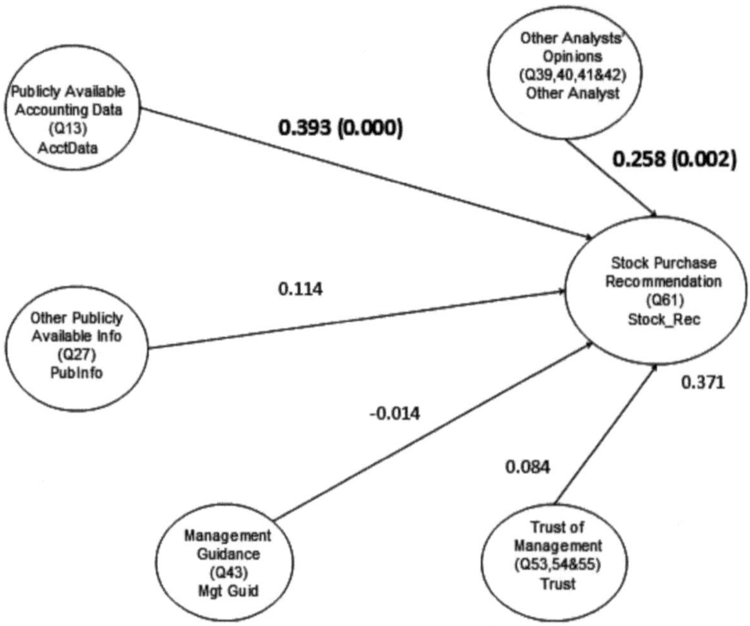

Fig. 8: Stock purchase recommendation model path coefficients

3.4.2 Model 2: Stock Purchase Recommendation with Interaction Effects

The r^2 of the structural model used to measure the Stock Purchase Recommendation Model with Interaction Effects has more explanatory power than the first model. The r^2 is 0.527 with an adjusted r^2 of 0.434. As can be in Fig. 9, three significant main effects and one interaction effect are found:

- Publicly available accounting datas path increases from 0.393 to 0.497 with a significance of 0.000;
- Degree of usage of publicly available accounting data has a path of 0.175 with a significance of 0.035;

- The other analysts opinions path coefficient decreases from 0.258 to 0.197 with a significance of 0.040 (compared to 0.002 in the main effects model); and,
- Trust of management becomes significant with a path coefficient of 0.172 and a significance of 0.097.

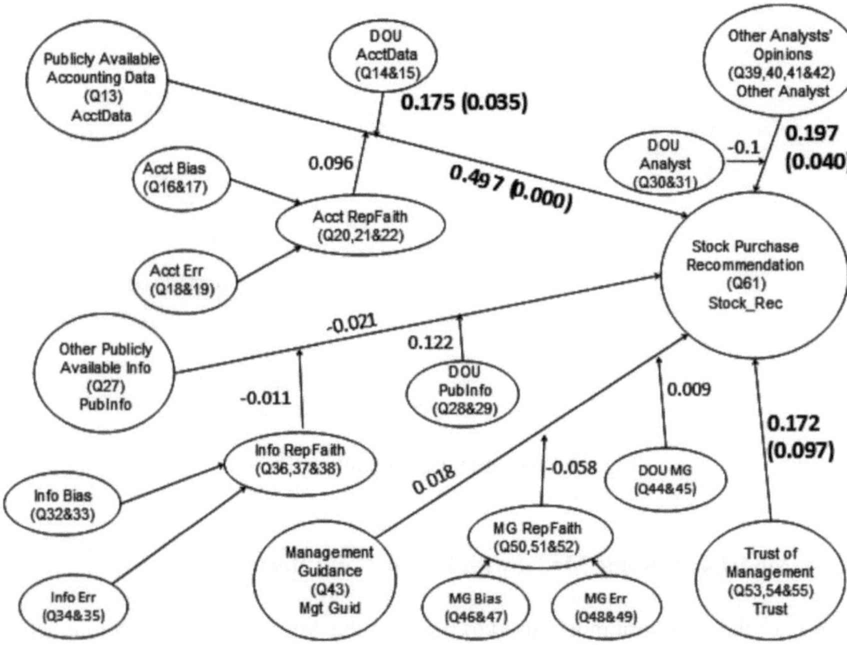

Fig. 9: Stock purchase recommendation model with interaction effects path coefficients

3.4.3 Model 3: Attitude Model

The r^2 of the structural model used to measure the Attitude Model is 0.642 with an adjusted r^2 of 0.626. As can be seen in Fig. 10, three significant main effects are found in the Attitude Model (main effects model), the first two of which are similar to the Recommendation Model (main effects model):

- Use of publicly available accounting data has a path of 0.450 with a significance <0.001;
- The opinions of other analysts has a path of 0.195 with a significance of 0.002; and,
- Trust of management is significant in this model with a path coefficient of 0.383 and a significance of <0.001.

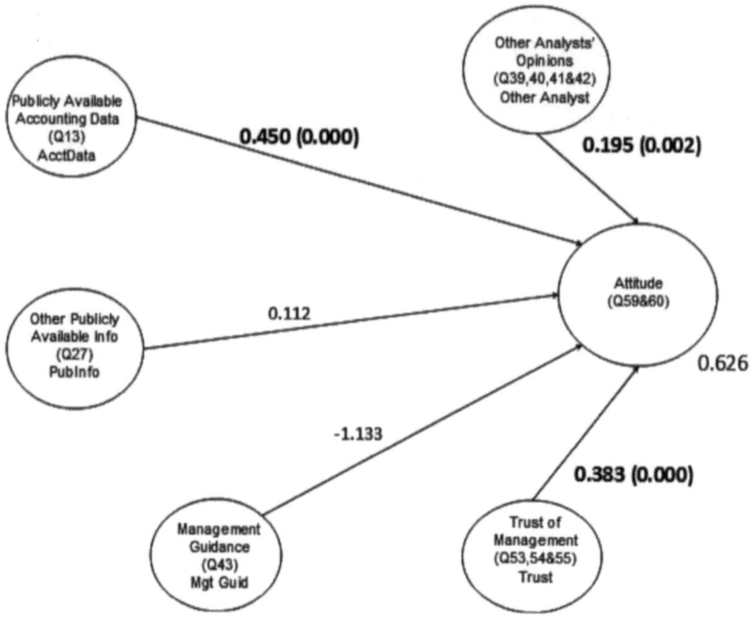

Fig. 10: Attitude model path coefficients

3.4.4 Model 4: Attitude Model with Interaction Effects

The r^2 of the structural model used to measure the Attitude Model with Interaction Effects indicates that it has slightly more explanatory power than the Attitude Model (main effects model) with an $r2$ of 0.700 (compared to 0.642) and an adjusted r^2 of 0.641 (compared to 0.626). As can be seen in Fig. 11, the three significant main effects found in the Attitude Model (main effects model) remain significant although the path coefficients have changed:

- Use of publicly available accounting data has a path of 0.437 with a significance <0.001;
- The opinions of other analysts has a path of 0.194 with a significance of 0.011 (versus 0.002); and,
- Trust of Management is significant in this model with a path coefficient of 0.334 and a significance <0.001.

Three additional interaction effects are found that are significant:

- The interaction of degree of usage of publicly available accounting data and use of publicly available accounting data has a path coefficient of 0.192 and a significance of 0.005;
- The interaction of degree of usage of management guidance and management guidance has a path coefficient of 0.151 and a significance of 0.072; and,

- The interaction of representational faithfulness of management guidance and management guidance has a path coefficient of −0.186 and a significance of 0.015.

The significant main effects and significant interaction effects are shown on Fig. 11.

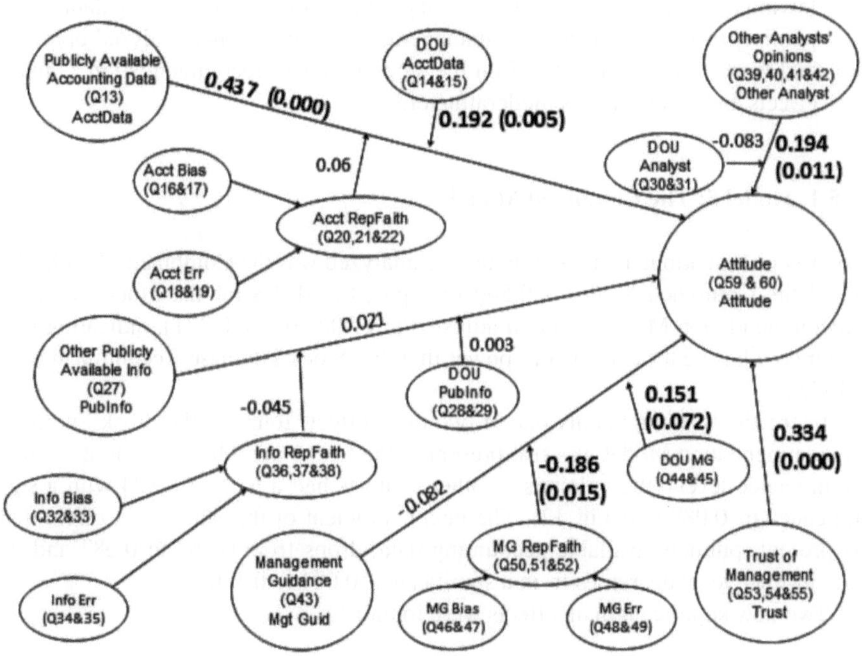

Fig. 11: Attitude model with interaction effects path coefficients

3.5 The Full and Combined Models

In order to provide better insight, an analysis of the effect that attitude has on the stock purchase recommendation is decomposed into three models:

1. The first model tested, the Combined Model, adds attitude to the Stock Purchase Recommendation Model as a variable to see what effect, if any, attitude would have on the stock purchase recommendation and on the influence of the other independent variables in the Stock Purchase Recommendation Model;
2. The second model tested, the Combined Model with Interaction Effects, adds attitude to the Stock Purchase Recommendation Model with Interaction Effects

as a main effect, again, with the intent of examining if a buy-side analyst's attitude has any significant effect on the stock purchase recommendation and, if so, what effect it would have on the other independent variables in the model; and,

3. The Full Model, which combined the Stock Purchase Recommendation Model with the Attitude Model, and treated attitude as the sixth independent variable. The variables influencing attitude were then treated as moderating variables, allowing us to investigate the effects that publicly available accounting data, other publicly available information, management guidance, other analysts' opinions and trust of management had on the stock purchase recommendation as main effects and on attitude as moderating effects.

3.5.1 Model 5: The Combined Model

As was just mentioned, the first model we analyzed was the Combined Model. The r^2 of the Combined Model is 0.569 (compared to 0.398 for the Stock Purchase Recommendation Model) with an adjusted r^2 of 0.546 (vs. 0.371), indicating that the model has better explanatory power than the Stock Purchase Recommendation Model.

Interestingly enough, only one of two main effects found in the Stock Purchase Recommendation Model was still present in the Combined Model albeit at a lower significance level—the opinions of other analysts had a path of 0.124 with a significance of 0.089 (see Fig. 12). The beta coefficient of the other main effect that is present, publicly available accounting data, drops from 0.383 to 0.083 and its significance level decreases from a significant <0.001 to 0.380.

Two new significant main effects were found:

- Trust of management has a negative path coefficient of −0.175 with a significance of 0.032; and,
- The main effect with the largest significance is attitude with a path coefficient of 0.685 and a significance level of <0.001.

The negative influence of trust of management on the stock purchase recommendation is interesting. One might infer that if management is truly trustworthy, there may be no opportunity to profit from information asymmetries. In other words, buy-side analysts might believe that trustworthy management would be too open with information (eliminating any asymmetry of information that could lead to an analyst finding private information that could result in abnormal profits) or perhaps fear that trustworthy management would be unwilling to make aggressive decisions that would result in higher reported earnings.

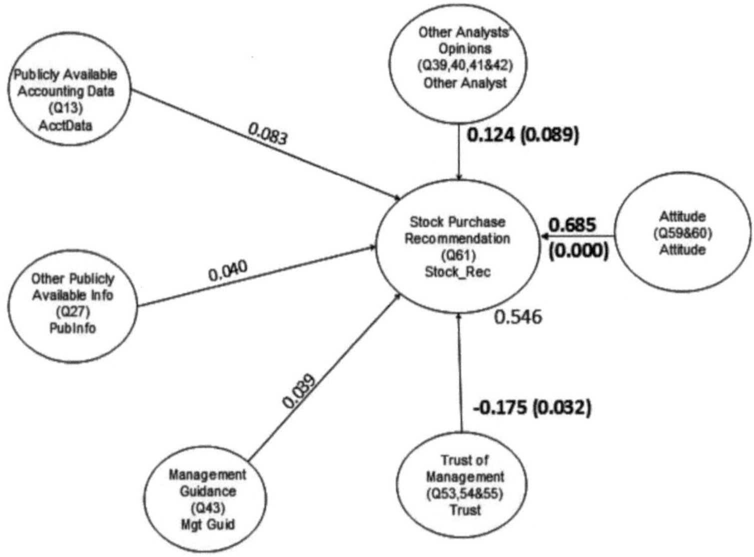

Fig. 12: Combined model path coefficients

3.5.2 Model 6: Combined Model with Interaction Effects

The next model we examined was the Combined Model with Interaction Effects (shown in Fig. 13). The r^2 of the structural model used to measure the Combined Model with Interaction Effects (the Stock Purchase Recommendation Model with Interaction Effects and Attitude added) is considerably higher (e.g., has much better explanatory power) than that of the original Stock Purchase Recommendation Model with Interaction Effects: 0.659 (compared to 0.527) with an adjusted r^2 of 0.587 (vs. 0.434).

There is one similarity and two differences in the significant main effects and interaction effects found between the Combined Model and the Combined Model with Interaction Effects (see Fig. 14).

- Attitude remains as the main effect with the largest significance with a path coefficient of 0.679 and a significance level <0.001;
- Publicly available accounting data becomes significant again (it has statistical significance in both of the Stock Purchase Recommendation Models and the two Attitude Models, but not in the Combination Model) with a path coefficient of 0.210 and a significance of 0.057; and,
- Degree of usage of publicly available accounting data becomes significant for the first time in any of our models with a path coefficient of 0.133 and a significance of 0.041.

It is interesting to note that trust of management loses significance with a negative path coefficient of -0.082 and a significance level of 0.401 in the Combination

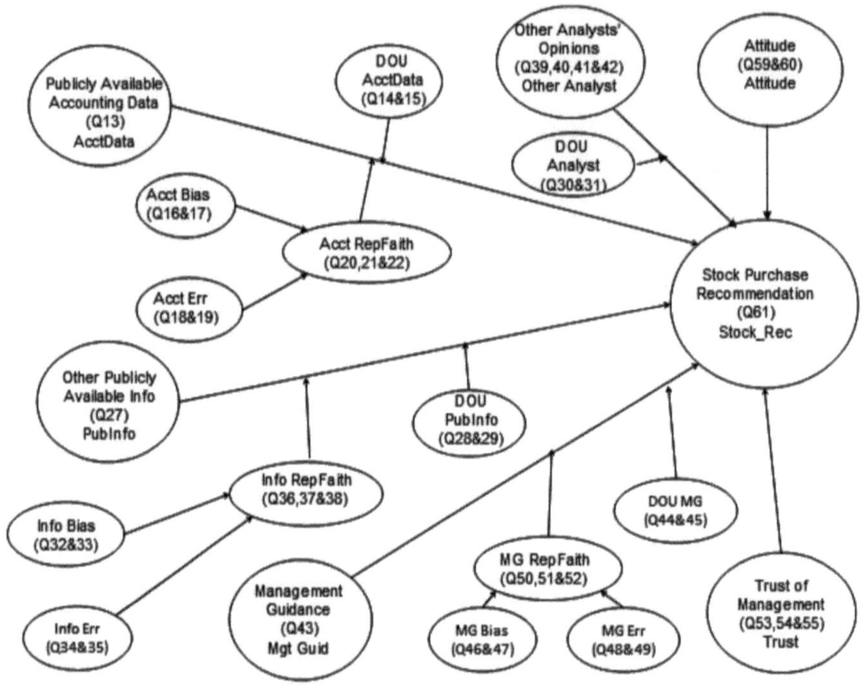

Fig. 13: Combined model with interaction effects

Model with Interactions Effects after being statistically significant in the Stock Purchase Recommendation Model with Interaction Effects, both Attitude Models and the Combined Model.

3.5.3 Model 7: The Full Model

We next performed an analysis of the Full Model. The r^2 of the Combined Main Effects Model is 0.565 (compared to 0.569 for the Combined Model and 0.659 for the Combined Model with Interaction Effects) with an adjusted r^2 of 0.541 (versus 0.546 and 0.587, respectively). Although its "goodness of fit" is less than the two Combination Models, it still has high explanatory power. Three significant main effects and three significant moderating effects are found in the Full Model (see Fig. 15):

- The largest significant effect is attitude with a 0.684 path coefficient and a significance <0.001 (this is consistent with the earlier Combination Models which have beta coefficients of 0.685 and 0.679, respectively, both at a significance level <0.001);
- Trust of management has two significant effects:

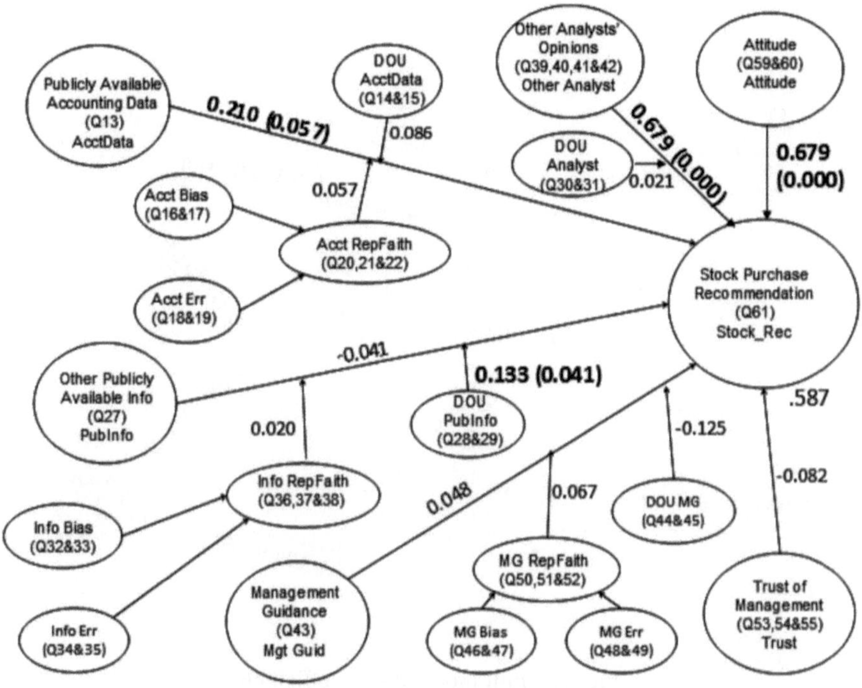

Fig. 14: Combined model with interaction effects path coefficients

1. A main effect negative path coefficient of −0.179 with a significance of 0.030 when regressed against stock purchase recommendation; and,
2. A moderating effect positive path coefficient of 0.383 with a significance <0.001 when regressed against attitude.

- Other analysts' opinions also has two significant effects:
  1. A main effect positive path coefficient of 0.124 with a significance of 0.091 when regressed against stock purchase recommendation; and,
  2. A moderating effect positive path coefficient of 0.195 with a significance of 0.002 when regressed against attitude.
- Publicly available accounting data does not have a significant main effect when regressed against the stock purchase recommendation dependent variable (0.083 with a 0.383 level of significance) but does have a moderating effect when regressed with attitude (with a path coefficient of 0.450 and a significance level <0.000).

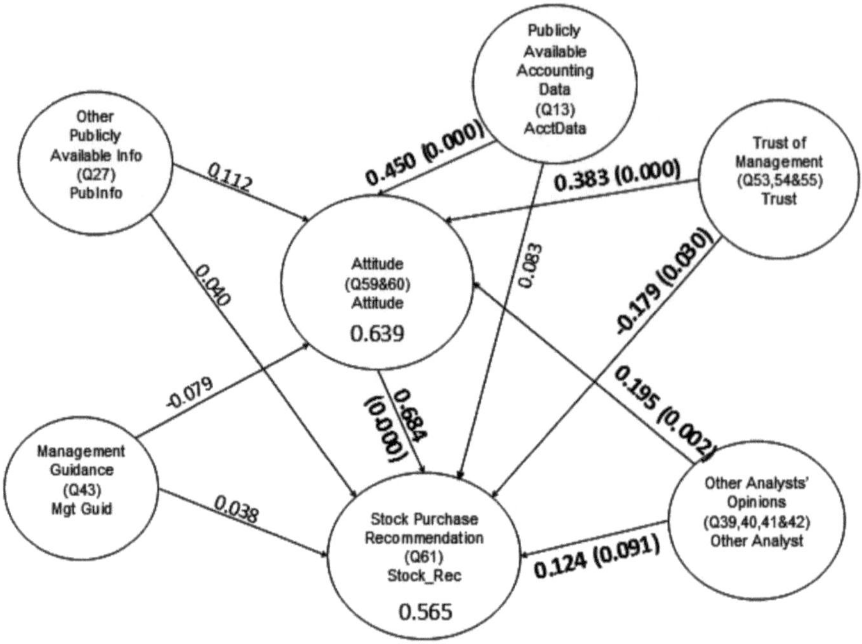

Fig. 15: Full model path coefficients

4 Discussion and Conclusion

The best interpretation of these findings is that a buy-side analyst bases his/her decisions to recommend purchase of a stock largely based on his/her attitude towards a company and that attitude is influenced largely by publicly available accounting data and trust of management, as well as what other analysts he/she respects say about the company. The empirical analysis of the data provides a number of interesting insights. Specifically:

- Our most interesting finding is that trust of management plays an important role in both a buy-side analyst's attitude towards an organization and in his/her decisions regarding purchasing, holding or selling a security. Specifically:
  - There is evidence that there is a positive, significant relationship between a buy-side analyst's attitude towards a stock and trust of management, and
  - A lesser weighted but still significant negative relationship between a buy-side analyst's stock purchase recommendation and trust of management.

This implies that while buy-side analysts may feel good about companies with management they trust they are cautious about investing in firms run by trustworthy management. An interesting side note is that while buy-side analysts do not consider management guidance in their decisions, their contact with management does seem to influence their trust of management.

- Our second most interesting finding is that the independent variable that has the greatest single effect on a buy-side analyst's stock purchase recommendation is the buy-side analyst's attitude towards that company. In other words, based on the findings of the models, all of which were found to have good explanatory power, it appears that buy-side analysts base their buy or sell decisions of a stock largely on their "gut feeling" ("attitude") towards a company and that this attitude is largely influenced by the publicly available accounting data, their trust of management and the opinions of other analysts they respect about the firm for which they are making a stock purchase recommendation.
- We also find that publicly available accounting data:
  - Plays an important role in the buy or sell recommendation a buy-side analysts makes when attitude is not part of the equation but loses some of its effect when attitude is added; and,
  - Has a huge effect on the buy-side analyst's attitude towards a company.

This leads us to conclude that while publicly available accounting data is considered in the stock purchase recommendation decision (direct impact), its largest influence is an indirect one (on a buy-side analyst's attitude towards the firm). It should be noted that the degree of use/usefulness of publicly available accounting data is also significant, leading us to conclude that buy-side analysts do use this information.

- Finally, we find that other analysts' opinions affects both the buy or sell recommendation and the attitude of buy-side analysts, and does so in a positive way, implying that buy-side analysts are more likely to recommend purchase of a stock if (other) analysts they trust are making similar recommendations (herd behavior).

References

[1] M. E. Barth and A. P. Hutton, *Information intermediaries and the pricing of accruals*, Stanford University, 2000.
[2] B. J. Bushee, "Do institutional investors prefer near-term earnings over long-run value?," *Contemporary Accounting Research*, **18**, 207–246, (2001).
[3] B. J. Bushee, D. A. Matsumoto, and G. S. Miller, "Managerial and investor responses to disclosure regulation: The case of reg fd and conference calls," *Accounting Review*, **79**, 617–643, (2004).
[4] B. J. Bushee and G. S. Miller, "Investor relations, firm visibility, and investor following," *Accounting Review*, **87**, 867–897, (2012).
[5] B. J. Bushee and C. F. Noe, "Corporate disclosure practices, institutional investors, and stock return volatility," *Journal of Accounting Research*, **38**, 171–202, (2000).
[6] W. W. Chin, "Issues and opinion on structural equation modeling," *MIS Quarterly*, **22**, vii–xvi, 1998.
[7] W. W. Chin, "Partial least squares approach to structural equation modeling," in G. A. Marcoulides, (ed.) *Modern Methods for Business Research*, Lawrence Erlbaum Associates, Mahwah, NJ, pp. 295–336, 1998.

[8] W. W. Chin, B. L. Marcolin, and P. R. Newsted, "A partial least squares latent variable modeling approach for measuring interaction effects: Results from a Monte Carlo simulation study and an electronic-mail emotion/adoption study," *Information Systems Research*, **14**, 189–217, (2003).

[9] W. W. Chin and P. Newsted, "Structural equation modeling: Analysis with small samples using partial least squares," in R. Hoyle (ed.) *Statistical Strategies for Small Sample Research*, SAGE, Thousand Oaks, CA, pp. 307–341, 1999.

[10] Lal C. Chugh and Joseph W. Meador, "The stock valuation process: The analysts' view," *Financial Analysts Journal*, **40**, pp. 41–48, (1984).

[11] D. DelGuercio, "The distorting effect of the prudent-man laws on institutional equity investments," *Journal of Financial Economics*, **40** 31–62, 1996.

[12] E. G. Falkenstein, "Preferences for stock characteristics as revealed by mutual fund portfolio holdings," *Journal of Finance*, **51**, 111–135, 1996.

[13] FASB, *Recognition and Measurement in Financial Statements of Business Enterprises*, 1986.

[14] T. J. Fogarty and R. K. Rogers, "Financial analysts' reports: an extended institutional theory evaluation," *Accounting Organizations and Society*, **30**, 331–356, 2005.

[15] C. Fornell, "A second generation of multivariate analysis: Classification of methods and implications for marketing research," in M. J. Houston (ed.) *Review of Marketing*, American Marketing Association, Chicago, IL, pp. 407 – 450, 1987.

[16] J. Francis and L. Soffer, "The relative informativeness of analysts' stock recommendations and earnings forecast revisions," *Journal of Accounting Research*, **35**, 193–211, 1997.

[17] D. Gefen and D. Straub, "A practical guide to factorial validity using pls-graph: tutorial and annotated example," *Communications of AIS*, **16**, 91–109, 2005.

[18] D. Givoly and J. Lakonishok, "Properties of analysts' forecast of earnings: A review and analysis of the research," *Journal of Accounting Literature*, **3**, 117–152, 1984.

[19] P. A. Gompers and A. Metrick, "Institutional investors and equity prices," *Quarterly Journal of Economics*, **116**, 229–259, 1984.

[20] J. F. Hair, R. E. Anderson, R. L. Tatham, and W. C. Black, *Multivariate data analysis with readings*, 4 ed., Pretice Hall, Englewood Cliffs, NJ, 1995.

[21] A. Hutton Healy, P. and K. Palepu, "Stock performance and intermediation changes surrounding sustained increases in disclosure," *Contemporary Accounting Research*, **16**, 485–520, 1999.

[22] C. Hessel and M. Norman, "Financial characteristics and neglected and institutionally held stocks," *Journal of Accounting, Auditing and Finance*, 313–334, 1992.

[23] H. Hong, J. D. Kubik, and A. Solomon, "Security analysts' career concerns and herding of earnings forecasts," *Rand Journal of Economics*, **31**, 121–144, 2000.

[24] C. Hongren, *Implications from accountants on the uses of financial statements by securities analysts*, Dissertation, University of Chicago, Chicago, IL, 1978.

[25] A. P. Hutton, G. S. Miller, and D. J. Skinner, "The role of supplementary statements with management earnings forecasts," *Journal of Accounting Research*, **41** , 867–890, 2003.

[26] C. D. Ittner and D. F. Larcker, "Are nonfinancial measures leading indicators of financial performance? an analysis of customer satisfaction," *Journal of Accounting Research*, **36** , 1–35, 1998.

[27] C. D. Ittner, D. F. Larcker, and M. V. Rajan, "The choice of performance measures in annual bonus contracts," *Accounting Review*, **72**, 231–255, 1997.

[28] M. D. Kaplowitz, T. D. Hadlock, and R. Levine, "A comparison of web and mail survey response rates," *Public Opinion Quarterly*, **68**, 94–101, 2004.

[29] S. P. Kothari, "Capital markets research in accounting," *Journal of Accounting & Economics*, **31**, 105–231, 2001.

[30] W. G. Lewellen, R. C. Lease, and G. G. Schlarbaum, "Patterns of investment strategy and behavior among individual investors," *Journal of Business*, **50**, 296–333, 1977.

[31] W. G. Lewellen, R. C. Lease, and G. G. Schlarbaum, "Personal investments of professional managers," *Financial Management*, **8**, 28–36, 1979.

[32] R. Libby, R. Bloomfield, and M. W. Nelson, "Experimental research in financial accounting," *Accounting Organizations and Society*, **27**, 777–812, 2002.

[33] T. Lys and S. K. Sohn, "The association between revisions of financial analysts earnings forecasts and security-price changes," *Journal of Accounting & Economics*, **13**, no. 4, 341–363, 1990.

[34] S. Mintz and R. Morris, *Ethical Obligations and Decision making in Accounting: Text and Cases*, McGraw-Hill/Irwin, New York, 2008.

[35] P. C. Obrien and R. Bhushan, "Analyst following and institutional ownership," *Journal of Accounting Research*, **28**, 55–76, 1990.

[36] L. Pankoff and R. Virgil, "Some preliminary findings from a laboratory experiment on the usefulness of financial accounting information to security analysts," *Journal of Accounting Research*, **8**, 1–48, 1970.

[37] R. Bricker T. Robinson Previts, G. and S. Young, "A content analysis of sell-side financial analyst company reports," *Accounting Horizons*, **8**, 55–70, 1994.

[38] S. Rivard, G. Poirier, L. Raymond, and F. Bergeron, "Development of a measure to assess the quality of user-developed applications," *Data Base for Advances in Information Systems*, **28**, 44–58, 1997.

[39] K. Schipper, "Commentary on earnings management," *Accounting Horizons*, **3**, 91–102, 1989.

[40] K. Schipper, "Commentary on analysts' forecasts," *Accounting Horizons*, **5**, 105–121, 1991.

[41] H. T. Tan, R. Libby, and J. E. Hunton, "Analysts' reactions to earnings preannouncement strategies," *Journal of Accounting Research*, **40**, 223–246, 2002.

Index

H. Abdi et al. (eds.), *New Perspectives in Partial Least Squares and Related Methods*, Springer Proceedings in Mathematics & Statistics 56, DOI 10.1007/978-1-4614-8283-3,

MIX
Papier aus verantwortungsvollen Quellen
Paper from responsible sources
FSC® C105338
FSC
www.fsc.org

If you have any concerns about our products,
you can contact us on
ProductSafety@springernature.com

In case Publisher is established outside the EU,
the EU authorized representative is:
Springer Nature Customer Service Center GmbH
Europaplatz 3, 69115 Heidelberg, Germany

Printed by Libri Plureos GmbH
in Hamburg, Germany